Geometry, Spectral Theory, Groups, and Dynamics

(Courtesy of Paul Urlieb, Photographer)

Robert Brooks
(1952–2002)

CONTEMPORARY MATHEMATICS

387

Israel Mathematical Conference Proceedings

Geometry, Spectral Theory, Groups, and Dynamics

Proceedings in Memory of Robert Brooks
December 29, 2003–January 2, 2004
January 5–9, 2004
Technion-Israel Institute of Technology
Haifa, Israel

Michael Entov
Yehuda Pinchover
Michah Sageev
Editors

American Mathematical Society
Providence, Rhode Island
Bar-Ilan University
Ramat Gan, Israel

2000 *Mathematics Subject Classification.* Primary 58–06, 53–06, 37–06; Secondary 11–06, 20–06, 55–06.

The figures on pages 5 and 6 of Peter Buser's article, "On the Mathematical Work of Robert Brooks," are reprinted by permission of Princeton University Press: Kra, Irwin; *Riemann Surfaces and Related Topics.* ©1981 Princeton University Press.

Library of Congress Cataloging-in-Publication Data

Robert Brooks Memorial Conference on Geometry and Spectral Theory (2003 : Haifa, Israel)
Geometry, spectral theory, groups, and dynamics : proceedings in memory of Robert Brooks, December 29, 2003–January 2, 2004 [and] January 5–9, 2004, Technion-Israel Institute of Technology, Haifa, Israel/Michael Entov, Yehuda Pinchover, Michah Sageev, editors.
p. cm. — (Israel mathematical conference proceedings) (Contemporary mathematics, ISSN 0271-4132 ; 387)
Includes bibliographical references.
ISBN 0-8218-3710-9 (alk. paper)
1. Geometry—Congresses. 2. Spectral theory (Mathematics)—Congresses. 3. Dynamics—Congresses. I. Brooks, Robert, 1952–2002. II. Entov, Michael, 1969– . III. Pinchover, Yehuda, 1953– . IV. Sageev, Michah, 1966– . V. Title. VI. Series. VII. Contemporary mathematics (American Mathematical Society); v. 387.

QA440.R63 2005
516—dc22   2005053006

∞ The paper used in this book is acid-free and falls within the guidelines established to ensure permanence and durability.
Visit the AMS home page at http://www.ams.org/

10 9 8 7 6 5 4 3 2 1   10 09 08 07 06 05

Contents

Preface

The Robert Brooks Memorial Conference on Geometry and Spectral Theory took place at the Technion - Israel Institute of Technology, Haifa, Israel, on December 29, 2003 - January 2, 2004 and was immediately followed by the Workshop on Groups, Geometry and Dynamics, January 5-9, 2004.

The conference and the workshop were held under the auspices of the Center for Mathematical Studies (CMS) at the Technion. We thank the CMS and its director, Professor Baruch Solel, for letting us organize this program. We are grateful to the Mallat Family Fund for Research in Mathematics that supported the program and to Professors Dov Aharonov, Alexander Ioffe, Vladimir Lin, Alexander Nepomnyashchy, Allan Pinkus, Jack Sonn and Uri Srebro who provided additional financial support for the conference from their chair funds. We owe a great debt to the CMS secretary Sylvia Schur (she has passed away to our great sorrow) for her excellent job in handling all the technical matters related to the conference. We also thank Hava Harel, Galia Khanin, Ruth Markevich and the rest of the staff at the Technion Department of Mathematics, for their professional technical and logistical support. We thank the Editorial Board of the Israel Mathematics Conference Proceedings for agreeing to publish this volume. We thank Professor Lawrence Zalcman for his editorial work on the manuscripts. We thank Maxim Iorsh who carefully prepared the volume for publication. Finally, we thank all our colleagues who advised and encouraged us in the course of organizing the program and preparing this volume for publication.

***

Robert Brooks was born on September 16, 1952 and passed away on September 5, 2002. He received his PhD from Harvard in 1977 under the guidance of Raoul Bott. Having completed his postdoctoral studies at the State University of New York at Stony Brook, Robert Brooks became an assistant professor at the University of Maryland (1979-84). He went on to become an associate (1985-88) and later a full professor (1988-1997) at the University of Southern California. In 1995 Brooks immigrated to Israel and became a professor at the Technion - Israel Institute of Technology.

Known to his friends as Bob, Robert Brooks was a dear and beloved colleague. He was an outstanding mathematician whose mathematics touched a number of different fields, such as Global Analysis, Spectral Theory, the theory of Riemann Surfaces, Riemannian and Discrete Geometry, and Number Theory. A survey of Brooks' mathematical work has been kindly written for this volume by his close colleague Peter Buser. The wide range of mathematical interests represented by those who attended Brooks' memorial conference and workshop is a testament to the breadth and depth of his mathematical work: in these proceedings you will find both a work on such analytic topics as Szegő's theorem and a geometric work on isoperimetric inequalities and symmetries of manifolds.

Bob Brooks will be sorely missed by the mathematical community. His works and ideas will live on for years to come.

Israel Mathematical Conference Proceedings (IMCP)

Israel Mathematical Conference Proceedings (IMCP) is a publication, part of the Contemporary Mathematics Series, devoted to the proceedings of conferences, symposia and seminars. Collections of papers focusing on a certain subject will also be published. Prospective volumes may be submitted to any member of the editorial board. Each volume has an editor (or editors) responsible for its preparation. In order to ensure inexpensive and timely distribution, authors are requested to submit to the Editor of the volume an electronic TEX file of their manuscript in AMS-LATEX, using the Contemporary Mathematics style file which can be downloaded at `http://www.ams.org/tex/author-info.html`. For further information, contact the Managing Editor, IMCP, Department of Mathematics, Bar-Ilan University, Ramat-Gan 52900, Israel; e-mail: `rowen@macs.biu.ac.il`.

List of Participants

Arav Marina
Georgia State University, Atlanta, GA, U.S.A.

Bader Uri
Technion-Israel Institute of Technology, Haifa, Israel

Biran Paul
Tel Aviv University, Tel Aviv, Israel

Birman Joan
Columbia University, New York, NY, U.S.A.

Buser Peter
L'Ecole Polytechnique Fédérale de Lausanne, Lausanne, Switzerland

Casson Andrew
Yale University, New Haven, CT, U.S.A.

Chavel Isaac
City College of New York, New York, NY, U.S.A.

Dodziuk Józef
City College of New York, New York, NY, U.S.A.

Entov Michael
Technion – Israel Institute of Technology, Haifa, Israel

Eremenko Alexandre
Purdue University, West Lafayette, IN, U.S.A.

Farkas Hershel
Hebrew University of Jerusalem, Jerusalem, Israel

Friedlander Leonid
University of Arizona, Tucson, AZ, U.S.A.

Fujiwara Koji
Tohoku University, Sendai, Japan

Gabai David
Princeton University, Princeton, NJ, U.S.A.

Gamburd Alexander
Stanford University, Stanford, CA, U.S.A.

Gelander Tsachik
Yale University, New Haven, CT, U.S.A.

Ginzburg Viktor
University of California, Santa Cruz, CA, U.S.A.

Givental Alexander
University of California, Berkeley, CA, U.S.A.

Hersonsky Saar
Ben Gurion University, Be'er Sheva, Israel

Kaimanovich Vadim
Université de Rennes 1, Rennes, France

Katz Mikhail
Bar Ilan University, Ramat Gan, Israel

Kotani Motoko
Tohoku University, Sendai, Japan

Lubotzky Alex
Hebrew University of Jerusalem, Jerusalem, Israel

Makover Eran
Connecticut College, New London, CT, U.S.A.

Mangoubi Dan
Technion – Israel Institute of Technology, Haifa, Israel

Minsky Yair
Yale University, New Haven, CT, U.S.A.

Monastyrsky Michael
Institute of Theoretical and Experimental Physics, Moscow, Russia

Nabutovsky Alexander
Pennsylvania State University, University Park, PA, U.S.A.

Narayanan E. K.
Bar Ilan University, Ramat Gan, Israel

Nevo Amos
Technion – Israel Institute of Technology, Haifa, Israel

Parfyonova Nataliya
Kharkiv State University, Kharkiv, Ukraine

Pinchover Yehuda
Technion – Israel Institute of Technology, Haifa, Israel

Pinsky Ross
Technion – Israel Institute of Technology, Haifa, Israel

Polterovich Iosif
Université de Montréal, Montréal, QC, Canada

Polterovich Leonid
Tel Aviv University, Tel Aviv, Israel

Potyagailo Leonid
Université de Lille 1, Lille, France

Rotman Regina
Pennsylvania State University, University Park, PA, U.S.A.

Sageev Michah
Technion – Israel Institute of Technology, Haifa, Israel

Sapir Mark
Vanderbilt University, Nashville, TN, U.S.A.

Shalen Peter
University of Illinois at Chicago, Chicago, IL, U.S.A.

Shubin Mikhail
Northeastern University, Boston, MA, U.S.A.

Sikorav Jean-Claude
Ecole Normale Supérieure de Lyon, Lyon, France

Simon Barry
California Institute of Technology, Pasadena, CA, U.S.A.

Smilansky Uzy
Weizmann Institute of Science, Rehovot, Israel

Sodin Mikhail
Tel Aviv University, Tel Aviv, Israel

Solomyak Michael
Weizmann Institute of Science, Rehovot, Israel

Sunada Toshikazu
Meiji University, Kawasaki, Japan

Vishne Uzi
Bar Ilan University, Ramat Gan, Israel

Weinberger Shmuel
University of Chicago, Chicago, IL, U.S.A.

Weinstein Gilbert
University of Alabama at Birmingham, Birmingham, AL, U.S.A.

Weiss Barak
Ben Gurion University, Be'er Sheva, Israel

Woess Wolfgang
Technische Universität Graz, Graz, Austria

Zalcman Lawrence
Bar Ilan University, Ramat Gan, Israel

Zelenko Igor
SISSA-ISAS, International School for Advanced Studies, Trieste, Italy

Zuk Andrzej
University of Chicago, Chicago, IL, U.S.A.

International Conference on Geometry and Spectral Theory

29 December, 2003 - 2 January, 2004
Technion, Haifa, Israel

PROGRAM

Monday, 29 December

Morning Session

08:30-09:00 Registration

09:00-09:30 Greetings and Opening Remarks

09:30-10:20 Peter Buser (Lausanne, Switzerland)
Special lecture on the mathematical works of Robert Brooks

11:00-11:50 Ross Pinsky (Haifa, Israel)
The behavior of the principal eigenvalue in a domain with a hole - Asymptotic behavior and domain monotonicity

12:00-12:50 Shmuel Weinberger (Chicago, Illinois, USA)
Manifolds with excessive symmetry

Afternoon Session

15:00-15:50 Mikhail Sodin (Tel Aviv, Israel)
Nodal domains of Laplace-Beltrami eigenfunctions on surfaces

16:00-16:50 Eran Makover (New London, Connecticut, USA)
Jacobians of isospectral Riemann surfaces

17:00 A reception

Tuesday, 30 December

Morning session

09:00-09:50 Isaac Chavel (New York, New York, USA)
Partial survey of isoperimetric inequalities

10:00-10:50 Iosif Polterovich (Montreal, Canada)
How large can the first eigenvalue be on a Klein bottle?

11:30-12:20 Jozef Dodziuk (New York, New York, USA)
Arithmetic properties of eigenvalues of generalized Harper operators on graphs

Afternoon Session

14:30-15:20 Hershel Farkas (Jerusalem, Israel)
A variant of the 3N+1 conjecture

15:30-16:20 Motoko Kotani (Tohoku, Japan)
Magnetic transition operators on a crystal lattice

17:00-17:50 Mikhail Shubin (Boston, Massachusetts, USA)
Two-sided estimates for the bottoms of spectra and essential spectra of Schrodinger operators

Wednesday, 31 December

Morning session

09:00-09:50 Alex Lubotzky (Jerusalem, Israel)
Property "tau" and its applications in combinatorics, computer science and geometry

10:00-10:50 Alexander Gamburd (Stanford, California, USA)
Discrete and continuous variations on the expanding theme

11:30-12:20 Barry Simon (Pasadena, California, USA)
The sharp form of the strong Szegö theorem

13:00 Excursion and banquet.

Thursday, 1 January

Morning session

09:00-09:50 Toshikazu Sunada (Kawasaki, Japan)
Spectral geometry of crystal lattices

10:00-10:50 Gilbert Weinstein (Birmingham, Alabama, USA)
On a Penrose inequality with charge

11:30-12:20 Uzi Vishne (Ramat Gan, Israel)
Ramanujan complexes, and isospectral manifolds, complexes and regular graphs

Afternoon Session

14:30-15:20 Andrzej Zuk (Chicago, Illinois, USA)
The Cheeger constant of modular surfaces

15:30-16:20 Andrey Reznikov (Ramat Gan, Israel)
L_p*-norms on automorphic representations*

17:00-17:50 Leonid Friedlander (Tucson, Arizona, USA)
On the density of states function in quasi-gaps

Friday, 2 January

Morning session

09:00-09:50 Vadim Kaimanovich (Rennes, France)
The spectrum of the foliated Laplacian and amenability

10:00-10:50 Michael Monastyrsky (Moscow, Russia)
Hecke surfaces and monodromial representation of braid groups

11:30-12:20 Peter Buser (Lausanne, Switzerland)
Isospectrality and spectral rigidity of surfaces with small topology

International Workshop on Groups, Geometry and Dynamics

5-9 January, 2004
Technion, Haifa, Israel

Program

Monday, 5 January

Morning session

08:30-09:00 Registration

09:00-09:10 Opening remarks

09:10-10:00 David Gabai (Princeton, New Jersey, USA)
D^2 limit laminations in 3-manifolds

10:10-11:00 Joan Birman (New York, New York, USA)
Distinct transversal knots with the same topological knot type and Bennequin number

11:40-12:30 Sa'ar Hersonsky (Beer-Sheva, Israel)
The Dirichlet-Neumann boundary value problem and tiling by squares

Afternoon Session

14:30-15:20 Mark Sapir (Nashville, Tennessee, USA)
Dynamics of polynomial maps over finite fields and residually finite groups

15:30-16:20 Viktor Ginzburg (Santa Cruz, California, USA)
Almost existence of periodic orbits of Hamiltonian systems and branching Floer homology

16:30-17:20 Alexander Nabutovsky (Toronto, Canada and State College, Pennsylvania, USA)
The length of the shortest closed geodesic and related problems

17:30 A reception

Tuesday, 6 January

Morning session

09:00-09:50 Peter Shalen (Chicago, Illinois, USA)
Hyperbolic volume and the lattices of subgroup of a free group

10:00-10:50 Koji Fujiwara (Sendai, Japan)
Cat(0)-dimension of groups

11:30-12:20 Dan Mangoubi (Haifa, Israel)
Riemann surfaces and 3-regular graphs

Afternoon Session

14:30-15:20 Leonid Potyagailo (Lille, France)
On right-angled reflection groups in hyperbolic spaces

15:30-16:20 Uri Bader (Haifa, Israel)
Property (T) and unitary representation on L_p

17:00-17:50 Tsachik Gelander (New Haven, Connecticut, USA)
A topological Tits alternative

Wednesday, 7 January

Morning session

09:00-09:50 Leonid Polterovich (Tel Aviv, Israel)
Quasimorphisms in symplectic topology

10:00-10:50 Jean-Claude Sikorav (Lyon, France)
Cohomology with bounds on manifolds and groups

11:00 An excursion and banquet

Thursday, 8 January

Morning session

09:00-09:50 Andrew Casson (New Haven, Connecticut, USA)
A remark on the stable Andrews-Curtis conjecture

10:00-10:50 Paul Biran (Tel Aviv, Israel)
Linear systems, algebraic families and Lagrangian submanifolds

11:30-12:20 Mikhail Katz (Ramat Gan, Israel)
Recent progress on Gromov's filling area conjecture

Afternoon Session

14:30-15:20 Alexander Givental (Berkeley, California, USA)
Quantum Hirzebruch-Riemann-Roch

15:30-16:20 Barak Weiss (Beer-Sheva, Israel)
Bounded trajectories on parameter spaces

Friday, 9 January

Morning session

09:00-09:50 Amos Nevo (Haifa, Israel)
Spectrum theory on the boundary, and invariants of discrete groups of isometries

10:00-10:50 Wolfgang Woess (Graz, Austria)
Transition operators on co-compact G-spaces

11:30-12:20 Yair Minsky (New Haven, Connecticut, USA)
Deformation spaces of hyperbolic 3-manifolds

Contemporary Mathematics
Volume **387**, 2005

On the Mathematical Work of Robert Brooks

Peter Buser

Dedicated to the memory of Robert Brooks

ABSTRACT. The aim of this article is to give an insight into the ideas and methods in Brooks' rich mathematical work. Each section covers an individual theme and is written in a way that should allow it to be read independently. In Section 5 we give a new proof of the monotonicity of Brooks' circle packing parameter. The complete list of published research articles by Robert Brooks is given in the first part of the bibliography.

1. Introduction

In the introduction to his article "Some geometric aspects of the work of Lars Ahlfors" [**68**], Bob Brooks wrote:

> "If I were to try to quantify Ahlfors' attitude towards geometry, I would put it this way: the proper way to do a calculation in complex analysis is to be guided primarily by geometric intuition. Indeed, a calculation is not properly understood, and probably not carried out properly, if it is not primarily guided by geometric intuition."

Certainly, this describes also Bob's own approach to mathematics. Geometric intuition is an ever-present element in his work, and wanting to understand and get insight a permanent motivation. I hope this will transpire in the following pages, where I have selected a number of subjects from Bob's impressive work, trying to give an overview of his ideas and methods.

At various places I found it worthwhile to go into considerable detail, and so this article will highlight a few subjects rather than give a systematic account. So several areas of Bob's research, as, e.g., his contributions to topology, L^p-spectral geometry and isoscattering are completely left out, and I apologize for this.

2000 *Mathematics Subject Classification.* Primary 05C25, 11F72, 30F40, 52C15, 58J50, 58J53; Secondary 05C80, 11R42, 20H15, 30F45.

Key words and phrases. Cheeger's inequality, circle packings, eigenvalues of graphs and manifolds, isospectrality, Jørgensen's inequality, Kleinian groups, random graphs and random Riemann surfaces.

Supported by the Swiss National Science Foundation, FNS-grant 20-101639.

The subjects then treated begin in Section 2, where we detail Brooks' proof of Jørgensen's inequality. In Sections 3 and 8 we give an account of his work on "λ_1", the first eigenvalue of the Laplacian; a central theme with contributions to Cheeger's isoperimetric constant, Kazhdan's Property T, Selberg's 3/16-Theorem, and random Riemann surfaces. The latter make crucial use of the Ahlfors-Schwarz Lemma, which Bob liked particularly, and we present his own version of it in Section 7.

The second main subject in Brooks' work is inverse spectral geometry, where he contributed to Sunada's method, constructed isospectral Riemann surfaces of small genus, compared isospectrality of graphs with isospectrality of manifolds, and also worked on isoscattering. Here, too, it was necessary to make a selection, and so we will focus in Section 6 on a single paper, "Mutually isospectral Riemann surfaces", which gives us an occasion to talk about the finite groups aspect of Brooks' work.

Sections 4 and 5, finally, are about circle packings and Kleinian groups. Here we see how the spectral geometry of the Apollonian packing gradually leads to the theorem about the density of packable Riemann surfaces, a cornerstone in the literature for the numerical uniformization of Riemann surfaces via circle packings. A tool for this is the circle packing parameter, a veritable "coup de génie", which we present separately in Section 5. In that section we will momentarily abandon the survey character of this article and give a new proof of the monotonicity.

The bibliography is in two parts. The first part comprises the published research articles of Bob Brooks. The second part contains the remaining references; these are numbered beginning with 101.

2. A Geometric Proof of Jørgensen's Inequality and the Brooks-Matelski Set

The first work of Bob Brooks to be described here is a rather early proceedings article, [**5**], together with Peter Matelski, dealing with a generalization of Jørgensen's inequality. Even though the paper is short and somewhat incomplete, it is remarkable for two reasons. First of all it provides – it seems – the only proof in the literature that approaches Jørgensen's inequality from a hyperbolic geometric point of view. And second, in this paper Brooks and Matelski discover a remarkable subset of the complex plane (Figure 2) which later became known as the Mandelbrot set, but which we will rather call the Brooks-Matelski set.

We shall first outline the hyperbolic geometric proof of Jørgensen's inequality given in [**5**], and then examine what it has to do with the strange set shown in Figure 2.

Jørgensen's inequality states that if X and Y are matrices from $\mathrm{PSL}(2,\mathbb{R})$, and if the group $\langle X, Y\rangle$ generated by X and Y is discrete and non-elementary, then

$$|\mathrm{tr}^2(X) - 4| + |\mathrm{tr}(XYX^{-1}Y^{-1}) - 2| \geq 1.$$

For what follows it is more suitable to rephrase Jørgensen's result in this form: if $|\mathrm{tr}^2(X) - 4| + |\mathrm{tr}(XYX^{-1}Y^{-1}) - 2| < 1$, and if X and Y do not commute, then $\langle X, Y\rangle$ is not discrete.

In order to avoid dealing with special cases we will make the *assumption* that X is neither elliptic nor parabolic and give a proof only for this case. As usual, we may interpret $\mathrm{PSL}(2,\mathbb{C})$ as the group of orientation preserving isometries of the upper half space $\mathbb{H}^3$, where the latter is endowed with the hyperbolic metric of constant curvature -1. Then X has a uniquely defined invariant geodesic, the *axis* of X, and the points on this axis are not kept fixed.

The main tool from hyperbolic geometry is a trigonometric formula for right-angled geodesic hexagons in $\mathbb{H}^3$:

Consider oriented geodesics ℓ_1, ℓ_2 in $\mathbb{H}^3$ and assume that the two have no common endpoint at infinity. Then there exists a unique geodesic γ perpendicular to both ℓ_1 and ℓ_2, and there is an isometry $X \in \mathrm{PSL}(2,\mathbb{C})$ with axis γ sending ℓ_1 to ℓ_2 in a way which respects the given orientations. The *complex distance* from ℓ_1 to ℓ_2, and at the same time the *complex translation length* of X, is defined as the complex number $\tau = \tau(X) = \delta(\ell_1, \ell_2)$, whose real part is the hyperbolic distance from ℓ_1 to ℓ_2, and whose imaginary part is the angle made by ℓ_1 and ℓ_2 along γ. This number can be expressed in terms of the trace of X:

$$\mathrm{tr}^2(X) = 4\cosh^2(\tau/2) = 2\cosh(\tau) + 2. \tag{2.1}$$

To complete this definition we let the distance of geodesics with a common endpoint at infinity be zero.

Let now ℓ_0, ℓ_1, ℓ_2 be three such geodesics and denote by ω the complex distance from ℓ_1 to ℓ_2; by τ_1, τ_2 the complex distances from ℓ_0 to ℓ_1, ℓ_2; and by α the complex distance from the perpendicular between ℓ_0 and ℓ_1 to the perpendicular between ℓ_0 and ℓ_2, the perpendiculars being oriented from ℓ_0 to ℓ_1, ℓ_2. For this configuration the trigonometric formula in question is the following (see e.g., Fenchel [**115**, formula (6), p. 83], with different sign conventions):

$$\cosh(\omega) = \cosh(\tau_1)\cosh(\tau_2) - \cosh(\alpha)\sinh(\tau_1)\sinh(\tau_2). \tag{2.2}$$

(In order to transpose this formula to $\mathbb{H}^2$, one has to take into account that the distances have imaginary parts either 0 or π.)

Following the usual approach to Jørgensen's inequality one defines inductively

$$Y_1 = YXY^{-1}, \quad Y_{i+1} = Y_iXY_i^{-1}, \quad i = 1, 2, \ldots.$$

Let $\tau = \tau(X)$, and denote by β_i the complex distance from the axis of X to the axis of Y_i. It is easy to check that

$$\mathrm{tr}((Y_iXY_i^{-1})X^{-1}) - 2 = -(1-\cosh(\tau))(1-\cosh(\beta_{i+1})), \tag{2.3}$$

for $i = 0, 1, 2, \ldots$, where we set $Y_0 = Y$.

The strategy is to show that if Jørgensen's inequality does not hold for X and Y, then β_i converges to zero, and to deduce from it that the group generated by X and Y is non-discrete or elementary.

Brooks and Matelski carry this out in the following way using formula (2.2). They take the configuration $\ell_1 = \mathrm{axis}(X)$, $\ell_0 = \mathrm{axis}(Y_i)$, $\ell_2 = \mathrm{axis}(Y_{i+1})$. It is then checked that ℓ_2 is the image of ℓ_1 under Y_i, where Y_i (being conjugate to X) has complex translation length τ. With the terminology of formula (2.2) this yields $\tau_1 = \tau_2 = \beta_i$, $\alpha = \tau$, $\omega = \beta_{i+1}$, and so they get

$$\cosh(\beta_{i+1}) = (1-\cosh(\tau))\cosh^2(\beta_i) + \cosh(\tau). \tag{2.4}$$

Setting

$$z_i = (1-\cosh(\tau))\cosh(\beta_i)$$

they write this in the striking form

$$z_{i+1} = z_i^2 + C, \quad \text{where} \quad C = (1-\cosh(\tau))\cosh(\tau). \tag{2.5}$$

In this way they reduce the study of the iteration $Y \mapsto YXY^{-1}$ to the study of the iteration $z \mapsto z^2 + C$ in the complex plane, which is well understood.

Let us first finish the proof of Jørgensen's inequality via equation (2.5), and then explore how the Brooks-Matelski set gradually comes into the picture via the same equation.

Abbreviate $p = 1 - \cosh(\tau)$, so that $C = p(1-p)$, $z_1 = \operatorname{tr}(XYX^{-1}Y^{-1}) - 2 + p$, $\operatorname{tr}^2(X) - 4 = -2p$. The hypothesis of Jørgensen's inequality then is

$$2|p| + |z_1 - p| = |\operatorname{tr}^2(X) - 4| + |\operatorname{tr}(XYX^{-1}Y^{-1}) - 2| < 1.$$

Now, if z is any complex number such that $2|p| + |z - p| < 1$, then $z' := z^2 + C$ satisfies

$$z' - p = z^2 + p(1-p) - p = (z-p)(z+p)$$
$$|z + p| = |z - p + 2p| \le |z - p| + 2|p| < 1.$$

From this we see that, either the z_i form an infinite sequence converging to p, or there exists k such that $z_k = p$.

In the first case the complex distances β_i converge to zero. (This is the geometric property of the Y_i revealed by this proof.) Using the fact that on its own axis X is a non-trivial translation, we find suitable powers m_i of X such that the shortest arc σ_i from the axis of X to the axis of $Y_i' = X^{-m_i} Y_i X^{m_i}$ remains in a bounded neighbourhood as i tends to infinity. Hence, a non-discrete set of Y_i''s in the group $\langle X, Y \rangle$.

In the second case the axes of X and Y_k have distance equal to 0, which means that they have a common endpoint at infinity. If the two axes are distinct, then either for $m \to \infty$ or for $m \to -\infty$ the family $X^m Y_k X^{-m}$ is non-discrete. If the axes coincide, then $Y_{k+1} = X$, and a well-known theorem (e.g., Beardon [**104**, Theorem 5.1.4]) says that in this case $\langle X, Y \rangle$ is elementary.

This finishes the proof of Jørgensen's inequality (under our initial assumption that X is neither elliptic nor parabolic).

The critical question for the next step is a typical feature in Brooks' work: *What are the precise conditions under which this proof works ?*

The central argument in the proof was the convergence of z to p under the iteration of the mapping

$$z \mapsto f(z) = z^2 + C,$$

where

$$p = -\frac{1}{2}(\operatorname{tr}^2(X) - 4), \quad C = p(1-p).$$

Here p is a fixed point of f, and it is well known that if $|\frac{d}{dz} f(p)| > 1$, such convergence can hold only in the trivial case $z = p$, whereas for p with $|\frac{d}{dz} f(p)| < 1$ (an attractive fixed point) there exists an entire open neighbourhood E_p of z's who converge to p. Hence, the "accessible" X, that is to say, those for which the above argument is capable to produce a Jørgensen type result, are the ones with $2|p| < 1$; in other words, the ones with $|\operatorname{tr}^2(X) - 4| < 1$.

The rest of the proof then was to show that if X is of this kind and if for

$$z_1 = \operatorname{tr}(XYX^{-1}Y^{-1}) - 2 + p$$

one has $|z_1 - p| < 1 - |2p|$, then z_1 belongs indeed to E_p.

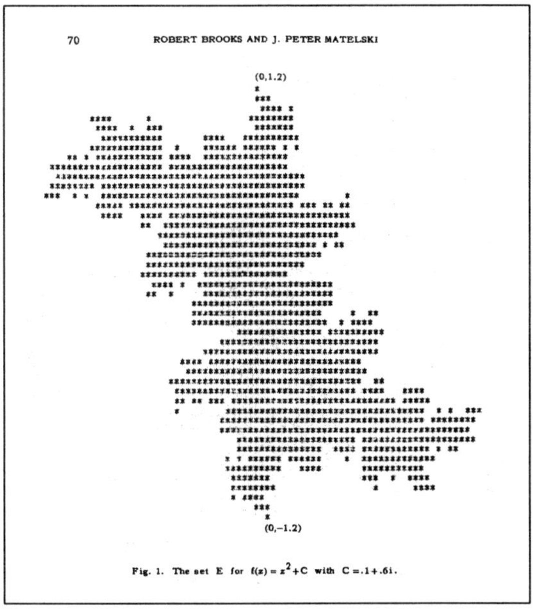

FIGURE 1. A filled Julia set

The set E_p is known as the (filled) *Julia set* corresponding to p (respectively, C). It is, in general, much larger than the set where this estimate holds. Figure 1 is the reproduction of an illustration in [**5**] showing E_p for $C = 0.1 + 0.6i$. For comparison: the set of all z_1 satisfying $|z_1 - p| < 1 - 2|p|$ is a disk of radius $0.0028\ldots$.

One has therefore a more general Jørgensen theorem saying that if X and Y are such that the corresponding values C, p and z_1 satisfy $C = p(1-p)$, $|p| < \frac{1}{2}$ and $z_1 \in E_p$, then $\langle X, Y\rangle$ is not discrete.

At this point, Brooks and Matelski go a crucial step further observing that the decisive property of the sequence z_i is not its convergence to p but rather its *boundedness*. And the latter is well understood: Let f^n denote the n-th iteration $f^n(z) = f \circ f \circ \cdots \circ f(z)$ (n times). A solution p to the polynomial equation $f^n(p) = p$ is called a *stable periodic point* (of f of period n) if $|\frac{d}{dz} f^n(p)| < 1$. It is then known that $f(p), f^2(p), \ldots, f^{n-1}(p)$ are also stable, and there is again a filled Julia set E_p, this time with the property that for $z_1 \in E_p$ the sequence z_i converges to the orbit $p, f(p), f^2(p), \ldots, f^{n-1}(p)$, whereas for z_1 in the exterior of E_p the sequence is unbounded.

A sufficient condition is then described in [**5**], under which z_1 belongs to E_p, hence, a further generalization of Jørgensen's theorem.

Of course, the authors must have been curious to know which are the "accessible" X in this new setting, and so they were led to consider the set $\mathcal{BM}$ of all C's for which $f(z) = z^2$ has a stable periodic orbit.

For C of the form $C = p(1-p)$ with $|p| < \frac{1}{2}$ (the case of the classical Jørgensen inequality) one has the stable fixed points, so these values certainly lie in this set. But $\mathcal{BM}$ is bigger and more complicated. Figure 2 reproduces a computer picture of $\mathcal{BM}$ from [**5**] (probably from around 1979). Its quality is in no comparison with the spectacular pictures that appeared in the literature not long afterwards, but

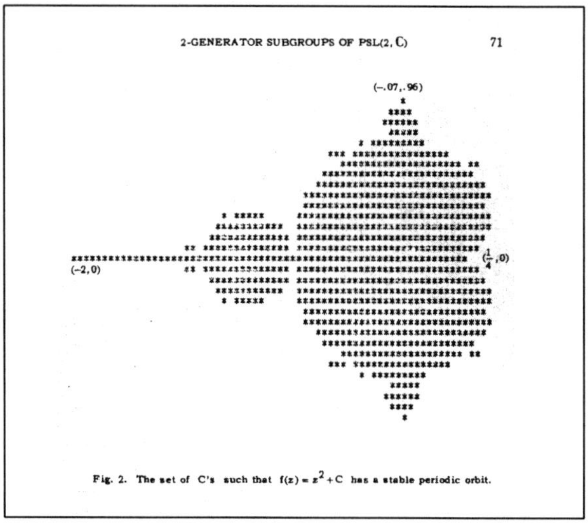

FIGURE 2. Early print of the Brooks-Matelski set

one may easily imagine their excitement when Brooks and Matelski saw it for the first time.

3. Towers of Coverings and Cheeger's Inequality

Let us now come to a central theme in Brooks' work, the behaviour of the bottom of the Laplace spectrum under large coverings and the attempt to understand Selberg's celebrated $\lambda_1 \geq 3/16$ theorem from a geometric and also a combinatorial point of view.

The investigation begins in Commentarii [**7**], 1981 and is motivated by a theorem of Milnor-Švarč comparing the growth rate of the fundamental group $\pi_1(M)$ of a compact Riemannian manifold M with the volume growth of distance balls in the universal covering $\tilde{M}$.

To this, Brooks adds the following beautiful theorem, where $\lambda_0(\tilde{M})$ is the greatest lower bound of the spectrum of the Laplace-Beltrami operator on the universal covering $\tilde{M}$ of M.

THEOREM 3.1 ([**7**]). *$\lambda_0(\tilde{M}) = 0$ if and only if $\pi_1(M)$ is an amenable group.*

By Følner's theorem, a group G is *amenable* if and only if for any $\mu \in (0,1)$ and any finite sequence $g_1, \dots, g_n \in G$ there exists a finite subset E of G such that $\#(E \cap g_i \cdot E) \geq \mu \#(E)$, for $i = 1, \dots, n$, $\#$ denoting the cardinality of a set. Brooks translates this first into an isoperimetric characterization of $\tilde{M}$:

Call a fundamental domain F for M in $\tilde{M}$ *admissible* if it is the union of inverse images (under the covering $\tilde{M} \to M$) of simplices of a smooth triangulation of M. The characterization is then as follows, where $\pi_1(M)$ is interpreted as the group of deck transformations.

PROPOSITION 3.2 ([**7**]). *$\pi_1(M)$ is amenable if and only if for every admissible fundamental domain F and for every $\varepsilon > 0$, there is a finite subset E of $\pi_1(M)$*

such that the union $A = \bigcup_{h \in E} h(F)$ *satisfies*

$$\frac{\mathrm{area}(\partial A)}{\mathrm{vol}(A)} < \varepsilon.$$

Here, $\mathrm{vol}(A)$ is the n-volume of A, where n is the dimension of M, and $\mathrm{area}(\partial A)$ is the $(n-1)$-volume of its boundary. The proof of the proposition is not too hard, but the step from this proposition to Theorem 3.1 is rather difficult and needs, among other things, results about minimizing currents.

We will not go into the details of the proof but rather advance to the next mile stone, the article in the Journal of Differential Geometry [**23**], 1986, where the ideas have been pushed forward and are easier to describe here. In this article finitely sheeted coverings

$$M_i \to M$$

of a given Riemannian manifold M are considered, where the number of sheets goes to infinity as $i \to \infty$. The coverings are not necessarily normal, but the M_i carry the lifted Riemannian metric. The main question is: *Under what conditions does the first non-zero eigenvalue* $\lambda_1(M_i)$ *of the Laplace operator of* M_i *remain bounded away from zero as* $i \to \infty$?

Brooks first reduces this problem to a combinatorial one in the following way. Consider the universal covering $\tilde{M} \to M$ and interpret the fundamental group $\pi_1(M)$ again as the group of deck transformations. The fundamental groups $\pi_1(M_i)$ are then finite index subgroups of $\pi_1(M)$ in a natural way. Pick a compact fundamental domain F for $\tilde{M} \to M$, and let $g_1, \ldots, g_k$ be the deck transformations sending F to its adjacent neighbours. Then M is equal to F modulo the side pairings given by $g_1, \ldots, g_k$, and so these elements generate $\pi_1(M)$. For each subgroup $H = \pi_1(M_i)$ we have the *Schreier graph* $\mathcal{G} = \mathcal{G}_i$ defined as follows:

The vertices of $\mathcal{G}$ are the cosets aH, $a \in \pi_1(M)$, and two vertices aH, bH are joined by an edge if and only if $bH = (g_j^{\pm 1} a)H$ for some generator g_j.

For a finite graph such as $\mathcal{G}$ the *Cheeger constant* $h(\mathcal{G})$ is defined as

$$h(\mathcal{G}) = \inf_E \frac{\#(E)}{\min\{\#(A), \#(B)\}}, \tag{3.1}$$

where E ranges over all sets of edges for which $\mathcal{G} - E$ (the graph with the edges of E removed) falls into two disconnected non-empty pieces A and B. For example, if $\mathcal{G}$ is the 1-skeleton of the ordinary cube, i.e., the 3-regular graph with 8 vertices, then $h(\mathcal{G}) = 1$.

The reduction of the eigenvalue problem to a combinatorial one is given by the following theorem stated for the present conditions.

THEOREM 3.3 ([**23**]). $\lambda_1(M_i) \to 0$ *as* $i \to \infty$ *if and only if* $h(\mathcal{G}_i) \to 0$.

A main ingredient in the proof is Cheeger's inequality [**111**],

$$\lambda_1(N) \geq \frac{1}{4} h^2(N), \tag{3.2}$$

for a compact or, more generally, a complete Riemannian manifold N with finite volume, where $h(N)$ is *Cheeger's constant* for manifolds:

$$h(N) = \inf_S \frac{\mathrm{area}(S)}{\min\{\mathrm{vol}(A), \mathrm{vol}(B)\}}. \tag{3.3}$$

In this definition, S runs through all codimension 1 submanifolds for which the complement $N \setminus S$ consists of two disjoint open subsets A, B with common boundary S. (In Section 4 we shall consider yet another version of this constant.)

The similarity between (3.1) and (3.3) is obvious. Brooks' achievement in [**7**] and [**23**] is to have shown that the two are equivalent:

LEMMA 3.4 ([**23**]). *For the above coverings $M_i \to M$ there exists a constant $c > 1$ (depending on M and the choice of the generators $g_1, \ldots, g_k$ of $\pi_1(M)$) such that for each i,*

$$\frac{1}{c} h(M_i) \leq h(\mathcal{G}_i) \leq c \cdot h(M_i).$$

The difficult part is the inequality on the right hand side. In [**23**] this inequality is not stated as a lemma, but it appears as a text statement at the end of Section 1.

The following application gives an idea of Brooks' deep insight into the subject.

THEOREM 3.5 ([**23**]). *Suppose $\pi_1(M)$ has Kazhdan's Property* T. *Then there exists a constant $C > 0$ with the following properties:*

(1) *If M' is a finite covering of M then $\lambda_1(M') > C$.*
(2) *If M'' is an infinite covering of M then $\lambda_0(M'') > C$.*

It is instructive to see how the argument goes. For simplicity we restrict ourselves to part (1).

Let us recall Property T for the present context: A discrete group G with generators $g_1, \ldots, g_k$ has *Property* T if there exists a constant $\varepsilon > 0$ such that for any unitary representation (denoted by $\cdot$) of G in a Hilbert space $\mathcal{H}$ the following holds: If there is a vector $X \in \mathcal{H}$ of norm $\|X\| = 1$ and such that $\|g \cdot X - X\| < \varepsilon$ for $g = g_1, \ldots, g_k$, then there exists $Y \in \mathcal{H}$ with $\|Y\| = 1$ and $g \cdot Y = Y$ for all $g \in G$. For example (see [**121**, Example 3.2.4]) $\mathrm{SL}(n, \mathbb{Z})$ has this property for all $n \geq 3$, but $\mathrm{SL}(2, \mathbb{Z})$ does not.

In view of Theorem 3.3, the strategy to prove (1) is to show that if there exists a tower of coverings with $h(\mathcal{G}_i) \to 0$, then $\pi_1(M)$ cannot have Property T. So let such a tower be given. Consider, for each i, the Hilbert space of functions on the vertex set, $u : \mathcal{G}_i \to \mathbb{R}$, with the inner product

$$\langle u, v \rangle = \sum_{x \in \mathcal{G}_i} u(x) v(x),$$

and let $\mathcal{H}_i$ be the subspace of all u satisfying $\langle u, 1 \rangle = 0$, i.e., the orthogonal complement of the constant functions.

$\pi_1(M)$ acts transitively on the vertex set of $\mathcal{G}_i$ (but not necessarily as a graph automorphism), each $g \in \pi_1(M)$ sending any coset $x = a\pi_1(M_i)$ to $g(x) := (ga)\pi_1(M_i)$, and we have the canonical unitary representation of $\pi_1(M)$ in $\mathcal{H}_i$ defined by $(g \cdot u)(x) := u(g(x))$, $x \in \mathcal{G}_i$, $u \in \mathcal{H}_i$.

Now take a subset E_i of edges of $\mathcal{G}_i$ separating $\mathcal{G}_i$ into two disconnected pieces A_i, B_i such that

$$\frac{\#(E_i)}{\min\{\#(A_i), \#(B_i)\}} = h(\mathcal{G}_i),$$

and consider the function

$$u_i(x) = \begin{cases} \alpha, & x \in A_i \\ \beta, & x \in B_i \end{cases}$$

with constants α, β such that $u_i \in \mathcal{H}_i$ and $\|u_i\| = 1$. Then the key argument is this: For each generator, in fact, for each $g \in \pi_1(M)$ we have

$(g \cdot u_i)(x) - u_i(x) \neq 0$ *only if* $g(x)$ *and* x *are connected by an edge of* E.

From this we easily compute

$$\|g \cdot u_i - u_i\|^2 \leq 2(\alpha - \beta)^2 \#(E_i) = 2\frac{\#(A_i) + \#(B_i)}{\#(A_i)\#(B_i)} \#(E_i) \leq 4h(\mathcal{G}_i).$$

Now assume that $\pi_1(M)$ has Property T with some constant ε. Taking i sufficiently large we then have $\|g_j \cdot u_i - u_i\| < \varepsilon$ for the generators $g_1, \ldots, g_k$ of $\pi_1(M)$, and so there exists a function u in $\mathcal{H}_i$ with $\|u\| = 1$ such that $g \cdot u - u = 0$ for all $g \in \pi_1(M)$. But this is impossible, because $\pi_1(M)$ acts transitively on the vertices of $\mathcal{G}_i$ so that for any $x \in A_i$ there exists $g \in \pi_1(M)$ with $g(x) \in B_i$.

Among the tower of coverings with λ_1 bounded away from zero, probably the most famous ones are those from Selberg's theorem:

THEOREM 3.6 (Selberg, [**128**]). *Let* $\Gamma = \mathrm{PSL}(2, \mathbb{Z})$*, and consider the congruence subgroups*

$$\Gamma_n = \left\{ \begin{pmatrix} a & b \\ c & d \end{pmatrix} \in \Gamma \;\middle|\; \begin{pmatrix} a & b \\ c & d \end{pmatrix} \equiv \pm \begin{pmatrix} 1 & 0 \\ 0 & 1 \end{pmatrix} \pmod{n} \right\}.$$

Then, for all n*,*

$$\lambda_1(\mathbb{H}^2/\Gamma_n) \geq \frac{3}{16}.$$

Selberg also conjectured that the constant $3/16$ may be improved to $1/4$ (which then would be the best possible constant).

In [**24**], [**41**] and also later on, Brooks tried to understand, from a geometric point of view, what exactly prevents $\lambda_1(\mathbb{H}^2/\Gamma_n)$ from becoming small as $n \to \infty$. Certainly, he also hoped to get hold of Selberg's conjecture itself. Theorem 3.5 (which can also be formulated for manifolds with finite volume) is of no help here, because $\mathrm{PSL}(2, \mathbb{Z})$ does not have Kazhdan Property T. (See, however, Lubotzky-Zimmer [**120**, p. 291].) Now $1/4$ is the bottom of the spectrum of the hyperbolic plane $\mathbb{H}^2$ for which Cheeger's inequality, modified for infinite volume manifolds, states that $\lambda_0(\mathbb{H}^2) \geq 1/4$ (this sharpness of Cheeger's inequality was first pointed out by Osserman [**125**]). It was therefore legitimate to hope to prove Selberg's conjecture via Cheeger's inequality.

In [**45**] Brooks showed that this cannot be done:

THEOREM 3.7 ([**45**]). *Let* p *be a prime number,* $p \equiv 1 \pmod{4}$*. Then*

$$h(\mathbb{H}^2/\Gamma_p) < \frac{3\log(3)}{2\pi} \cdot \frac{p-1}{p+1}.$$

This gives us an upper bound $h^2/4 < \frac{1}{16} \cdot 1.11$, and so Cheeger's inequality is not even strong enough to prove Theorem 3.6.

In the remainder of this paragraph we will sketch the proof of Theorem 3.7 given in [**45**], since it provides a nice picture of the geometry of the surfaces $\mathbb{H}^2/\Gamma_p$.

$$F = \{z \in \mathbb{H}^2 \mid -\tfrac{1}{2} \leq \Re(z) \leq \tfrac{1}{2},\ |z| \geq 1\}$$

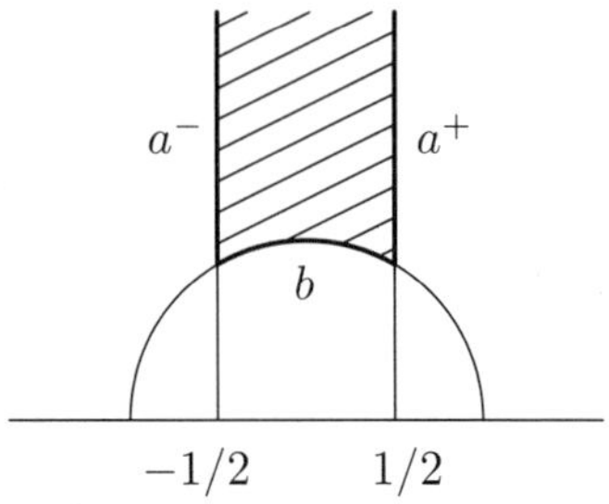

FIGURE 3. The standard fundamental domain of the modular group

We let F be the standard fundamental domain for the modular group $\Gamma = \mathrm{PSL}(2,\mathbb{Z})$ and use the following generators of Γ,

$$U = \begin{pmatrix} 1 & 1 \\ 0 & 1 \end{pmatrix}, \quad V = \begin{pmatrix} 0 & -1 \\ 1 & 0 \end{pmatrix}.$$

$U^{\pm 1}$ maps F to a copy of F adjacent along side $a^{\pm} = \{z \in F \mid \Re(z) = \pm\frac{1}{2}\}$, and V maps F to a copy adjacent along side $b = \{z \in F \mid |z| = 1\}$. The combinatorial structure of the tiling of $\mathbb{H}^2$ given by the images of F under the action of Γ is a 3-regular graph, where each tile is a vertex, and two vertices are joined by an edge whenever as tiles they are adjacent along a side.

For given (so far arbitrary) n, Γ_n is a normal subgroup of Γ, and there is a natural isomorphism $\Gamma/\Gamma_n \simeq \mathrm{PSL}(2,n) = \mathrm{PSL}(2,\mathbb{Z}/n\mathbb{Z})$. We may then interpret U and V (by abuse of notation) as generators of $\mathrm{PSL}(2,n)$, and obtain the Schreier graph of the cosets Γ_n in Γ as the *Cayley graph* $\mathcal{G}_n$ of $\mathrm{PSL}(2,n)$ with respect to the generators U, V: the vertices of $\mathcal{G}_n$ are the elements of $\mathrm{PSL}(2,n)$, and two vertices X, Y are joined by an edge whenever $Y = SX$ with $S \in \{U, U^{-1}, V\}$.

Since $\mathcal{G}_n$ is the Schreier graph of the cosets Γ_n in Γ we have the following *geometric picture* of $\mathcal{G}_n$ on the surface $\mathbb{H}^2/\Gamma_n$: projecting the tiling of $\mathbb{H}^2$ onto this surface we get a tiling of $\mathbb{H}^2/\Gamma_n$ with copies of F, where each tile is naturally identified with a vertex of $\mathcal{G}_n$, and two tiles are adjacent along a side if and only if as vertices they are joined by an edge.

Figure 4 shows part of $\mathcal{G}_n$ for $n = 5$. The vertices on the circle in the center of the figure represent the elements $U^k = \left(\begin{smallmatrix} 1 & k \\ 0 & 1 \end{smallmatrix}\right)$, $k = 0, \ldots, n-1$. From each U^k a radial edge leads to vertex VU^k lying on another circle, composed of all U^iVU^k,

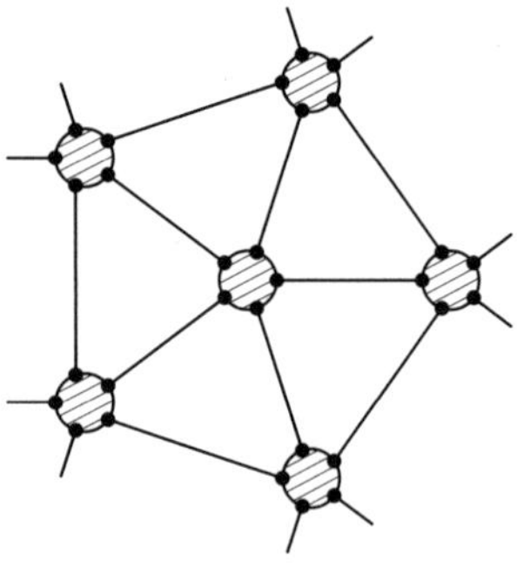

FIGURE 4. Part of a Cayley graph of PSL(2, 5)

$i = 0, \ldots, n-1$. It is easy to see that for $k \neq l$ an U^iVU^k is joined with an U^jVU^l by an edge in exactly two cases: $l = k-1$, $i = 1$, $j = -1$, and $l = k+1$, $i = -1$, $j = 1$ (all equalities (mod n)). To see this, one simply computes (in PSL)

$$VU^iVU^k = \pm \begin{pmatrix} -1 & -k \\ i & ik-1 \end{pmatrix}, \quad U^jVU^l = \pm \begin{pmatrix} j & jl-1 \\ 1 & l \end{pmatrix}.$$

Since Γ acts (by right multiplication) transitively on its own Cayley graph, the drawn picture remains correct if we interpret the circle in the center of the figure as an arbitrary circle of vertices U^kX, $k = 0, \ldots, n-1$.

As Brooks remarks in [**45**], one gets a better picture by passing from $\mathcal{G}_n$ to the n-regular graph $\mathcal{G}'_n$ obtained by defining each circle to be a single vertex with n edges leading from it. From the given description of $\mathcal{G}_n$ we see that $\mathcal{G}'_n$ is the 1-skeleton of a regularly triangulated closed surface. The vertices of $\mathcal{G}_n$

$$U^k \begin{pmatrix} a & b \\ c & d \end{pmatrix} = \begin{pmatrix} a+kc & b+kd \\ c & d \end{pmatrix}, \quad k = 0, \ldots, n-1,$$

in each circle have the same bottom row, and so one may use the rows $(c,d) (\mathrm{mod}\ \pm 1)$ (i.e., (c,d) and $(-c,-d)$ are considered equivalent) to label the vertices of $\mathcal{G}'_n$. Since $V\left(\begin{smallmatrix} a & b \\ c & d \end{smallmatrix}\right) = \left(\begin{smallmatrix} -c & -d \\ a & b \end{smallmatrix}\right)$, two vertices (c,d), (c',d') of $\mathcal{G}'_n$ are joined by an edge if and only if $\det \left(\begin{smallmatrix} c & d \\ c' & d' \end{smallmatrix}\right) = \pm 1 \pmod n$.

For simplicity, Brooks now restricts the values of n to prime numbers p with $p \equiv 1 \pmod 4$. This allows one to work with quadratic residues, and one has the following advantages:

1. $\mathrm{PSL}(2,p)$ is a group of order $p(p^2-1)/2$.
2. Precisely $\frac{p-1}{2}$ elements in $\mathbb{Z}/p\mathbb{Z}$ are non-zero squares.
3. (-1) is a square in $\mathbb{Z}/p\mathbb{Z}$.

The task is now to find a small upper bound for Cheeger's constant. By cutting away just one vertex we get the trivial estimate $h(\mathcal{G}'_p) \leq p$. A better way is to take the set of vertices

$$A = \{\text{all } (0,x^2) \text{ and } (x^2,y) \text{ with } x = 1, \ldots, \tfrac{p-1}{2}\},$$

and to let B be its complement in $\mathcal{G}'_p$. Then $\#(A) = \#(B) = \frac{1}{2}\#(\mathcal{G}'_p) = \frac{1}{4}(p^2-1)$. For the size of the set E of edges leading from A to B we observe that if $(0,x^2)$ is joined to (c,d) then $\det\left(\begin{smallmatrix} 0 & x^2 \\ c & d \end{smallmatrix}\right) = \pm 1 = -cx^2$ which implies that c is a square and hence $(c,d) \in A$. On the other hand, if (x^2,y) with $x \neq 0$ is joined with (c,d), then it is exactly joined with all $(c+kx^2, d+ky)$, $k = 0, \ldots, p-1$, and exactly half of these belong to A. This yields $\#(E) = \frac{p(p-1)}{4} \cdot \frac{p-1}{2}$ and

$$h(\mathcal{G}'_p) \leq \frac{p}{2} \cdot \frac{p-1}{p+1}.$$

(Brooks also shows that this cannot be improved by more than a factor $\frac{1}{2}$.) For the original graph we get $h(\mathcal{G}_p) \leq \frac{1}{2} \cdot \frac{p-1}{p+1}$ because the edges of E also disconnect $\mathcal{G}_p$. On the surface $\mathbb{H}^2/\Gamma_p$ each vertex of $\mathcal{G}_p$ is a copy of F with area $\pi/3$. Each edge of E corresponds to a copy of side b of F with length $\log(3)$, and the $\#(E)$ copies together disconnect the surface. This yields

$$h(\mathbb{H}^2/\Gamma_p) \leq \frac{\log(3)}{\pi/3} h(\mathcal{G}_p),$$

and the theorem follows.

The estimate has been improved in the article [**73**] on asymptotic isoperimetric constants with Andrzej Żuk, in 2002. There the arguments with the quadratic residues have been replaced by probabilistic ones, and the following improvement of Theorem 3.7 is obtained.

THEOREM 3.8 ([**73**]). *For all integers k sufficiently large one has*

$$h(\mathbb{H}^2/\Gamma_k) \leq 0.4402.$$

In [**122**] Luo, Rudnick and Sarnak improved Selberg's Theorem 3.6 obtaining the bound

$$\lambda_1(\mathbb{H}^2/\Gamma_k) \geq \frac{171}{784}.$$

An interesting history of Selberg's eigenvalue conjecture until 1995 may be found in Sarnak [**127**].

4. Circle Packings

In a series of four papers [**19, 21, 25, 42**] Brooks deals with a subject completely different from his main stream: Kleinian groups and circle packings. One particular result, Theorem 4.3, was later to play an important rôle in the computational approach to the uniformization of Riemann surfaces via circle packings ([**108, 109, 130**]), and we shall study it in some detail. But let us first cite the main results, which are from Inventiones [**25**], 1986, and let us also look at how Brooks became interested in the subject.

Let Γ be a Kleinian group acting as discrete group of isometries on $\mathbb{H}^3$. We say that Γ is *co-compact*, respectively, *co-finite* if $\mathbb{H}^3/\Gamma$ is compact, respectively, non-compact with finite volume. By Mostow's rigidity theorem, there are only countably many pairwise non-conjugate co-compact and co-finite Kleinian groups. In contrast to this, if Γ is geometrically finite and such that $\mathbb{H}^3/\Gamma$ has infinite volume, then with few exceptions, Γ has a non-trivial deformation space. Such groups do therefore not appear as subgroups of co-compact or co-finite Kleinian groups in general. *Where in the deformation space lie those who do appear?*

Answering to this, Brooks gets the following remarkable results.

THEOREM 4.1 ([**25**]). *Let Γ be a geometrically finite Kleinian group without cusps. Then there exist arbitrarily small quasi-conformal deformations Γ_ε of Γ, such that Γ_ε admits a co-compact extension.*

THEOREM 4.2 ([**25**]). *Let Γ be a geometrically finite Kleinian group with cusps. Then there exist arbitrarily small quasi-conformal deformations Γ_ε of Γ, such that Γ_ε admits a co-finite extension.*

For the third result we need some terminology from circle packings. For this we let for a moment $\mathbf{S}$ be any compact Riemann surface endowed with a conformally equivalent Riemannian metric of constant curvature.

A *configuration of circles* on $\mathbf{S}$ is a finite collection of embedded geodesic circles bounding compact disks such that the interiors of these disks are pairwise disjoint. The configuration is a *packing* of $\mathbf{S}$ if the region complementary to the interiors of the disks consists only of curvilinear triangles.

There are, of course variants of this definition for infinite configurations and/or non compact surfaces, and also for surfaces, where no metric but still an intrinsic notion of round circle is given.

Now let us consider closed hyperbolic surfaces (i.e., with constant curvature -1). By the theorem of Andreev-Koebe-Thurston [**103**], [**119**], [**134**, Chapter 13], for each topological type of packing there is, up to isometry, exactly one hyperbolic surface carrying this packing. So we have a similar situation as before, and Brooks' third theorem is the following, equally remarkable result.

THEOREM 4.3 ([**25**]). *Let C be a configuration of circles on the closed hyperbolic surface $\mathbf{S}$. Then there are hyperbolic surfaces $\mathbf{S}_\varepsilon$ arbitrarily close to $\mathbf{S}$, and configurations C_ε on $\mathbf{S}_\varepsilon$ of the same combinatorial type as C, such that C_ε can be completed, by adding finitely many circles, to a packing.*

Let us now first examine how Brooks was led to the subject. It begins with the "Spectral geometry of the Apollonian packing" in Comm. Pure Appl. Math. [**19**], 1985, where the following problem is considered.

Take a configuration of circles (in the above sense) on the sphere $\mathbb{S}^2$ and group them into unordered pairs in some way. (A pair is allowed to consist of one circle only.) Select for each pair a Möbius transformation of $\mathbb{S}^2$ taking the interior of one circle onto the exterior of the other. The group Γ generated by these transformations is a classical Schottky group. *What is the Hausdorff dimension of its limit set* $\Lambda(\Gamma)$?

In [**126**] Phillips and Sarnak had shown that for the corresponding problem on $\mathbb{S}^n$, $n \geq 3$, there exists a constant $c_n > 0$ such that $\dim(\Lambda(\Gamma)) < n - c_n$. They conjectured that the same should hold for $n = 2$. Brooks proved the following partial result:

THEOREM 4.4 ([**19**]). *There is a constant $c_2 > 0$ such that, for any finite collection taken from the Apollonian packing,* $\dim(\Lambda(\Gamma)) < 2 - c_2$.

His proof gave strong evidence in favor of the conjecture, and a complete proof was eventually found by Peter Doyle, [**113**], in 1994.

The connection with the Laplacian comes from Sullivan [**131, 132**] who defines, for any discrete group G of isometries of $\mathbb{H}^{n+1}$, an invariant $\delta(G)$ with the following three properties. First, if $\delta(G) > \frac{n}{2}$, then the bottom $\lambda_0(G)$ of the spectrum of the Laplacian on $\mathbb{H}^{n+1}/G$ is given by

$$\lambda_0(G) = \delta(G)(n - \delta(G)).$$

Second, if G is a union of subgroups $G = \bigcup_\alpha G_\alpha$, then

$$\delta(G) = \sup_\alpha \delta(G_\alpha).$$

Third, if G is geometrically finite, then the Hausdorff dimension of the limit set $\Lambda(G)$ (on $\mathbb{S}^n$) is given by

$$\dim(\Lambda(G)) = \delta(G).$$

The task is thus to find a lower bound of $\lambda_0(\Gamma)$ for the above Schottky groups.

For this it is practical to take the upper half space model of $\mathbb{H}^3$ so that Γ is a Schottky group on $\mathbb{R}^2 \cup \{\infty\}$. Each circle of the given collection (placed such that none passes through ∞) is then the boundary of a hemisphere in $\mathbb{H}^3$. A fundamental domain for Γ is given by the region $\Omega = \Omega(\Gamma)$ formed by the points of $\mathbb{H}^3$ lying above all the hemispheres, and it is well known that $\lambda_0(\Gamma)$ is bounded from below by the bottom of the spectrum $\lambda_0^{(N)}(\Omega)$ of the Laplacian on Ω with respect to Neumann boundary conditions. For this quantity one has Cheeger's inequality $\lambda_0^{(N)}(\Omega) \geq \frac{1}{4}h_N^2(\Omega)$, with the isoperimetric constant

$$h_N(\Omega) = \inf_S \frac{\text{area}(S)}{\text{vol}(\text{int}_\Omega(S))}, \tag{4.1}$$

where S runs through the hypersurfaces separating Ω into a bounded part $\text{int}_\Omega(S)$ and an unbounded part. One is now left with the geometric problem of estimating $h_N(\Omega)$.

The intuition in [**19**] is that the cases with small $h_N(\Omega)$ are those where the circles are tightly together.

A very tight, albeit infinite case is the *Apollonian packing*: start with three mutually tangent circles (in $\mathbb{R}^2 \cup \{\infty\}$). There are two new circles which are each tangent to the three circles. When we add these to our collection, then there are six new circles, each tangent to some triple of mutually tangent circles. We add them too, and continue, adding at the k-th stage the $2 \cdot 3^{k-1}$ circles tangent to three mutually tangent circles from the previous stage.

The infinite configuration $\mathcal{A}$ obtained in this way – the Apollonian packing – is unique up to Möbius transformations. Figure 5 shows a picture of it, redrawn after Figure 2 of [**19**]. The packing is placed such that two circles become horizontal straight lines mutually tangent at infinity.

FIGURE 5. "This infinite configuration is the Apollonian packing"

The properties of Sullivan's δ allow one to consider the Apollonian packing itself, even though the corresponding Schottky group $G_\mathcal{A}$ is not geometrically finite: If G_α runs through all Schottky groups arising from a finite subcollection of $\mathcal{A}$, then $G_\mathcal{A} = \bigcup_\alpha G_\alpha$. Thus, any estimate of $\lambda_0(G_\mathcal{A})$ leads to a universal bound $\dim(\Lambda(G_\alpha)) \leq 2 - c_2$ for all G_α.

To prove Theorem 4.4, Brooks is thus left with the geometric problem of finding a lower bound for the Cheeger constant $h_N(\Omega_\mathcal{A})$, where $\Omega_\mathcal{A}$ is the region of all points lying above the hemispheres belonging to the circles of the Apollonian packing.

For this he shows a great deal of three-dimensional intuition. The following tiling of $\mathbb{H}^3$ allows him to reduce the problem to a combinatorial one: consider the seven circles (in $\mathbb{R}^2 \cup \{\infty\}$) drawn in Figure 6 (a cutting of Figure 4 of [**19**]) and introduce the four additional circles (dotted lines) orthogonal to the triples$(\mathbf{1}, \mathbf{2}, \mathbf{3})$, $(\mathbf{1}, \mathbf{2}, \mathbf{4})$, $(\mathbf{2}, \mathbf{3}, \mathbf{4})$, $(\mathbf{1}, \mathbf{3}, \mathbf{4})$.

Interpret $\mathbb{R}^2 \cup \{\infty\}$ as the boundary of the upper half space $\mathbb{H}^3$ and place over each of the $7 + 4$ circles the corresponding hemisphere. This defines us the domain F of all points of $\mathbb{H}^3$ lying below the hemisphere corresponding to $\mathbf{8}$ but above the hemispheres corresponding to the circles $\mathbf{1}-\mathbf{4}$ and $\mathbf{9}-\mathbf{11}$. The domain is of finite volume with six cusps, labelled by the points a–f. Its boundary consists of eight ideal hyperbolic triangles, one for each of the hemispheres over the circles $\mathbf{1}-\mathbf{4}$,

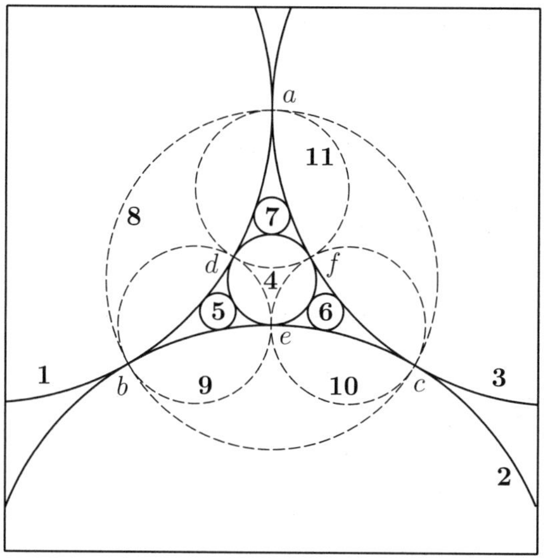

FIGURE 6. Hemispheres seen from above

8−**11**. For instance, the part of the boundary of F lying on hemisphere **1** is the ideal triangle whose vertices are the points a, b, d.

Returning to the region $\Omega_{\mathcal{A}}$, we place one copy of F for each configuration of four circles in the Apollonian packing consisting of a "center" circle (such as **4**) surrounded by three mutually tangent circles (such as **1**, **2**, **3**). It is then a pretty exercise to prove that these copies form a tiling of $\Omega_{\mathcal{A}}$ whose combinatorial structure is the 4-regular tree (graph without loops) $\mathcal{T}_4$.

With the methods developed in [**7**] and [**23**], the estimate of $h_N(\Omega_{\mathcal{A}})$ is reduced to the estimate of Cheeger's constant (adapted to infinite graphs) of $\mathcal{T}_4$ for which one easily shows that $h(\mathcal{T}_4) = 2$.

Let us return to the circle packings. The article "On the deformation theory of classical Schottky groups" [**21**], 1985, studies the deformation space $\mathcal{S}$ of a configuration $\mathcal{C}$ of circles on $\mathbb{S}^2$ which is such that the complement in $\mathbb{S}^2$ of the disks belonging to $\mathcal{C}$ consists only of triangles and quadrilaterals.

These quadrilaterals are bounded by circular arcs, and the interior angles at the four vertices are all equal to zero. With Brooks we call such quadrilaterals *rectangles*. Figures 7, 8, 9 show examples, filled with additional circles.

Denote by R the number of rectangles corresponding to $\mathcal{C}$. It was known that the deformation space $\mathcal{S}$ is homeomorphic to $\mathbb{R}^R$, and Brooks provides in [**21**] a new proof of this fact by combining the measurable Riemann mapping theorem of Ahlfors-Bers [**102**] with Sullivan's results on quasi-conformal deformations [**132**] of discrete groups of hyperbolic motions. The coup de génie in this proof is the introduction of a new Möbius invariant for any rectangle Q, the *packing parameter* $r(Q)$, which is more powerful than the cross ratio of its vertices, and which later led to the above-cited theorems from [**25**].

The parameter will be described in the next section; here we restrict ourselves to state two of its properties. The first property is a consequence of the definition. It clearly indicates the strength of the new invariant.

LEMMA 4.5. *If $r(Q)$ is a rational number, then Q admits a circle packing.*

For the second property we enumerate the rectangles corresponding to $\mathcal{C} \in \mathcal{S}$ in a coherent way: $Q_1, \ldots, Q_R$, and define $\bar{r}(\mathcal{C}) := (r(Q_1), \ldots, r(Q_R))$. We then have

THEOREM 4.6 ([**21**]). *The mapping $\bar{r} : \mathcal{S} \to (0, \infty)^R$ is a homeomorphism.*

The topology of $\mathcal{S}$ is the topology induced by quasi-conformal mappings:

Let $\mathcal{C}$ and $\mathcal{C}'$ be two configurations on $\mathbb{S}^2$ of the same combinatorial type (thus, $\mathcal{C}, \mathcal{C}' \in \mathcal{S}$) and let Γ and Γ' be the Schottky groups generated by the reflections in the circles of $\mathcal{C}$ and $\mathcal{C}'$ respectively. The two configurations are called *ε-close* to each other if there exists an ε-quasi-conformal mapping $\phi : \mathbb{S}^2 \to \mathbb{S}^2$ such that $\Gamma' = \phi\Gamma\phi^{-1}$.

To show Brooks' beautiful idea with this parameter we will sketch a proof of Theorem 4.3. Let $\mathbf{S} = \mathbb{D}/\Gamma$ be a compact Riemann surface, where now Γ is a Fuchsian group acting on the unit disk model $\mathbb{D}$ of the hyperbolic plane. To fit Γ into the preceding context we let it act on the entire Riemann sphere $\mathbb{R}^2 \cup \{\infty\}$. The limit set of Γ is $\partial\mathbb{D}$, the boundary of $\mathbb{D}$.

Fill $\mathbf{S}$ with circles to obtain a configuration on $\mathbf{S}$ such that the complementary region consists of triangles and rectangles, and let R be the number of rectangles. Lift the circles to the universal covering $\mathbb{D}$ of $\mathbf{S}$ and let Γ' be the group of Möbius transformations of $\mathbb{R}^2 \cup \{\infty\}$ generated by: Γ and the reflection in $\partial\mathbb{D}$ and the reflections in all the lifted circles. One can then show using Poincaré's Polyhedron Theorem (e.g., [**104**, Theorem 9.8.4], [**124**, Theorem H.11]) that Γ' is finitely generated and has a deformation space $\mathcal{S}(\Gamma')$ of dimension R. Furthermore, the methods used to prove Theorem 4.6 carry over to the present situation, showing that the map $\mathcal{S}(\Gamma') \to (0, \infty)^R$ given by sending the rectangles $Q_1, \ldots, Q_R$ on $\mathbf{S}$ to $r(Q_1), \ldots, r(Q_R)$ is a local homeomorphism.

A glance at Lemma 4.5 lets us guess what comes next: We deform Γ' into $\Gamma'_\varepsilon = \phi\Gamma'\phi^{-1}$ with an ε-quasiconformal mapping ϕ of arbitrarily small ε, such that all $r(Q)$ become rational numbers. Any rectangle Q on the deformed surface $S_\varepsilon = \phi(\mathbb{D})/\Gamma_\varepsilon$ with $\Gamma_\varepsilon = \phi\Gamma\phi^{-1}$ can now be circle-packed and $\mathbf{S}_\varepsilon$ too.

Suppressing all the difficulties we may also give a rough idea of the proofs of Theorems 4.1 and 4.2. If Γ is a geometrically finite Kleinian group with infinite co-volume, then Γ has a non-empty region of discontinuity $\mathcal{D}$ on $\partial\mathbb{H}^3$. An arbitrarily small quasi-conformal deformation produces a Kleinian group Γ_ε nearby, whose region of discontinuity $\mathcal{D}_\varepsilon$ admits a circle packing. Adding to Γ_ε the reflections in the circles of this packing we get the desired extension of Γ_ε.

Bowers and Stephenson generalized Theorem 4.3 in [**108**] and [**109**] to Riemann surfaces with geodesic boundary for the purpose of numerical uniformization via circle packings. Their proof is again based on the parameter $r(Q)$ but otherwise proceeds in a more elementary way avoiding the quasi-conformal deformation theory.

A rather recent treatment of Brooks' theorems on circle packings is given in Chapter 13 of Kapovich's book [**118**]. A tale about circle packings and their application to numerical uniformization is given by Stephenson in [**130**].

5. The Brooks Circle Packing Parameter

This section gives a short account of the circle packing parameter. We also provide a proof (of Proposition 5.1), whose presentation in [**21**] looks incomplete.

Consider a configuration of four circles A, B, C, D in $\mathbb{R}^2 \cup \{\infty\}$ bounding a rectangle Q as defined in the preceding section. We understand Q as being *marked* by the labelling of its sides. With Brooks we call A, B, C and D, respectively, the *bottom*, *right*, *top* and *left side* (or circle) of Q.

We fill Q with additional circles in the following way. First, there is a unique circle K inscribed in Q tangent to the top and left sides and also tangent to one of the sides A, B. If K is tangent to A we call it *horizontal*, if it is tangent to B we call it *vertical*. If K is tangent to all four sides of Q we stop the procedure. If the latter is not the case, then Q consists of four regions: the disk bounded by K, two triangles, and a new rectangle Q'. This new rectangle has sides A, B, C, K, if K is horizontal, and A, B, K, D, if K is vertical. We may then repeat the procedure by inscribing in Q' the unique circle K' tangent to its top and left sides and also tangent to one of the sides A and B. In this way we continue, either ad infinitum or until at some step the new inscribed circle is tangent to all four sides of the current rectangle, in which case we stop.

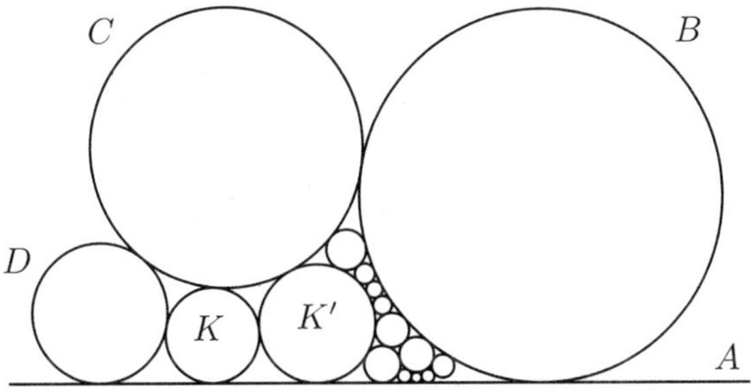

FIGURE 7. Rectangle filled with additional circles

To get a real number out of this, let n_1 be the number of horizontal circles required before a vertical circle is needed for the first time, then n_2 the number of vertical circles before again a horizontal circle is needed, and so on. In the example of Figure 7, e.g., we have $n_1, n_2, \ldots = 2, 5, 1, 1, 3, 1, \ldots$. The sequence will be encoded as a single real number using the continued fraction

$$r(Q) = n_1 + \cfrac{1}{n_2 + \frac{1}{n_3 + \ldots}} .$$

This is the *Brooks parameter* or, as he calls it, the *continued fraction parameter*. It is a rational number if and only if the sequence $n_1, n_2, \ldots$ is finite. Hence, rectangles with rational $r(Q)$ admit a circle packing (more precisely, a packing obtained by the above procedure).

It is fairly easy to convince oneself that $r(Q)$ varies continuously when A, B, C, D are varied continuously. But the following (crucial) monotonicity needs some effort.

Consider a family Q_t, of rectangles, where three of the circles, say A, C and D are kept fixed and the remaining circle, B_t, varies continuously for t in some interval I. Assume, moreover, that the variation is such that if $t < t'$, then Q_t is a proper subset of $Q_{t'}$.

PROPOSITION 5.1 (Key Proposition). *For the family Q_t the function $r(Q_t)$ is strictly monotone increasing.*

PROOF. We follow Brooks [**21**] but use a different geometric argument at a certain moment.

First we bring the family into a more suitable position, using a Möbius transformation, such that circles A and C are concentric forming an annulus, where A is the inner and C the outer boundary as in Figure 8 and 9. (In the limit case where A and C are mutually tangent, we choose the position such that A and C are parallel straight lines.) Then D and all B_t have the same size. Speaking in terms of the drawn figures, the condition of the Key Proposition is then simply that B_t is an inscribed circle that rolls to the right, away from D.

Denote by $\mathbf{C}_k$ and $\mathbf{C}'_k$ the $k-th$ circles in the fillings of Q_t and $Q_{t'}$ respectively. Set $\mathbf{S}(k) = -1$ if $\mathbf{C}_k$ is vertical, $\mathbf{S}(k) = 1$ if $\mathbf{C}_k$ is horizontal, and $\mathbf{S}(k) = 0$ if $\mathbf{C}_k$ is both. Define $\mathbf{S}'(k)$ similarly with $\mathbf{C}'_k$. As an example, in the second figure, $B_{t'}$ is drawn slightly further to the right, and we have $\mathbf{S}(k) = \mathbf{S}'(k)$ for $k = 1, \ldots, 7$, while $\mathbf{S}(8) = -1$ and $\mathbf{S}'(8) = 1$.

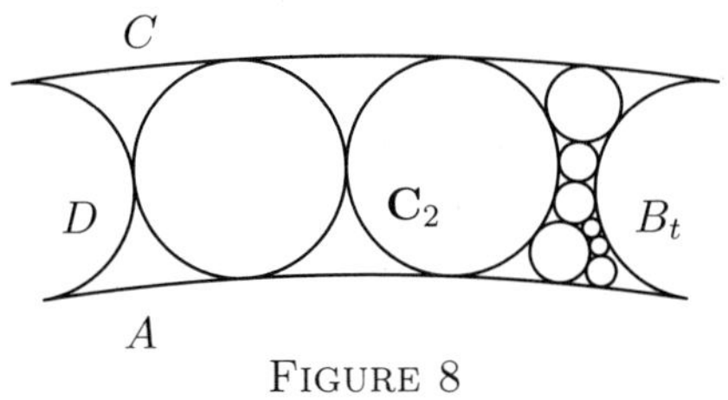

FIGURE 8

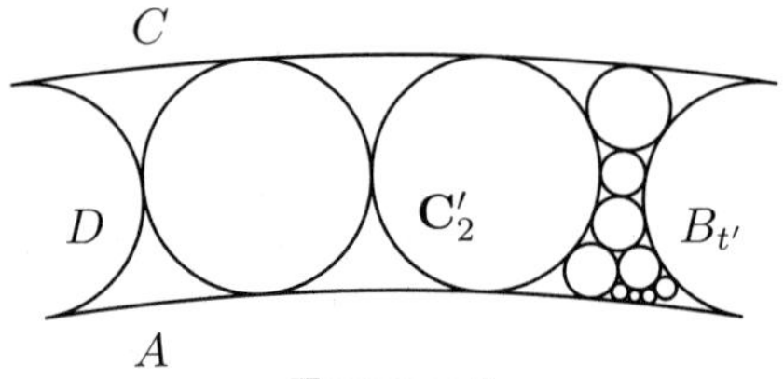

FIGURE 9

The strategy of proof is to look at the movements of the filled-in circles as B_t rolls to the right. For this we prove the Claim below, in which L, K, M, G, and similarly L', K', M', G' are Euclidean circles as in Figure 10 with disjoint interiors and L tangent to K, K tangent to M, and G tangent to L, K, M. The radii are denoted by ρ_L, ρ_K, etc. The *angle* $\sphericalangle(K, L, M)$ is defined as the value in $(0, 2\pi)$ such that $\rho_K \cdot \sphericalangle(K, L, M)$ is the length of the arc on K going from the point of contact with L via the point of contact with G to the point of contact with M.

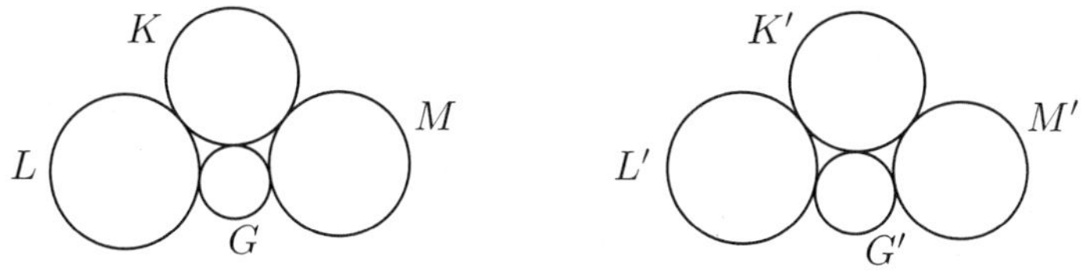

FIGURE 10

Angle $\sphericalangle(G, L, M)$ is defined as the value in $(0, 2\pi)$ such that $\rho_G \cdot \sphericalangle(G, L, M)$ is the length of the arc on G going from the point of contact with L to the point of contact with M *without meeting* K.

CLAIM. *Assume that* $\rho_L \geq \rho_{L'}$ *and* $\rho_M \geq \rho_{M'}$. *If* $\rho_K \leq \rho_{K'}$ *and* $\sphericalangle(K, L, M) \leq \sphericalangle(K', L', M')$ *and if at least one of the inequalities is strict, then* $\rho_G < \rho_{G'}$ *and* $\sphericalangle(G, L, M) < \sphericalangle(G', L', M')$.

To see this we use arguments as in [**21**]. First we increase the entire figure $LKMG$ (without changing the names) using a similarity transformation such that K assumes the size of K'. Then we shrink circles L and M individually, keeping the points of tangency with K fixed such that they assume the sizes of L' and M' respectively. Angle $\sphericalangle(K, L, M)$ has not changed so far. We now roll L and M (further) away from G until $\sphericalangle(K, L, M)$ becomes equal to $\sphericalangle(K', L', M')$. Circle G, which so far has been increased by a factor $\rho_{K'}/\rho_K$, is no longer tangent to L and M and has to be increased further (and properly rolled along K) to become tangent to L and M again, at which moment we have obtained figure $L'K'M'G'$. This argument shows that the radii in the claim satisfy

$$(*) \qquad \frac{\rho_{G'}}{\rho_G} > \frac{\rho_{K'}}{\rho_K} \geq 1.$$

Hence, the first conclusion of the claim. For the second conclusion we apply a variant of the procedure. We increase the original figure $LKMG$ using a similarity such that this time G assumes the size of G'. By (*), the three circles L, K, M are now momentarily too big. So we shrink them individually, keeping the points of tangency with G fixed. Angle $\sphericalangle(K, L, M)$ has not been affected so far, but L and K, respectively K and M are no longer mutually tangent. We may thus roll them together along G until this is again the case, at which moment we will obtain figure $L'K'M'G'$. This increases $\sphericalangle(G, L, M)$ and finishes the proof of the claim.

To finish the proof of the proposition we consult again Figures 8 and 9. Fix some value k_0 for the moment such that $\mathbf{S}(k) = \mathbf{S}'(k)$ for $k < k_0$ (but possibly $\mathbf{S}(k_0) \neq \mathbf{S}'(k_0)$). How do the $\mathbf{C}_k$ behave for $k < k_0$ as t gets bigger?

The horizontal circles $\mathbf{C}_k$, $k = 1, \ldots, n_1$, remain invariant as B_t rolls towards $B_{t'}$. The first vertical circle $\mathbf{C}_{n_1+1}$ clearly becomes bigger as t is increased, and angle $\sphericalangle(\mathbf{C}_{n_1+1}, \mathbf{C}_{n_1}, B_t)$ becomes bigger as well. Using our claim, we see that the same is true for all the vertical circles $\mathbf{C}_k$, $k = n_1 + 1, \ldots, n_2$. These circles move therefore downwards along $\mathbf{C}_{n_1}$ towards A. As a consequence, the next horizontal circle, $\mathbf{C}_{n_2+1}$ becomes smaller. Using the claim again, we see that also $\sphericalangle(\mathbf{C}_{n_2+1}, A, \mathbf{C}_{n_2})$ becomes smaller, and therefore all horizontal circles $\mathbf{C}_k$, $k = n_2 + 1, \ldots, n_3$, have decreasing radii and decreasing angles $\sphericalangle(\mathbf{C}_k, A, \mathbf{C}_{n_2})$.

Since the radius of $\mathbf{C}_{n_2}$ and $\sphericalangle(\mathbf{C}_{n_2}, \mathbf{C}_{n_1}, B_t)$ become bigger, and since furthermore all circles $\mathbf{C}_k$, $k = n_2 + 1, \ldots, n_3$, become smaller, we see that *also angle* $\sphericalangle(\mathbf{C}_{n_2}, \mathbf{C}_{n_3}, B_t)$ *becomes bigger.* Hence, the claim may also be applied to the configuration $\mathbf{C}_{n_3}\mathbf{C}_{n_2}B_t\mathbf{C}_{n_3+1}$, where $\mathbf{C}_{n_3+1}$ is the next vertical circle. In this way we may continue to get the following result: For $k < k_0$ all horizontal circles (except for the first generation $\mathbf{C}_1, \ldots, \mathbf{C}_{n_1}$) become smaller and move to the left, while all vertical circles become bigger and move down.

Now let us see what happens if we take k_0 such that, in addition, $\mathbf{S}(k_0) \neq \mathbf{S}'(k_0)$. Let s and j be the positive integers for which we have $k_0 = n_1 + \cdots + n_s + j$, $j \leq n_{s+1}$. From the behaviour of the $\mathbf{C}_k$ with $k < k_0$ follows that $\mathbf{C}_{k_0}$ must be a vertical circle, and thus s is an *odd* integer. (If $\mathbf{C}_{k_0}$ is vertical and horizontal at the same time we agree to write $k_0 = n_1 + \cdots + n_s + 1$ and $n_{s+1} = 1$.) If $\mathbf{C}_{k_0-1}$ is horizontal, then $n'_s > n_s$ because $\mathbf{C}'_{k_0}$ has become horizontal. If $\mathbf{C}_{k_0-1}$ is vertical, then n_s remains invariant but $n'_{s+1} < n_{s+1}$ for the same reason. Since s is odd we conclude, in either case, that $r(Q_t) < r(Q_{t'})$.

It remains to show that k_0 with $\mathbf{S}(k_0) \neq \mathbf{S}'(k_0)$ exists. For this we copy the arguments from [**21**].

If $r(Q_t)$ is rational and thus the sequence of $\mathbf{C}_k$ is finite, this is clear, because the last circle looses its status of being horizontal and vertical at the same time. If $r(Q_t)$ is irrational, we observe (see below) that the horizontal circles converge to the intersection point of B_t and A. If for all k we would have $\mathbf{S}(k) = \mathbf{S}'(k)$, then by the above claim, all horizontal circles move to the left whereas B_t moves to the right. This cannot happen because in $Q_{t'}$ the horizontal circles converge to the intersection point of $B_{t'}$ and A.

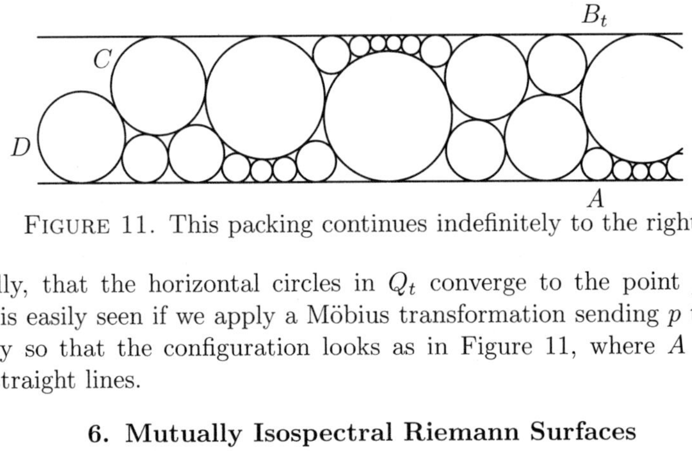

FIGURE 11. This packing continues indefinitely to the right

Finally, that the horizontal circles in Q_t converge to the point p where B_t meets A is easily seen if we apply a Möbius transformation sending p to the point at infinity so that the configuration looks as in Figure 11, where A and B_t are parallel straight lines. □

6. Mutually Isospectral Riemann Surfaces

Over twenty articles in the bibliography of Bob Brooks are in the field of inverse spectral geometry, with subjects such as the construction of isospectral paper models [**32, 39**], isospectral Riemann surfaces [**29, 45, 60**], accounts of Sunada's method on graphs and manifolds [**58, 59, 63, 64**], compactness properties [**35, 43, 44, 48**], and isoscattering [**69, 71, 74**].

Out of these we pick here the joint work [**60**] of 1998, with Ruth Gornet and William Gustafson, which unites several aspects of Brooks' work and gives insight into his combinatorial and group theoretic ideas.

The paper is based, so to say, on a "funny idea": *Apply to finite groups a strategy that has been designed for Lie groups.*

The authors address the following question: Given a natural number g, how many Riemann surfaces $S_1, \ldots, S_k$ of genus g can there be such that they are pairwise isospectral and non-isometric? *Isospectral* means that they have the same spectrum of the Laplacian.

It was known from [**110**] that the least upper bound $N(g)$ to the size of such sets has the upper estimate $N(g) \leq \exp(720g^2)$, and from the constructions of Tse in [**135**] that there exists a sequence of $g_i \to \infty$ such that $N(g_i) \geq \text{const}\sqrt{g_i}$. This is improved by giving more than polynomially many examples. The statement is the following.

THEOREM 6.1 ([**60**]). *For each natural number $n > 2$ and prime p, the number $N(g)$ of mutually isospectral non-isometric Riemann surfaces of genus*

$$g = 1 + (n-1)p^{2n}$$

is at least

$$N(g) \geq p^{n(n-1)}.$$

For completeness, we also state the two other theorems in this article. The first theorem is proved independently from the above using the technique of Seidel switching (for this technique see the article "Non Sunada graphs" [**63**]).

THEOREM 6.2 ([**60**]). *For any n, there are 2^{n-1} isospectral graphs that are 6-regular with $4n$ vertices.*

The second theorem arises from the considerations about the groups used to prove Theorem 6.1.

THEOREM 6.3 ([**60**]). *For all natural numbers n and odd primes p, there exist $p^{n(n-1)}$ non-isomorphic number fields of degree p^{2n} over the rational numbers that have the same zeta function.*

Let us first describe what is meant by the "funny" idea of the proof and then go into some details. In [**116**] Gordon and Wilson constructed continuous families of Lie groups which all have the same spectrum of the Laplacian (see also [**112**] for an overview of this and related examples). This is quite in contrast to the case of a Riemann surface, where such families are always finite. So one may ask oneself the following question: *Are there finite groups that look like the Lie groups in the isospectral deformations and "therefore" produce a large number of pairwise isospectral Riemann surfaces?*

The idea of the proof of Theorem 6.1 is that the answer is "Yes", and now let's look more closely at how this works.

The tool is Sunada's theorem [**133**] which we state in the following form. Let G be a finite group and H_1, H_2 two subgroups of G. We say that H_1 and H_2 are *almost conjugate* in G if they satisfy:

$$\#([g] \cap H_1) = \#([g] \cap H_2), \quad \forall g \in G.$$

Here $\#$ denotes the cardinality of a set, and $[g]$ is the conjugacy class of g in G. Now let M be a Riemannian manifold and let $\phi : \pi_1(M) \to G$ be a surjective homomorphism from the fundamental group of M onto G. Then there exists a Riemannian covering $\tilde{M}$ on which G acts freely by isometries such that $\tilde{M}/G = M$, and we may define two manifolds

$$M^{H_1} = \tilde{M}/H_1, \quad M^{H_2} = \tilde{M}/H_2.$$

Sunada's theorem then says that if H_1 and H_2 are almost conjugate, then M^{H_1} and M^{H_2} are isospectral (for both, the spectrum of the Laplacian and the spectrum of the lengths of closed geodesics).

Sunada also shows that if H_1 and H_2 are non-conjugate in G, then in a generic sense, the manifolds M^{H_1} and M^{H_2} are non-isometric.

The goal is thus to find G with a large number of pairwise almost conjugate non-conjugate subgroups $H_1, \ldots, H_l$.

The following concept is used: A group automorphism $\phi : G \to G$ is an *almost inner automorphism* if for all $g \in G$, $\phi(g)$ and g are conjugate. Clearly, if ϕ is such an automorphism and H a subgroup of G, then H and $\phi(H)$ are almost conjugate. The strategy is then to start with a suitable pair G, H and to take the images of H under a large number of almost inner automorphisms.

The group G that looks almost like a nilpotent Lie group is the three dimensional Heisenberg group

$$G = \left\{ \left(\begin{matrix} 1 & x & z \\ 0 & 1 & y \\ 0 & 0 & 1 \end{matrix} \right) \middle| \ x, y, z \in R \right\},$$

where R is a finite commutative ring with unity, and the distinguished subgroup H will be

$$H = \left\{ \left(\begin{matrix} 1 & x & 0 \\ 0 & 1 & 0 \\ 0 & 0 & 1 \end{matrix} \right) \middle| \ x \in R \right\}.$$

It is useful to denote the elements of G by (x, y, z) and to write out the group operations:

$$(x, y, z) * (x', y', z') = (x + x', y + y', z + z' + xy')$$
$$(x, y, z)^{-1} = (-x, -y, -z + xy).$$

Conjugacy may be worded as follows: (x, y, z) and (x', y', z') are conjugate in G if and only if

$$x = x', \quad y = y', \quad z - z' \in Rx + Ry.$$

Now let $f : R \times R \to R$ be an arbitrary map. Setting

$$\psi_f(x, y, z) := (x, y, z + f(x, y)), \quad (x, y, z) \in G,$$

we get a one-to-one mapping from G onto itself, and we immediately see that ψ_f becomes a group automorphism if and only if $f : R \times R \to R$ is an additive group homomorphism.

The question is: when will ψ_f be an almost inner automorphism? For this the following definition is made. A function $f : R \times \cdots \times R \to R$ (n factors) is called *almost linear* if it has the following form

$$f(x_1, \ldots, x_n) = a_1(x_1) + \cdots + a_n(x_n)$$

with suitable functions $a_i : R \to R$ which must be such that f becomes an additive group homomorphism.

Of course, if we take the functions a_i to be multiplications by constants, then f is linear. The condition for f to be a homomorphism between the additive groups $R \times \cdots \times R$ and R may readily be given explicitly: f is an additive group homomorphism if and only if each of the functions $a = a_i$ satisfies the functional relation

$$(*) \qquad a(x+y)(x+y) = a(x)x + a(y)y.$$

Taking again $n = 2$ and bringing all definitions together we have the following lemma, which will not be applied as it stands, but clears the picture:

LEMMA 6.4. *Let $f : R \times R \to R$ be an arbitrary function. Then*

(1) *ψ_f is an inner automorphism of G if and only if f is linear.*
(2) *ψ_f is an almost inner automorphism of G if and only if f is almost linear.*

In order to apply this to the above subgroup H, we now restrict ourselves to almost linear mappings f which depend only on x, i.e.,

$$f(x,y) = f(x) = a(x)x, \quad x, y \in R,$$

where a satisfies the functional relation (*). For any such f we have the subgroup

$$H_f = \psi_f(H) = \left\{ \begin{pmatrix} 1 & x & f(x) \\ 0 & 1 & 0 \\ 0 & 0 & 1 \end{pmatrix} \middle| \, x \in R \right\}.$$

The main property is the following lemma which nicely gathers all the previous definitions.

LEMMA 6.5. *Let $f_1, f_2 : R \to R$ be almost linear. Then*

(1) *H_{f_1} and H_{f_2} are almost conjugate.*
(2) *H_{f_1} and H_{f_2} are conjugate if and only if $f_1 - f_2$ is linear.*

PROOF. $\psi_{f_i}(x, 0, z) = (x, 0, z + a_i(x)x)$ is conjugate to $(x, 0, z)$ for any $(x, 0, z) \in H$, and so H_{f_i} is almost conjugate to H, $i = 1, 2$, and also H_{f_1} and H_{f_2} are almost conjugate. This is the first claim. If $(a, b, c) \in G$ and $(x, 0, f_1(x)) \in H_{f_1}$, then

$$(a,b,c)^{-1} * (x, 0, f_1(x)) * (a, b, c) = (x, 0, f_1(x) + bx).$$

The right hand side is equal to $(x, 0, f_2(x))$ if and only if $f_2(x) - f_1(x) = bx$. This gives the second claim. □

It's time for an example! The goal is to have a ring R which admits a large number of almost linear maps. In [**60**], three such examples are given, and we pick out the following one, in which almost linearity appears in a particularly nice way.

EXAMPLE 6.6. Let $R = \mathbb{F}_{p^n}$, the unique field with p^n elements. This field is at the same time an n-dimensional vector space over $\mathbb{F}_p$.

The two properties together imply that a map $f : R \to R$ is almost linear in the above sense if and only if it is a linear map from the $\mathbb{F}_p$-vector space R to itself in the usual sense. As the dimension is n, there are precisely p^{n^2} such maps.

On the other hand, the maps which are R-linear in the above sense are just those of the form $f(x) = ax$ for some constant $a \in R = \mathbb{F}_{p^n}$, and their cardinality is p^n. Hence, there exist $p^{n^2}/p^n = p^{n(n-1)}$ almost linear maps whose pairwise differences are not R-linear. By Lemma 6.5, this means that taking G and H over the ring $\mathbb{F}_{p^n}$ we obtain $p^{n(n-1)}$ subgroups $H_{f_1}, H_{f_2}, \ldots$, which are pairwise almost conjugate non-conjugate.

We finish the example by noting that if $\zeta_1, \ldots, \zeta_n$ form a basis of the vector space $\mathbb{F}_{p^n}$ over $\mathbb{F}_p$, then the $2n$ elements $(\zeta_i, 0, 0)$ and $(0, \zeta_i, 0)$ generate G. In fact, they certainly generate all $(x,0,0)$ and all $(0,x,0)$, $x \in \mathbb{F}_{p^n}$, and any other element of G is then obtained as follows,

$$(1,0,0) * (0,z,0) * (-1,0,0) * (0,-z,0) = (0,0,z)$$
$$(0,y,0) * (x,0,0) * (0,0,z) = (x,y,z).$$

It remains to construct the surfaces out of the above groups. To this end we let, for a moment, S be any topological closed surface of genus

$$g = n, \text{ if } n \text{ is even}, \qquad g = n+1, \text{ if } n \text{ is odd}.$$

Its fundamental group has the canonical presentation

$$\pi_1(S) = \left\langle A_1, B_1, \ldots, A_g, B_g \;;\; \textstyle\prod_i A_i B_i A_i^{-1} B_i^{-1} = 1 \right\rangle.$$

In order to get a surjective homomorphism from $\pi_1(S)$ to G as required by Sunada's theorem, it suffices to send the generators $A_1, B_1, \cdots, A_{g/2}, B_{g/2}$ of $\pi_1(S)$ to the generators $(\zeta_i, 0, 0)$ of G, and $A_{g/2+1}, B_{g/2+1}, \cdots, A_g, B_g$ to the $(0, \zeta_i, 0)$ in such a way that each generator of G occurs as an image. The relation is then clearly preserved, since all commutators involved are 0.

Sunada has shown that if $S = \mathbb{H}^2/\Gamma$ is a compact Riemann surface of genus $g \geq 3$ with trivial isometry group and Γ a non-arithmetic Fuchsian group, then the examples obtained by his construction are non-isometric [**133**, p. 173]. By Singerman [**129**] and Margulis [**123**], the surfaces S with these properties are dense in the Teichmüller space of a Riemann surface of genus $g \geq 3$. Thus, taking any such surface concludes the proof.

7. The Ahlfors-Schwarz Lemma

One of Brooks' favorite results from classical conformal analysis was the Ahlfors-Schwarz Lemma. In a survey [**68**] presented in the "Lectures in Memory of Lars Ahlfors", Haifa, 1996, he gave the Lemma a geometrically more complete form by adding just two lines of arguments to the original proof. The Lemma is an indispensable tool in Brooks' investigation on random Riemann surfaces (see the next section), and the simplicity of the proof contrasts highly with the powerfulness of its applications.

In what follows, $d\sigma = 2|dz|/(1 - x^2 - y^2)$ denotes the hyperbolic metric on the unit disk $\mathbb{D} = \{z = x + iy \in \mathbb{C} \mid x^2 + y^2 < 1\}$. A Riemannian metric ds on $\mathbb{D}$ is called a *conformal metric* if it is of the form $ds = e^{u(z)}|dz|$ for some smooth function u on $\mathbb{D}$. The lemma is:

THEOREM 7.1 (Ahlfors [**101**]). *Let ds be a conformal metric on $\mathbb{D}$ with Gauss curvature $\kappa \leq -1$, then everywhere $ds \leq d\sigma$.*

The classical Schwarz Lemma is obtained from this theorem in the following way. Let $f : \mathbb{D} \to \mathbb{D}$ be a holomorphic function with $f(0) = 0$. Then the pulled-back metric $ds = f_* d\sigma$ is conformal and has curvature -1. By the theorem, $ds \leq d\sigma$. Since $ds = |f'(z)| d\sigma$, $(f' = \frac{df}{dz})$, we conclude that $|f'(z)| \leq 1$, and so $|f(z)| \leq |z|$ for all $z \in \mathbb{D}$.

In [**68**] Brooks comments: "It was Ahlfors' insight to see that the essential point here [Pick's version of the Schwarz Lemma] is that the curvature of the target metric be ≤ -1, rather than just constant curvature". Brooks' own insight is that neither metric needs constant curvature:

THEOREM 7.2 (Brooks' version of the Ahlfors-Schwarz Lemma [**68**]). *Let ds_1 and ds_2 be conformal metrics on $\mathbb{D}$, with ds_2 complete. If the Gauss curvatures satisfy* $\sup(\kappa_1) \leq \inf(\kappa_2) < 0$, *then $ds_1 \leq ds_2$.*

Here is the proof from [**68**], which in turn follows closely Ahlfors [**101**]. The curvature κ of any conformal metric $ds = e^{u(z)}|dz|$ for z on some open domain in $\mathbb{C}$ is given by the classical formula

$$\kappa(z) = \frac{1}{e^{2u(z)}}\Delta u(z),$$

where $\Delta = -(\frac{\partial^2}{\partial x^2} + \frac{\partial^2}{\partial y^2})$ is the Euclidean Laplacian. Now assume, for a moment, that for the conformal metrics $ds_i = e^{u_i(z)}|dz|$ as in the theorem, the difference $u_2 - u_1$ has a global minimum at some point z_0. Then $\Delta(u_2 - u_1)(z_0) \leq 0$, and therefore

$$\kappa_2(z_0)e^{2u_2(z_0)} \leq \kappa_1(z_0)e^{2u_1(z_0)}.$$

The curvature assumptions then imply that $e^{u_1(z_0)} \leq e^{u_2(z_0)}$ and, since $u_2 - u_1$ is minimal at z_0, we get $u_1(z) \leq u_2(z)$ for all z, as claimed in the theorem.

Of course, it is not known whether or not $u_2 - u_1$ has a global minimum. So one applies a trick from conformal analysis: For r with $0 < r < 1$ (and later $r \to 1$) we let $ds_{r,2}$ be the pulled-back of ds_2 to the disk $|z| < r$ under the mapping $z \mapsto z/r$. We have $ds_{r,2} = e^{u_{r,2}(z)}|dz|$ with $u_{r,2}(z) = \frac{1}{r}u_2(z/r)$. Since $ds_{r,2}$ is a complete metric on this disk which, moreover, has a lower curvature bound, the function $u_{r,2}$ blows up on the circle of radius r in the sense that for any $M > 0$ there exists $\delta > 0$ such that $u_{r,2}(z) > M$ whenever $|z| > r - \delta$. But u_1 is bounded on $|z| < r$ and so $u_{2,r} - u_1$ has a global minimum. For $|z| < r$ we now have $u_1(z) \leq u_{r,2}(z) = \frac{1}{r}u_2(z/r)$, and the theorem follows by taking $r \to 1$.

In the introduction to [**68**] Brooks adds that "what I learned about how to use the Ahlfors-Schwarz Lemma comes from Scott Wolpert, who was using it to great effect at the time we first met". Wolpert's version of the Lemma is restricted to compact surfaces, but for these it is more general in the sense that the curvature assumptions are pointwise. In the following statement, "conformally equivalent" means that $d\sigma_1 = e^u \cdot d\sigma_2$ for some smooth function u.

THEOREM 7.3 (Wolpert, [**136**]). *Let $d\sigma_1$ and $d\sigma_2$ be conformally equivalent metrics on a compact Riemann surface S with negative curvatures κ_1 and κ_2 respectively. If $\kappa_1 \leq \kappa_2$ (pointwise), then $d\sigma_1 \leq d\sigma_2$.*

8. Plato, Belyĭ, and Random Riemann Surfaces

The Ahlfors-Schwarz Lemma comes into play much like a "deus ex machina" in a geometric construction which Brooks uses in [**61, 70, 72, 76**] and also in some other articles concerning Riemann surfaces with large λ_1.

The construction is based on the intuition that on Riemann surfaces *large cusps look like large disks.*

We will first extract the precise statement for this from [**61**] and [**70**] and then describe its applications to graphs and ideal triangles which led from the Platonic surfaces in [**61**] and the Belyĭ surfaces in [**72**] to Brooks' last concept: graph-randomness for Riemann surfaces, yet to appear in J. Differential Geom. [**76**] (jointly with Eran Makover).

In what follows, S (and S^{op}, S^{cl}, etc.) will be a Riemann surface of genus $g \geq 2$, compact or non-compact, endowed with a complete metric of finite area and constant curvature -1. In the non-compact case, S has a finite number of cusps. We make two definitions.

An open subset $\mathcal{C} \subset S$ is called a *cusp neighbourhood of size* l, if it is isometric to the cylinder

$$\mathcal{C}_l = \{z \in \mathbb{H}^2 \mid \Im(z) > 1\}/(z \sim z + l)\}.$$

An open subset $\mathcal{B} \subset S$ is called a *disk neighbourhood of size* l, if it is isometric to the interior $\mathcal{B}_l$ of a circle of hyperbolic length l in $\mathbb{H}^2$.

Introducing horocyclic coordinates for $\mathcal{C}$ or $\mathcal{C}_l$ and polar coordinates for $\mathcal{B}$ or $\mathcal{B}_l$ (e.g., [**110**, pp. 3,5]), we may describe these neighbourhoods and their Riemannian metrics, respectively, as follows,

$$(8.1)\qquad \begin{array}{cc} \{(\rho,\vartheta) \mid -\infty < \pi e^{\rho} < l,\ \vartheta \in \mathbb{S}^1\}, & \{(\rho,\vartheta) \mid 0 \leq 2\pi \sinh(\rho) < l,\ \vartheta \in \mathbb{S}^1\} \\ ds^2_{\mathcal{C}} = d\rho^2 + \frac{1}{4}e^{2\rho}d\vartheta^2 & ds^2_{\mathcal{B}} = d\rho^2 + \sinh^2(\rho)\,d\vartheta^2. \end{array}$$

The simple observation underlying everything that follows is that for large $l < L$ the regions $\mathcal{C}_L \smallsetminus \mathcal{C}_l$ and $\mathcal{B}_L \smallsetminus \mathcal{B}_l$ are almost isometric to one another. The less simple observation (Theorem 8.2) will then be that the first eigenvalue λ_1 of the Laplacian can hardly distinguish the difference between $\mathcal{C}_L$ and $\mathcal{B}_L$.

The construction, used in [**61**] and [**70**] in order to obtain compact examples out of Selberg's 3/16 Theorem 3.6, and which was to become the underlying element for the randomness concept in [**76**] is as follows.

Let S^{op} be an open, i.e., non-compact Riemann surface of genus $g \geq 2$. Let m be the number of cusps of S^{op} and choose pairwise disjoint cusp neighbourhoods $\mathcal{C}^1, \dots, \mathcal{C}^m$. We will say that S^{op} *is of cusps* $\geq l$ if we can choose these neighbourhoods such that each has size $\geq l$.

For each $\mathcal{C}^i$ there is an open disk $U^i \subset \mathbb{C}$, a point $q^i \in U^i$, and a conformal homeomorphism $\varphi^i : \mathcal{C}^i \to U^i \smallsetminus \{q^i\}$. Adding, for $i = 1, \dots, m$, a "point at infinity" p^i to $\mathcal{C}^i$ and extending φ^i to a mapping $\varphi^i : \mathcal{C}^i \cup \{p^i\} \to U^i$ sending p^i to q^i, we obtain a surface S^{cl} with a conformal structure such that $S^{cl} \smallsetminus \{p^1, \dots, p^m\} = S^{op}$.

At this step, S^{cl} is a compact Riemann surface of genus $g \geq 2$ only in the sense that it carries a conformal structure. But, by the Koebe-Poincaré Uniformization Theorem (e.g., [**114**]), there exists on S^{cl} a unique metric of constant curvature -1 which is conformally compatible with this structure, and we will then understand S^{cl} as being endowed with this metric.

Brooks' insight is that the metrics on S^{op} and S^{cl} may be compared quantitatively with each other despite the notorious in-explicitness of the Uniformization Theorem. For this he works with variable curvature. A bare hands approximation [**61**, pp. 158–162] starting from (8.1) shows that, given $\varepsilon > 0$, and assuming that the size l of our cusp neighbourhoods is sufficiently large, there exists on each $\mathcal{C}^i$ with initial metric $ds_{\mathcal{C}}$ a comparison metric $ds_{\mathcal{C},\varepsilon}$ with variable curvature $\kappa_{\mathcal{C},\varepsilon}$ such that

1. $ds_{\mathcal{C},\varepsilon} = ds_{\mathcal{C}}$ *in a neighbourhood of the boundary of* $\mathcal{C}^i$.
2. $ds_{\mathcal{C},\varepsilon}$ *is a conformally equivalent metric on* $\mathcal{C}^i$.
3. $ds_{\mathcal{C},\varepsilon}$ *extends to a smooth conformally equivalent metric on* $\mathcal{C}^i \cup \{p^i\}$.
4. $-1 - \varepsilon \leq \kappa_{\mathcal{C},\varepsilon} \leq -1 + \varepsilon$.

Replacing $ds_{\mathcal{C}}$ by $ds_{\mathcal{C},\varepsilon}$ in each $\mathcal{C}^i$, we end up with three metrics on the closed surface S^{cl}: the original metric on $S^{op} = S^{cl} \smallsetminus \{p^1, \ldots, p^m\}$, which we now denote by ds_{op}; then the smooth metric $ds_{cl,\varepsilon}$ just constructed on S^{cl}, which agrees with ds_{op} on $S^{op} \smallsetminus (\mathcal{C}^1 \cup \cdots \cup \mathcal{C}^m)$; and finally the metric ds_{cl} on S^{cl} given by the Uniformization Theorem. Both, $ds_{cl,\varepsilon}$ and ds_{cl} are conformally compatible with the conformal structure given on S^{cl}, and so they are conformally equivalent.

It might yet look as though nothing is gained, for ds_{cl} is as inexplicit as before. But now the deus ex machina comes in: ds_{cl} and $ds_{cl,\varepsilon}$ have almost the same curvature, and so by Brooks' as well as Wolpert's version of the Ahlfors-Schwarz Lemma, these metrics are almost identical on S^{cl}. As a consequence, ds_{cl} is almost identical with ds_{op} on $S^{op} \smallsetminus (\mathcal{C}^1 \cup \cdots \cup \mathcal{C}^m)$.

Thus, after re-adapting the arithmetic for ε and l:

THEOREM 8.1 ([**61**]). *For every $\varepsilon > 0$ there exists $l > 0$ such that if S^{cl} and $S^{op} = S^{cl} \smallsetminus \{p^1, \ldots, p^m\}$ with pairwise disjoint cusp neighbourhoods $\mathcal{C}^1, \ldots, \mathcal{C}^m$ of size l are as above, then on the subset $S^{op} \smallsetminus (\mathcal{C}^1 \cup \cdots \cup \mathcal{C}^m)$ of S^{cl}, we have*

$$\left(\tfrac{1}{1+\varepsilon}\right) ds_{op} \leq ds_{cl} \leq (1+\varepsilon)\, ds_{op}.$$

Analyzing the behavior of the first eigenfunction inside the cusp neighbourhoods, Brooks and Makover obtain the following application of Theorem 8.1.

THEOREM 8.2 (Cusp-Closing Lemma,[**70**]). *For every $\varepsilon > 0$ there exists $l > 0$ such that if S^{cl} and $S^{op} = S^{cl} \smallsetminus \{p^1, \ldots, p^m\}$ with pairwise disjoint cusp neighbourhoods $\mathcal{C}^1, \ldots, \mathcal{C}^m$ of size l are as above, then*

$$\lambda_1(S^{cl}) \geq \left(\tfrac{1}{1+\varepsilon}\right) \min\{1/4,\ \lambda_1(S^{op})\}.$$

They also prove the analogous theorems in the opposite direction, where the initial surface, S^{cl}, has large pairwise disjoint disk neighbourhoods and S^{op} is obtained from S^{cl} by removing their centers. The difficult part is to show that then S^{op} has large pairwise disjoint cusp neighbourhoods. This is done with Brooks' version (Theorem 7.2) of the Ahlfors-Schwarz Lemma.

EXAMPLE 8.3. A very nice explicit case for the above compactification comes from the congruence subgroups Γ_n of the modular group Γ. The rôle of S^{op} is played by the congruence surfaces $S_n := \mathbb{H}^2/\Gamma_n$, and we denote the corresponding closed surfaces S^{cl} by P_n.

In Section 3 we saw that S_n may be obtained by a graph construction: one replaces each vertex of the 3-regular graph $\mathcal{G}_n$ (Figure 4 for $n = 5$) by a copy of the standard fundamental domain F of the modular group.

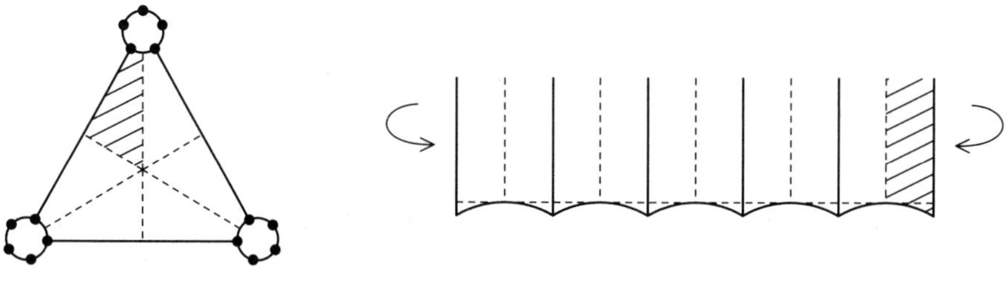

FIGURE 12. Constructing a Platonic surface

For each circle of n vertices, the corresponding n copies of F are pasted together as illustrated in Figure 12. This produces a cusp of size n (the horizontal dotted line is the horocycle of length n in the cusp). Denoting by ω_n the order of $\Gamma/\Gamma_n \simeq \mathrm{PSL}(2,n)$, we have ω_n/n such cusps, and the genus of S_n becomes

$$g_n = 1 + \frac{\omega_n}{2}\left(\frac{1}{6} - \frac{1}{n}\right).$$

For each triangle in the n-regular graph $\mathcal{G}'_n$ (whose vertices are the circles of $\mathcal{G}_n$), six copies of the part $F_+ = \{z \in F \mid \Re(z) \geq 0\}$ of F come together on S_n as illustrated schematically in Figure 12, forming and *ideal hyperbolic triangle.* Thus, S_n may also be described as a surface obtained by pasting together $\omega_n/3$ copies of an ideal triangle.

Knowing that the conformal and anticonformal automorphisms of S^{op} act by isometries on both, S^{op} and S^{cl}, and observing that for each side of an ideal triangle in $S^{op} = S_n$ there is an anticonformal automorphism fixing this side pointwise, we see that the ideal triangles of S_n become equilateral geodesic triangles in P_n. As n triangles meet at every vertex, P_n is a Platonic surface of genus g_n with Schlaefli symbol $\{3,n\}$.

Theorem 8.2 yields the following nice comparison:

$$\lambda_1(P_n) \geq c_n \cdot \lambda_1(\mathbb{H}^2/\Gamma_n), \quad n = 7, 8, \ldots$$

with constants $c_n \in (0,1)$ satisfying $c_n \to 1$ as $n \to \infty$.

The surfaces P_n are rather special. How does a *typical* compact Riemann surface look like when the genus gets large?

In [**76**] and, more informally in [**72**], Brooks and Makover approach this question in a combinatorial way which leads to a new concept that one may call *graph-randomness* for Riemann surfaces. The goal behind it can be described in the following way.

Suppose you find yourself scanning through the moduli space $\mathcal{M}_g$ of a compact Riemann surface of genus g looking for interesting examples, say with an exceptionally large injectivity radius, or an exceptionally small diameter, or λ_1 very close to $1/4$, i.e., properties that are not strongly affected by small moves in $\mathcal{M}_g$ but for which examples seem difficult to find. Then it would be nice to have a machine producing candidates at random, and if the probability (relatively to the machine's random process) for the desired property is positive, such an example should eventually be found.

The discovery is that Belyĭ's theorem provides such a process.

A compact Riemann surface S is called a *Belyĭ surface* if there exists a holomorphic mapping $\phi : S \to \hat{\mathbb{C}} = \mathbb{C} \cup \{\infty\}$ with at most 3 critical values in $\hat{\mathbb{C}}$. In [**105**] Belyĭ showed that this is equivalent to saying that S is conformally equivalent to an algebraic curve $P(x,y) = 0$ in $\mathbb{C}P^2$, where the coefficients of P lie in some number field K. As a consequence, the Belyĭ surfaces are dense in the union of all moduli spaces $\mathcal{M}_g$. A thorough account on Belyĭ surfaces is given by Jones and Singerman in [**117**].

Observing that $\phi : S \to \hat{\mathbb{C}}$ may be chosen such that the critical values are among $0, 1, \infty$, and knowing that $\hat{\mathbb{C}} \smallsetminus \{0,1,\infty\}$ is a six-fold branched cover of $\mathbb{H}^2/\mathrm{PSL}(2,\mathbb{Z})$, one obtains that S is Belyĭ if and only if there are finitely many points $p_1, \ldots, p_k \in S$ such that $S \smallsetminus \{p_1, \ldots, p_k\}$ is conformally equivalent to a quotient $\mathbb{H}^2/\Lambda$, where Λ is a finite-index torsion-free subgroup of $\Gamma = \mathrm{PSL}(2,\mathbb{Z})$. One has therefore the following geometric construction which yields precisely the Belyĭ surfaces.

Let $\mathcal{G}$ be a connected 3-regular vertex-oriented graph on $2k$ vertices. *Vertex-oriented* means that for each vertex v an enumeration v^a, v^b, v^c of the emanating half-edges is given, and that two enumerations are equivalent if they differ by a cyclic permutation.

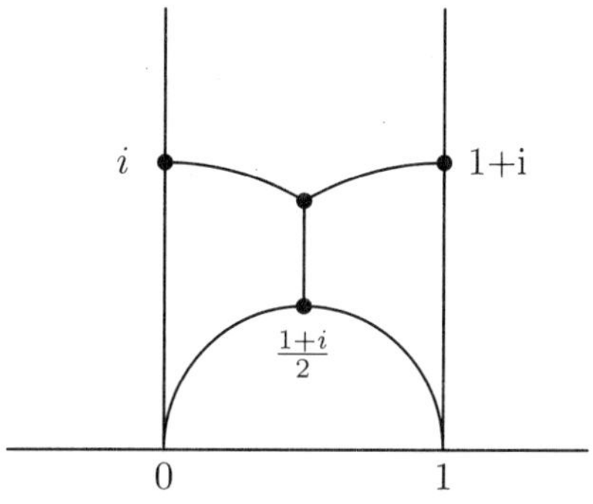

FIGURE 13. Ideal triangle with vertices in $0, 1, \infty$

Now take for each vertex of $\mathcal{G}$ a copy of the ideal hyperbolic geodesic triangle T, shown in Figure 13, and glue the triangles together in the following way: First note that T has a unique center, $(1+i\sqrt{3})/2$, and that dropping the perpendiculars from the center to the sides we obtain three geometrically well-determined "midpoints" of the sides. Then label the sides of each copy of T counterclockwise by a, b, c and glue the copies together in the following way:

If two half-edges v^x and w^y form an edge of $\mathcal{G}$, then paste side x of the corresponding copy T_v to side y of T_w such that the midpoints are glued together and such that the orientations of the copies are preserved.

This yields an orientable surface $S^{op}(\mathcal{G})$ with a complete metric of finite area and constant curvature -1, and the Belyĭ surface is then the conformally compactified $S^{cl}(\mathcal{G})$. It is easy to see that the result is not affected if we cyclically permute the enumeration of the vertices, so the surface really just depends on $\mathcal{G}$.

Individual changes of orientations of vertices, however, have strong effects as may be seen by the following way of computing the genus g of $S^{cl}(\mathcal{G})$. Call a closed edge path in $\mathcal{G}$ a *left hand turn path* if an observer walking along the path in the proper direction is turning left at each vertex (with respect to the orientation of the vertex). The number of such paths, say L, is equal to the number of cusps of $S^{op}(\mathcal{G})$, and so the genus is computed by the formula

$$g = 1 + \frac{k-L}{2}.$$

We note that the Platonic surfaces are a special case of this: $\mathcal{G}$ is the pattern of the triangulation of P_n, respectively S_n, with $2k = \omega_n/3$ triangles, and there are $L = \omega_n/n$ left hand turn paths, each of path length n.

In view of the fact that the above construction sweeps out the Belyĭ surfaces which are known to be dense in the union of the moduli spaces $\mathcal{M}_g$, Brooks and Makover ([**66, 72, 76**]) base their concept of random Riemann surface on the above construction, and we call it here *graph-randomness.*

The random process is an adaption of Bollobás' model [**106, 107**] to vertex-oriented graphs and may be described as follows. Consider $2k$ vertices represented by points in the plane, numbered $1, 2, \ldots, (2k)$, each with three emanating straight line segments, called "half-edges" and labelled counterclockwise $1^a, 1^b, 1^c, \ldots, (2k)^a, (2k)^b, (2k)^c$. Join the half-edges into edges stepwise in altogether $3k$ steps, by picking at each step at random a pair of free half-edges, and gluing them together to obtain an edge. This procedure defines a probability measure on the set $\mathcal{F}_k^*$ of all (connected and non-connected) 3-regular vertex-oriented graphs on $2k$ vertices.

If Q is a property of graphs in $\mathcal{F}_k^*$, we denote by $Prob_k[Q]$ the probability that a graph $\mathcal{G}$ picked from $\mathcal{F}_k^*$ by the above random process has property Q. For instance:

THEOREM 8.4 (Bollobás,[**106**]). *There exists a constant $C > 0$ such the Cheeger constant satisfies*

$$\mathrm{Prob}_k[h(\mathcal{G}) > C] \to 1, \text{ as } k \to \infty.$$

In particular, the probability that the graph is connected converges to 1 as $k \to \infty$.

The important property is the following, where "to be of cusps $\geq L$" means that the cusp neighbourhoods can be chosen pairwise disjoint and each of size $\geq L$.

THEOREM 8.5 ([**76**]). *For any $L > 0$ one has*

$$\mathrm{Prob}_k[S^{op}(\mathcal{G}) \text{ is of cusps } \geq L] \to 1, \text{ as } k \to \infty.$$

In connection with Theorems 8.1 and 8.2, Brooks and Makover obtain the following result about the first eigenvalue of the Laplacian of a "typical" compact Riemann surface of large genus:

THEOREM 8.6 ([**76**]). *There exists a constant $C_1 > 0$ such that*

$$\mathrm{Prob}_k[\lambda_1(S^{cl}(\mathcal{G})) > C_1] \to 1, \text{ as } k \to \infty.$$

Another result in [**76**] is that the expected value $E(\text{genus}(S^{cl}(\mathcal{G})))$ of the genus is of order $k/2 - (3/4)\log(k)$ as $k \to \infty$. It would be interesting to know whether one can also say something about an individual $\mathcal{M}_g$, for instance, what kind of distribution one might get if in the above random process one sorts out the examples of a given genus.

9. Conclusion

I would like to conclude with an anecdote.

Many years ago, in March 1987, we had organized a workshop in Ste-Croix-Les Rasses, a small place, about three quarters of an hour from Yverdon-Les Bains, way up in the Swiss Jura. The subject was "Graphs and Riemannian Geometry"; Bob Brooks was among the lecturers, and there are actually a number of his ideas that can be traced back to that workshop.

The evening before the workshop began, I waited at the train station in Yverdon for the small train that would bring me up to Ste-Croix. Nobody familiar was in sight, and I was beginning to wonder whether the other participants had found their way, when I observed a person in heavy, North-American-looking winter outfit. Certainly someone going for adventure. The person turned around, and I recognized the well-known face, smiling at me, but saying nothing. Then suddenly: "I'm safe".

I found this characteristic of Bob's way of thinking and his approach to things. And a bit it is characteristic of all of us. A few years later, Bob and I were sitting in a café in Grenoble, the day after Hubert Pesce's habilitation. Drawing on a napkin, Bob explained to me his ideas about random Riemann surfaces and Belyĭ's theorem. But I was suspicious all the time, because of a conformal argument in it which seemed geometrically out of control. Then Bob told me about the Ahlfors-Schwarz Lemma and suddenly I saw: it's safe.

The napkin reappeared in a stack of preprints when I prepared the lecture for the meeting in Haifa, and I copy it here.

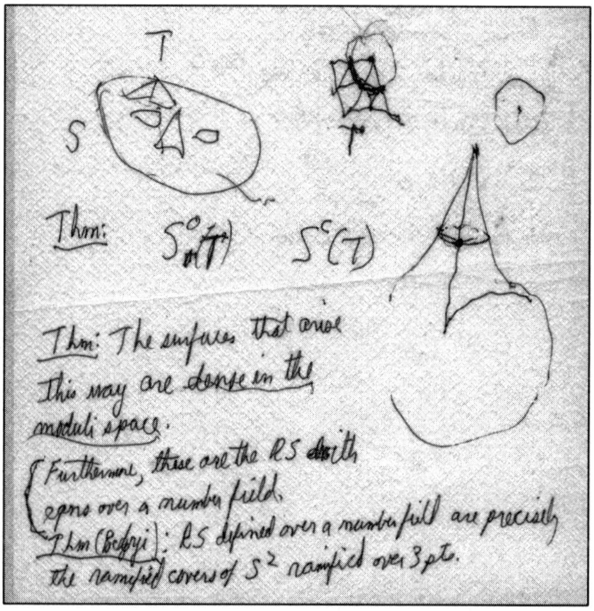

FIGURE 14. From a café in Grenoble, February 21, 1997

Acknowledgements. I would like to thank Aline Aigon, Grégoire Aubry, Yves de Cornulier, Robert Dalang, Thomas Gauglhofer, Hugo Parlier and Klaus-Dieter Semmler for many animated discussions about Riemann surfaces, Kazhdan's Property T, and philosophic aspects of randomness.

a) Articles by Robert Brooks

1. *Appendix* to "On some formulas for the characteristic classes of group actions", by R. Bott, Lecture Notes in Math., vol. 652, Springer, Berlin, 1978, pp. 57–60.
2. (with P. Trauber) *The van Est theorem for groups of diffeomorphisms*, Hadronic J. **1** (1978), 916–934.
3. *Volumes and characteristic classes of foliations*, Topology **18** (1979), 295–304.
4. *Some remarks on bounded cohomology*, Riemann Surfaces and Related Topics: Proceedings of the 1978 Stony Brook Conference (State Univ. New York, Stony Brook, N.Y., 1978), Ann. of Math. Stud., vol. 97, Princeton Univ. Press, Princeton, N.J., 1981, pp. 53–63.
5. (with J. P. Matelski) *The dynamics of* 2*-generator subgroups of* PSL(2, **C**), Riemann Surfaces and Related Topics: Proceedings of the 1978 Stony Brook Conference (State Univ. New York, Stony Brook, N.Y., 1978), Ann. of Math. Stud., vol. 97, Princeton Univ. Press, Princeton, N.J., 1981, pp. 65–71.
6. *Exponential growth and the spectrum of the Laplacian*, Proc. Amer. Math. Soc. **82** (1981), 473–477.
7. *The fundamental group and the spectrum of the Laplacian*, Comment. Math. Helv. **56** (1981), 581–598.
8. *A relation between growth and the spectrum of the Laplacian*, Math. Z. **178** (1981), 501–508.
9. *Amenability and the spectrum of the Laplacian*, Bull. Amer. Math. Soc. (N.S.) **6** (1982), 87–89.
10. (with J. P. Matelski) *Collars in Kleinian groups*, Duke Math. J. **49** (1982), 163–182.
11. *The $\hat{A}$-genus of complex hypersurfaces and complete intersections*, Proc. Amer. Math. Soc. **87** (1983), 528–532.
12. *Some Riemannian and dynamical invariants of foliations*, Differential Geometry (College Park, Md., 1981/1982), Progr. Math., vol. 32, Birkhäuser Boston, Mass., 1983, pp. 56–72.
13. *The spectral geometry of foliations*, Amer. J. Math. **106** (1984), 1001–1012.
14. (with C. Series) *Bounded cohomology for surface groups*, Topology **23** (1984), 29–36.
15. (with W. Goldman) *The Godbillon-Vey invariant of a transversely homogeneous foliation*, Trans. Amer. Math. Soc. **286** (1984), 651–664.
16. (with W. Goldman) *Volumes in Seifert space*, Duke Math. J. **51** (1984), 529–545.
17. *On the spectrum of noncompact manifolds with finite volume*, Math. Z. **187** (1984), 425–432.
18. *On branched coverings of* 3*-manifolds which fiber over the circle*, J. Reine Angew. Math. **362** (1985), 87–101.
19. *The spectral geometry of the Apollonian packing*, Comm. Pure Appl. Math. **38** (1985), 359–366.
20. *The bottom of the spectrum of a Riemannian covering*, J. Reine Angew. Math. **357** (1985), 101–114.
21. *On the deformation theory of classical Schottky groups*, Duke Math. J. **52** (1985), 1009–1024.
22. *The first eigenvalue in a tower of coverings*, Bull. Amer. Math. Soc. (N.S.) **13** (1985), 137–140.
23. *The spectral geometry of a tower of coverings*, J. Differential Geom. **23** (1986), 97–107.
24. *Combinatorial problems in spectral geometry*, Curvature and Topology of Riemannian Manifolds (Katata, 1985), Lecture Notes in Math., vol. 1201, Springer, Berlin, 1986, pp. 14–32.
25. *Circle packings and co-compact extensions of Kleinian groups*, Invent. Math. **86** (1986), 461–469.
26. (with P. Waksman) *The first eigenvalue of a scalene triangle*, Proc. Amer. Math. Soc. **100** (1987), 175–182.
27. *On the angles between certain arithmetically defined subspaces of* $\mathbf{C}^n$, Ann. Inst. Fourier (Grenoble) **37** (1987), 175–185.
28. *On manifolds of negative curvature with isospectral potentials*, Topology **26** (1987), 63–66.
29. (with R. Tse) *Isospectral surfaces of small genus*, Nagoya Math. J. **107** (1987), 13–24, (Correction: ibid **117** (1990), p. 227).
30. *Some remarks on volume and diameter of Riemannian manifolds*, J. Differential Geom. **27** (1988), 81–86.
31. *Isospectral potentials on a surface of genus* 3, Holomorphic Functions and Moduli, Vol. I (Berkeley, CA, 1986), Math. Sci. Res. Inst. Publ., vol. 10, Springer, New York, 1988, pp. 203–208.

32. *Constructing isospectral manifolds*, Amer. Math. Monthly **95** (1988), 823–839.
33. *Injectivity radius and low eigenvalues of hyperbolic manifolds*, J. Reine Angew. Math. **390** (1988), 117–129.
34. *A construction of metrics of negative Ricci curvature*, J. Differential Geom. **29** (1989), 85–94.
35. (with P. Perry and P. Yang) *Isospectral sets of conformally equivalent metrics*, Duke Math. J. **58** (1989), 131–150.
36. *Designer metrics on Riemannian manifolds*, Recent Developments in Geometry (Los Angeles, CA, 1987), Contemp. Math., vol. 101, Amer. Math. Soc., Providence, RI, 1989, pp. 31–46.
37. *Isospectral problems in conformal geometry*, Geometry and Topology of Submanifolds, II (Avignon, 1988), World Sci. Publishing, Teaneck, NJ, 1990, pp. 50–64.
38. (with C. Gordon) *Isospectral families of conformally equivalent Riemannian metrics*, Bull. Amer. Math. Soc. (N.S.) **23** (1990), 433–436.
39. (with Y. Eliashberg and C. McMullen) *The spectral geometry of flat disks*, Duke Math. J. **61** (1990), 119–131.
40. *The spectral geometry of k-regular graphs*, J. Anal. Math. **57** (1991), 120–151.
41. *Some relations between spectral geometry and number theory*, Topology '90 (Columbus, OH, 1990), Ohio State Univ. Math. Res. Inst. Publ., vol. 1, de Gruyter, Berlin, 1992, pp. 61–75.
42. *The continued fraction parameter in the deformation theory of classical Schottky groups*, Curves, Jacobians, and Abelian Varieties (Amherst, MA, 1990), Contemp. Math., vol. 136, Amer. Math. Soc., Providence, RI, 1992, pp. 41–54.
43. (with P. Perry and P. V. Petersen) *Compactness and finiteness theorems for isospectral manifolds*, J. Reine Angew. Math. **426** (1992), 67–89.
44. (with P. Perry and P. V. Petersen) *Finiteness of diffeomorphism types of isospectral manifolds*, Differential Geometry: Riemannian Geometry (Los Angeles, CA, 1990), Proc. Sympos. Pure Math., vol. 54, Amer. Math. Soc., Providence, RI, 1993, pp. 89–94.
45. (with P. Perry and P. V. Petersen) *On Cheeger's inequality*, Comment. Math. Helv. **68** (1993), 599–621.
46. *Spectral geometry and the Cheeger constant*, Expanding Graphs (Princeton, NJ, 1992), DIMACS Ser. Discrete Math. Theoret. Comput. Sci., vol. 10, Amer. Math. Soc., Providence, RI, 1993, pp. 5–19.
47. (with P. Perry and P. V. Petersen) *Some examples in L^p spectral geometry*, J. Geom. Anal. **3** (1993), 293–313.
48. (with P. Perry and P. V. Petersen) *Spectral geometry in dimension* 3, Acta Math. **173** (1994), 283–305.
49. *L^p spectral geometry*, Geometry of the Spectrum (Seattle, WA, 1993), Contemp. Math., vol. 173, Amer. Math. Soc., Providence, RI, 1994, pp. 71–87.
50. (with P. Glezen) *An L^p spectral bootstrap theorem*, Geometry of the Spectrum (Seattle, WA, 1993), Contemp. Math., vol. 173, Amer. Math. Soc., Providence, RI, 1994, pp. 89–97.
51. *Trace-formula methods in spectral geometry*, Proceedings of the Conference in Honor of Jean-Pierre Kahane (Orsay, 1993), J. Fourier Anal. Appl., Special Issue (1995), 87–95.
52. (with H. M. Farkas) *Factorization of N^2+1 and the modular group*, unpublished manuscript (1995).
53. *Cheng's theorem in L^p spectral geometry*, Algebra i Analiz **8** (1996), 98–102.
54. *Reflections on the first eigenvalue*, Texas Tech U. Math. Series **9** (1997), 71–84.
55. *Inverse spectral geometry*, Progress in Inverse Spectral Geometry, Trends Math., Birkhäuser, Basel, 1997, pp. 115–132.
56. (with Y. Kopeliovich) *Uniformization of some quotients of modular curves*, Extremal Riemann Surfaces (San Francisco, CA, 1995), Contemp. Math., vol. 201, Amer. Math. Soc., Providence, RI, 1997, pp. 155–164.
57. (with H. M. Farkas and I. Kra) *Number theory, theta identities, and modular curves*, Extremal Riemann Surfaces (San Francisco, CA, 1995), Contemp. Math., vol. 201, Amer. Math. Soc., Providence, RI, 1997, pp. 125–154.
58. *Some relations between graph theory and Riemann surfaces*, Proceedings of the Ashkelon Workshop on Complex Function Theory (1996), Israel Math. Conf. Proc., vol. 11, Bar-Ilan Univ., Ramat Gan, 1997, pp. 61–73.
59. *Isospectral graphs and isospectral surfaces*, Séminaire de Théorie Spectrale et Géométrie, No. 15, Année 1996–1997, Sémin. Théor. Spectr. Géom., vol. 15, Univ. Grenoble I, Saint Martin d'Hères, 1997, pp. 105–113.

60. (with R. Gornet and W. H. Gustafson) *Mutually isospectral Riemann surfaces*, Adv. Math. **138** (1998), 306–322.
61. *Platonic surfaces*, Comment. Math. Helv. **74** (1999), 156–170.
62. *Twist surfaces*, Random Walks and Discrete Potential Theory (Cortona, 1997), Sympos. Math., vol. XXXIX, Cambridge Univ. Press, Cambridge, 1999, pp. 85–103.
63. *Non-Sunada graphs*, Ann. Inst. Fourier (Grenoble) **49** (1999), 707–725.
64. *The Sunada method*, Tel Aviv Topology Conference: Rothenberg Festschrift (1998), Contemp. Math., vol. 231, Amer. Math. Soc., Providence, RI, 1999, pp. 25–35.
65. (with E. Makover) *The first eigenvalue of a Riemann surface*, Electron. Res. Announc. Amer. Math. Soc. **5** (1999), 76–81 (electronic).
66. *Typical surfaces and random graphs*, Séminaire de Théorie Spectrale et Géométrie, Année 1998–1999, Sémin. Théor. Spectr. Géom., vol. 17, Univ. Grenoble I, Saint Martin d'Hères, 1999, pp. 99–103.
67. *Lars Ahlfors* (1907–1996), Lectures in Memory of Lars Ahlfors (Haifa, 1996), Israel Math. Conf. Proc., vol. 14, Bar-Ilan Univ., Ramat Gan, 2000, pp. 1–2.
68. *Some geometric aspects of the work of Lars Ahlfors*, Lectures in Memory of Lars Ahlfors (Haifa, 1996), Israel Math. Conf. Proc., vol. 14, Bar-Ilan Univ., Ramat Gan, 2000, pp. 31–39.
69. (with R. Gornet and P. Perry) *Isoscattering Schottky manifolds*, Geom. Funct. Anal. **10** (2000), 307–326.
70. (with E. Makover) *Riemann surfaces with large first eigenvalue*, J. Anal. Math. **83** (2001), 243–258.
71. (with P. Perry) *Isophasal scattering manifolds in two dimensions*, Comm. Math. Phys. **223** (2001), 465–474.
72. (with E. Makover) *Belyi surfaces*, Entire Functions in Modern Analysis (Tel-Aviv, 1997), Israel Math. Conf. Proc., vol. 15, Bar-Ilan Univ., Ramat Gan, 2001, pp. 37–46.
73. (with A. Żuk) *On the asymptotic isoperimetric constants for Riemann surfaces and graphs*, J. Differential Geom. **62** (2002), 49–78.
74. (with O. Davidovich) *Isoscattering on surfaces*, J. Geom. Anal. **13** (2003), 39–53.
75. *A statistical model of Riemann surfaces*, Complex Analysis and Dynamical Systems, Contemp. Math., vol. 364, Amer. Math. Soc., Providence, RI, 2004, pp. 15–25.
76. (with E. Makover) *Random construction of Riemann surfaces*, to appear in J. Differential Geom. (2005).

b) Further References

101. L. Ahlfors, *An extension of Schwarz's lemma*, Trans. Amer. Math. Soc. **43** (1938), 359–364.
102. L. Ahlfors and L. Bers, *Riemann's mapping theorem for variable metrics*, Ann. of Math. (2) **72** (1960), 385–404.
103. E. M. Andreev, *Convex polyhedra of finite volume in Lobačevskiĭ space*, Mat. Sb. (N.S.) **83** (1970), 256–260; English transl. in Math. USSR-Sb **12** (1970), 255–259.
104. A. F. Beardon, *The geometry of discrete groups*, Graduate Texts in Mathematics, vol. 91, 2nd ed., Springer-Verlag, New York, 1995.
105. G. V. Belyĭ, *Galois extensions of a maximal cyclotomic field*, Izv. Akad. Nauk SSSR Ser. Mat. **43** (1979), 267–276, 479; English transl. in Math. USSR-Izv **14** (1980), 247–256.
106. B. Bollobás, *The isoperimetric number of random regular graphs*, European J. Combin. **9** (1988), 241–244.
107. B. Bollobás, *Random Graphs*, 2nd ed., Cambridge University Press, Cambridge, 2001.
108. Ph. Bowers and K. Stephenson, *The set of circle packing points in the Teichmüller space of a surface of finite conformal type is dense*, Math. Proc. Cambridge Philos. Soc. **111** (1992), 487–513.
109. Ph. Bowers and K. Stephenson, *Circle packings in surfaces of finite type: an in situ approach with applications to moduli*, Topology **32** (1993), 157–183.
110. P. Buser, *Geometry and spectra of compact Riemann surfaces*, Progress in Mathematics, vol. 106, Birkhäuser, Boston, 1992.
111. J. Cheeger, *A lower bound for the smallest eigenvalue of the Laplacian*, Problems in Analysis (Papers dedicated to Salomon Bochner, 1969), Princeton Univ. Press, Princeton, N.J., 1970, pp. 195–199.

112. D. M. DeTurck and C. S. Gordon, *Isospectral deformations. I. Riemannian structures on two-step nilspaces*, Comm. Pure Appl. Math. **40** (1987), 367–387.
113. P. G. Doyle, *On the bass note of a Schottky group*, Online document, August 1994, available at `http://www.math.dartmouth.edu/~ doyle/`.
114. H. M. Farkas and I. Kra, *Riemann surfaces*, Graduate Texts in Mathematics, vol. 71, 2nd ed., Springer-Verlag, New York, 1992.
115. W. Fenchel, *Elementary geometry in hyperbolic space*, de Gruyter Studies in Mathematics, vol. 11, Walter de Gruyter & Co., Berlin, 1989.
116. C. S. Gordon and E. N. Wilson, *Isospectral deformations of compact solvmanifolds*, J. Differential Geom. **19** (1984), 241–256.
117. G. Jones and D. Singerman, *Belyĭ functions, hypermaps and Galois groups*, Bull. London Math. Soc. **28** (1996), 561–590.
118. M. Kapovich, *Hyperbolic manifolds and discrete groups*, Progress in Mathematics, vol. 183, Birkhäuser, Boston, Basel, Berlin, 2001.
119. P. Koebe, *Kontaktprobleme der Konformen Abbildung*, Ber. Sachs. Akad. Wiss. Leipzig, Math.-Phys. Klasse **88** (1936), 141–164.
120. A. Lubotzky and R. J. Zimmer, *Variants of Kazhdan's property for subgroups of semi-simple groups*, Israel J. Math. **66** (1989), 289–299.
121. A. Lubotzky, *Discrete groups, expanding graphs and invariant measures*, Progress in Mathematics, vol. 125, Birkhäuser, Boston, Basel, Berlin, 1994.
122. W. Luo, Z. Rudnick, and P. Sarnak, *On Selberg's eigenvalue conjecture*, Geom. Funct. Anal. **5** (1995), 387–401.
123. G. A. Margulis, *Discrete groups of motions of manifolds of nonpositive curvature*, Proceedings of the International Congress of Mathematicians (Vancouver, B.C., 1974), Vol. 2, Canad. Math. Congress, Montreal, Que., 1975, pp. 21–34.
124. B. Maskit, *Kleinian groups*, Grundlehren der mathematischen Wissenschaften, vol. 287, Springer-Verlag, New York, 1988.
125. R. Osserman, *The isoperimetric inequality*, Bull. Amer. Math. Soc. **84** (1978), 1182–1238.
126. R. Phillips and P. Sarnak, *The Laplacian for domains in hyperbolic space and limit sets of Kleinian groups*, Acta Math. **155** (1985), 173–241.
127. P. Sarnak, *Selberg's eigenvalue conjecture*, Notices Amer. Math. Soc. **42** (1995), 1272–1277.
128. A. Selberg, *On the estimation of Fourier coefficients of modular forms*, Proc. Sympos. Pure Math., Vol. VIII, Amer. Math. Soc., Providence, RI, 1965, pp. 1–15.
129. D. Singerman, *Finitely maximal Fuchsian groups*, J. London Math. Soc. (2) **6** (1972), 29–38.
130. K. Stephenson, *Circle packing: a mathematical tale*, Notices Amer. Math. Soc. **50** (2003), 1376–1388.
131. D. Sullivan, *The density at infinity of a discrete group of hyperbolic motions*, Inst. Hautes Études Sci. Publ. Math. **50** (1979), 171–202.
132. D. Sullivan, *On the ergodic theory at infinity of an arbitrary discrete group of hyperbolic motions*, Ann. Math. Stud. **97** (1981), 465–496.
133. T. Sunada, *Riemannian coverings and isospectral manifolds*, Ann. of Math. (2) **121** (1985), 169–186.
134. W. P. Thurston, *The geometry and topology of three-manifolds*, Lecture Notes, Princeton University, 1980; Electronic version 1.1 March 2002, `http://www.msri.org/publica-tions/books/gt3m/`.
135. R. Tse, *A lower bound for the number of isospectral surfaces*, Recent Developments in Geometry (Los Angeles, CA, 1987), Contemp. Math., vol. 101, Amer. Math. Soc., Providence, RI, 1989, pp. 161–164.
136. S. Wolpert, *A generalization of the Ahlfors-Schwarz lemma*, Proc. Amer. Math. Soc. **84** (1982), 377–378.

SECTION DE MATHÉMATIQUES, ECOLE POLYTECHNIQUE FÉDÉRALE DE LAUSANNE, SB-IGAT-GEOM STATION 8, CH-1015 LAUSANNE, SWITZERLAND
E-mail address: `peter.buser@epfl.ch`

Contemporary Mathematics
Volume **387**, 2005

Moduli Spaces of Homotopy Theory

David Blanc

This article is dedicated to the memory of Bob Brooks, colleague and friend

ABSTRACT. The moduli spaces referred to are topological spaces whose path components parametrize homotopy types. Such objects have been studied in two separate contexts: rational homotopy types, in the work of several authors in the late 1970's; and general homotopy types, in the work of Dwyer-Kan and their collaborators. We here explain the two approaches, and show how they may be related to each other.

1. Introduction

The concept of "moduli" for a mathematical object goes back to Riemann, who used it to describe a set of parameters determining the isomorphism class of a Riemann surface of a given topological type. He also recognized that the set of all conformal equivalence classes of such surfaces can itself be given a complex structure, although this was only made precise by Teichmüller, Ahlfors, Bers, and others in the twentieth century. Similarly, the family of birational equivalence classes of algebraic curves of each genus (over $\mathbb{C}$, say) are parametrized by an algebraic variety of moduli, as Mumford showed, and there are analogues in higher dimensions.

If we fix an oriented surface S of genus g, a marked Riemann surface is a choice of an orientation-preserving diffeomorphism of S into a compact Riemann surface X, and the equivalence classes of such marked surfaces form Teichmüller space $\mathcal{T}_g$, a complex analytic variety. The mapping class group Γ_g of orientation preserving diffeomorphisms of S acts on $\mathcal{T}_g$, and the quotient is isomorphic to the *moduli space* $\mathcal{M}_g$ of isomorphism classes of complex structures on S – or equivalently, of smooth projective curves over $\mathbb{C}$. Since the action of Γ_g is virtually free, and $\mathcal{T}_g$ is contractible, $\mathcal{M}_g$ has the same rational homology as the classifying space $B\Gamma_g$.

Note that the set of all Riemann surfaces of a given genus can be obtained by deforming a given conformal (or equivalently, complex) structure on a fixed sample

2000 *Mathematics Subject Classification.* Primary 55P15; Secondary 55Q99, 55P62, 18G30, 14J10.

surface Σ_g – or the corresponding Fuchsian group Γ. Such deformations are typically governed by an appropriate cohomology group.

Note also that there are several levels of possible structures on a compact surface: topological, differentiable, real or complex analytic, hyperbolic, conformal, metric, and so on, which define different types of moduli space. See [**31**] or [**51**] for more details and references on the classical theory.

1.1. Moduli spaces. Thus, there are a number of common themes in the moduli problems arising in various areas of mathematics:

(a) Finding a set of parameters to describe appropriate equivalence classes of objects in a certain geometric category. These may occur at various levels, and we distinguish between discrete parameters (such as the genus of a surface) and the finer (continuous) *moduli*.
(b) Giving the set of such equivalence classes the structure of an object in the same category, which we may call a *moduli space*.
(c) The *deformation* of a given structure, controlled by cohomology, as a means of obtaining all other possible structures.
(d) To pass between levels of structure, one can often quotient by a group action.

As we shall see, all these themes will find expression in the context of homotopy theory. One aspect of the usual moduli spaces for which we have no analogue in our setting is the important role played by compactifications.

1.2. Moduli spaces in homotopy theory. The homotopy theory of the classical moduli spaces of Riemann surfaces and the corresponding mapping class groups has been studied extensively – see, for example, [**29, 35, 53**]. However, this is not our subject; we shall be concerned here rather with the analogues of such moduli spaces in the category of topological spaces, answering to the general description given above.

The simplest examples of this approach are provided by various mapping spaces, starting with the loop space ΩX of all pointed maps from the circle into a topological space X. Here the set $\pi_0\Omega X$ of components is isomorphic to the fundamental group of X, and the higher homotopy groups are again those of X (re-indexed). More significantly, the components of space of maps from X to $B\mathrm{O}$ or $B\mathrm{U}$, the classifying spaces of the infinite orthogonal or unitary groups, correspond to equivalence classes of (real or complex) vector bundles over X, and the higher homotopy groups correspond to various reductions of structure, by Bott periodicity. Further examples of simple "moduli spaces" of topological spaces are provided by the configuration spaces of n distinct points in a manifold M, which have also been studied extensively (see [**21**] for a comprehensive survey).

1.3. Moduli spaces and homotopy types. Our focus here will be more specifically on moduli spaces of all homotopy types (suitably interpreted), which have been investigated from at least two points of view over the past twenty-five years:

- In work of Lemaire-Sigrist, Félix and Halperin-Stasheff, simply-connected rational homotopy types correspond to the components of a certain (infinite dimensional) algebraic variety over $\mathbb{Q}$ – or alternatively, the quotient of such a variety by appropriate group actions (Sec. 2).

- The "classification complex" approach of Dwyer-Kan yields a space whose components parametrize the homotopy types of all CW complexes (Sec. 6).

Of course, neither statement has much content, as stated; the point is that both approaches offer (surprisingly similar) tools for inductively analyzing the collection of all relevant homotopy types:

In each case we begin with a coarse classification of homotopy types by algebraic invariants, such as the cohomology algebra H of a space, or the corresponding structure on homotopy groups, called a Π-algebra (which reduces rationally to a graded Lie algebra over $\mathbb{Q}$ – see Sec. 3). One might also refine this initial classification using both homotopy and cohomology groups (Sec. 5).

(a) For each H, we have homological classification of spaces with this cohomology algebra. This is usually presented as an obstruction theory, stated in terms of (algebraically defined) cohomology groups of H. It can be reinterpreted as expressing the moduli space as the limit of a tower of fibrations (Sec. 7), with successive fibers described in terms of these cohomology groups (Sec. 8).

(b) In the rational case, we have a deformation theory approach, which describes all homotopy types with $H^*(X,\mathbb{Q}) \cong H$ by perturbing a canonical model. In fact, this was the main focus of the rational homotopy theory approach mentioned above.

Our main goal here is to bring out the connection between the integral and rational cases. From the first point of view, this can be made explicit at the algebraic level, comparing the two cohomology theories through appropriate spectral sequences (Sec. 9).

On the other hand, the rational deformation theory has no integral analogue, since it uses differential graded models, which exist only for simply-connected rational homotopy types. However, it turns out that the deformations can be described geometrically in terms of higher homotopy operations (Sec. 4), and these do have integral versions.

1.4. Remark. One could – with some justice – argue that the analogy between the rational and integral "spaces of homotopy types" we describe here and classical moduli spaces is rather tenuous (although it is merely incidental to the point we want to make). One aspect in particular for which we can offer no real analogue is the fact that the classical moduli spaces have the same structure (algebraic or analytic variety, etc.) as the objects being classified.

In some sense, however, a homotopy type *is* just the collection of all its homotopy invariants, suitably interpreted (see [**41**, I,4.9]). In fact, one goal of homotopy theory is to produce a manageable set of such invariants, sufficient for distinguishing between type. A starting point for such a set of invariants is the Postnikov system of a space (and its invariants); so it is perhaps fitting that our analysis of the space of all homotopy types is carried out by means of its Postnikov system, and the associated collections of homotopy invariants (higher homotopy operations, or cohomology classes) which appear in our obstruction theory. Thus one could choose to view these collections of invariants themselves – rather than the space of homotopy types – as the true "moduli object". This is certainly the best way to understand the connections between the rational and integral cases, and it also might make (somewhat far-fetched) sense of the claim that the moduli object is itself of the same kind as the things it classifies. We leave the philosophically-inclined reader to pursue this line of thought at his or her discretion.

1.5. ACKNOWLEDGEMENTS. I would like to thank the referee for his or her comments, and Bill Dwyer, Paul Goerss, Jim Stasheff, and Yves Félix for helpful conversations and remarks on various aspects of the subject.

2. Rational homotopy and deformation theory

We first describe the deformation-theoretic approach to rational homotopy types, starting with some basic background material:

2.1. Rational homotopy. Rational homotopy theory deals with rational algebraic invariants of homotopy types – that is, it disregards all torsion in the homology and homotopy groups. Quillen and Sullivan, respectively, proposed two main algebraic models for the rational homotopy type of a simply-connected space X:

(i) A differential graded-commutative algebra, or DGA (A^*_X, d), where $(A^i_X)_{i=0}^\infty$ is a graded-commutative algebra over $\mathbb{Q}$, equipped with a differential d of degree $+1$ (which is a graded derivation with respect to the product), and $H^*(A_X, d) \cong H^*(X, \mathbb{Q})$. Since X is simply-connected, we may assume (A^*_X, d), is, too – that is, $A^0_X = \mathbb{Q}$, $A^1_X = 0$.

(ii) A differential graded Lie algebra, or DGL (L^X, ∂), with $L^X = (L^X_i)_{i=1}^\infty$ a positively-graded Lie algebra over $\mathbb{Q}$, and ∂ a graded Lie derivation of degree -1, such that $H_i(L^X, \partial) \cong \pi_i \Omega X \otimes \mathbb{Q}$ for $i \geq 1$, with the Samelson products as Lie brackets.

For more information see the survey [**25**], and the original sources [**42, 50**].

2.2. DEFINITION. A DGA of the form $(\Lambda V, d)$, where ΛV is the free graded-commutative algebra generated by the graded vector space $V = (V^i)_{i=0}^\infty$ is called *cofibrant.* A cofibrant DGA $(\Lambda V, d)$ such that $\mathrm{Im}\,(d) \subseteq \Lambda^+ V \cdot \Lambda^+ V$ (where $\Lambda^+ V$ is the sub-algebra of elements in positive degree), is called a *minimal model.*

A DGA or DGL map which induces an isomorphism in (co)homology is called a *quasi-isomorphism.*

2.3. PROPOSITION (see [**1**]). *Any simply-connected DGA A has a minimal model with a quasi-isomorphism $q : (\Lambda V, d) \to (A, d_A)$, and this is unique up to isomorphism.*

One can also define cofibrant and minimal models for DGLs, and show that they have similar properties (cf. [**25**, §21]).

2.4. REMARK. Both the category $\mathcal{D}GA_1$ of simply-connected DGAs, and the category $\mathcal{D}GL_0$ of connected DGLs have model category structures (see [**41**]); thus each has a concept of homotopy, and as Quillen showed, the corresponding homotopy categories $\mathrm{ho}\,\mathcal{D}GA_1$ and $\mathrm{ho}\,\mathcal{D}GL_0$ are both equivalent to the rational homotopy category $\mathrm{ho}\,\mathcal{T}_1^{\mathbb{Q}}$ of simply-connected topological spaces (cf. [**42**]). The weak equivalences in $\mathcal{T}_1^{\mathbb{Q}}$ correspond of course to quasi-isomorphisms for DGAs or DGLs.

Note that every simply-connected graded-commutative algebra over $\mathbb{Q}$ is realizable as the rational cohomology ring of some (1-connected) space, and similarly, every positively graded Lie algebra is realizable as the rational homotopy groups of a space.

2.5. The space of rational homotopy types. We may therefore identify the collection $\mathcal{M} = \mathcal{M}^{\mathcal{T}_1^{\mathbb{Q}}}$ of all rational weak homotopy types with the set of weak homotopy types of 1-connected DGAs (over $\mathbb{Q}$) – i.e., equivalence classes under the equivalence relation generated by the quasi-isomorphisms – and similarly for DGLs. Moreover, there is a coarse classification of such types, with the discrete parametrization provided by the cohomology rings; we denote the sub-collection of $\mathcal{M}$ consisting of all DGAs with a given graded cohomology ring H by $\mathcal{M}_H$. This has a distinguished element:

2.6. Definition. A DGA A is called *formal* if it is quasi-isomorphic to $(H^*A, 0)$ (the cohomology ring of A thought of as a DGA with zero differential). Dually, a DGL L is *coformal* if it is quasi-isomorphic to $(H_*L, 0)$.

$\mathcal{M}_H$ was studied by Jean-Michel Lemaire and François Sigrist in [**34**], by Yves Félix in [**23**], and independently by Steve Halperin and Jim Stasheff in [**27**]. They showed that every rational homotopy type in $\mathcal{M}_H$ can be obtained by suitable deformations of a fixed model of the unique formal object:

2.7. Definition. The *bigraded model* (B, d) for H is the minimal model $B = \Lambda Z$ for $(H, 0)$ of Proposition 2.3, where $Z = Z_*^*$, the graded vector space of generators, is equipped with an additional homological grading (indicated by the lower index).

The differential takes Z_n^i to $\Lambda(\oplus_{k=1}^{n-1} Z_k)^{i+1}$, so one still has a homological grading on the cohomology groups of B. The vector spaces Z_n^* are defined (almost canonically) by induction on n, so that at the n-th stage we kill the cohomology in homological dimension k for $0 < k < n$. $(\Lambda Z_*, d)$ is essentially the Tate-Jozefiak resolution of a graded commutative algebra H (cf. [**52, 32**]) – i.e., a minimal "cellular" resolution. See [**27**, §3] for the details.

The deformation consists of a perturbation of the differential d to $D = d + D'$ – of course, this is done in such a way that the cohomology is unchanged, and $D \circ D = 0$. We can no longer expect D to respect both gradings, but if we think of B as being only *filtered* by $\mathcal{F}_n B := \oplus_{k-0}^{n} (\Lambda Z^*)_k$, then the perturbation is such that D' takes Z_n into $\mathcal{F}_{n-2}B$.

2.8. Theorem (cf. [**27**, Theorem 4.4]). *If* (A, d_A) *is any DGA with* $H^*(A, d_A) \cong H$, *one can deform the bigraded model* (B, d) *for* $(H, 0)$ *into a filtered model* (B, D) *quasi-isomorphic to* (A, d_A), *and this is unique up to isomorphism.*

2.9. The variety of deformations. The deformations D of d are determined by successive choices of linear transformations $D'_n : Z_n \to \mathcal{F}_{n-2}\Lambda Z^*_{<n}$, satisfying a quadratic algebraic condition (to ensure that $D^2 = 0$). Thus the collection of all possible perturbations of the given bigraded model (B, d) constitute an (infinite dimensional) algebraic variety over $\mathbb{Q}$, which we denote by $\mathcal{V}_H$. Every simply-connected rational homotopy type is represented by at least one point of $\mathcal{V}_H$; each such point actually represents an "augmented" DGA X, equipped with a specific isomorphism $\phi : H^*(X; \mathbb{Q}) \to H$.

However, one can have more than one point corresponding to each augmented rational homotopy type: the points of $\mathcal{M}_H$ correspond bijectively with the orbits of V_H under the action of an appropriate group, as in [**34**, Thm. 4] or [**23**, Ch. 1] (note that these two superficially different descriptions are in fact the same, by [**24**]):

In the approach described in [**47**], we first divide by the action of a certain pro-unipotent algebraic group G_H to obtain the set V_H/G_H of augmented rational homotopy types as above. We must then further mod out by $\mathrm{Aut}(H)$ to get the plain (i.e., unaugmented) rational homotopy types, so that $\mathcal{M}_H \cong \mathrm{Aut}(H)\backslash V_H/G_H$. In [**47**, §6] it is then shown that the quotient set V_H/G_H can also be described as the set of path components of the Cartan-Chevalley-Eilenberg "standard construction" $C(L)$ (cf. §5.2) on a certain Lie algebra L (namely, the algebra of derivations of the bigraded model for H).

Note that though $\mathcal{M}_H$ can be uncountable, it can also be small: a DGA (A, d) is called *intrinsically formal* if $\mathcal{M}_{H^*A}$ consists of a single point.

2.10. Obstruction theory. We have seen that the collection $\mathcal{M}^{\mathcal{T}_1^{\mathbb{Q}}}$ of all simply-connected rational homotopy types indeed exhibits the basic characteristics of a moduli space listed in §1.1. However, to make this useful, we need further information about its overall structure.

This is provided by Halperin and Stasheff in the form of an obstruction theory for realizing a given isomorphism of rational cohomology algebras $\phi : H^*(A, d_A) \to H^*(B, d_B)$ by a DGA map (or equivalently, by a map of the corresponding rational spaces). Since such a map induces a homotopy equivalence between the respective cofibrant models, this can be thought of as a method for distinguishing between the homotopy types in $\mathcal{M}_H$.

The obstruction theory, which only works in full generality if the DGAs in question are of finite type, was presented in [**27**, §5] in terms of a sequence of elements $\mathcal{O}_n(f) \in \mathrm{Hom}(Z_{n+1}, H(B))$ – or rather, in a quotient of this group. In [**49**, §4], these are re-interpreted as elements in $\bigoplus_p H^{p+1}(B; \pi_p(H(A)))$.

2.11. Remark. By analogy with classical deformation theory, we would expect the deformations of the bigraded model to be "controlled" by a suitable differential graded Lie algebra (L, d), or perhaps by its cohomology. In fact, Félix observed that the Halperin-Stasheff obstructions $\mathcal{O}_n(f)$ can be thought of as lying in the F-cohomology group ${}_FH^1_n(B, d)$ of the bigraded model (B, d) for $(H, 0)$ (see [**23**, Ch. 5, Prop. 7]). Moreover, these groups can be identified with the tangent (or Harrison) cohomology of the graded algebra H (cf. [**23**, Ch. 3, Prop. 4]), which is the same as the André-Quillen cohomology, and this can indeed be calculated as the cohomology of a suitable DGL (cf. [**46**, §4]).

Moreover, both Schlessinger and Stasheff (see [**49**, §5]) and Félix (cf. [**23**, Ch. 4-5]) obtain results connecting formality and intrinsic formality of a DGA with its tangent cohomology, though the relation does not work both ways (see [**23**, Ch. 4, Ex. 2]).

We shall give a uniform treatment of this cohomological description in a more general context in Section 8. However, for technical reasons (inter alia, convergence of second- versus first-quadrant spectral sequences), this is done for the Eckmann-Hilton dual – that is, in terms of Π-algebras (the integral version of graded Lie algebras over $\mathbb{Q}$). The mod p version – that is, an obstruction theory for realizing unstable (co)algebras over the Steenrod algebra – appears in [**6**].

3. Moduli spaces of graded Lie algebras

The description of the moduli space of rational homotopy types in terms of DGLs, rather than DGAs, has received less attention. However, as noted above, it has certain technical advantages, especially in the integral case (where we need no longer restrict to simply-connected spaces, incidentally). The dualization to DGLs is quite straightforward, but we shall describe it here from a slightly different point of view.

3.1. Deformations of DGLs. The deformation theory for DGLs proceeds precisely along the lines indicated in §2.5*ff.*: given a positively-graded Lie algebra $P = P_*$ over $\mathbb{Q}$, consider the collection $\mathcal{M}_P$ of all rational homotopy types of simply-connected spaces X – or equivalently, of DGLs (L, ∂) – such that $\pi_*\Omega X \otimes \mathbb{Q} = H_*(L, \partial) \cong P$ (as Lie algebras).

Again there is a distinguished coformal element $(P, 0)$ in $\mathcal{M}_P$ (Def. 2.6). Its DGL minimal model (B, ∂) can be given an additional homological grading, and the differential in this bigraded model can be deformed so as to obtain a filtered model for any X in $\mathcal{M}_P$. See the sketch in [**28**]; the details appear in the unpublished thesis [**40**].

3.2. The resolution model category. Since the category $\mathcal{DGL}_0$ is not abelian, from the point of view of homotopical algebra it is more natural to consider simplicial resolutions of objects in this category, rather than "chain complex" resolutions (which is what the homological grading of the bigraded model gives us). Moreover, this generalizes more naturally to the integral setting, where differential graded models are no longer available (although see [**36**] for a way around this).

Recall that a *simplicial object* $X_\bullet$ over a category $\mathcal{C}$ is a functor $X : \Delta^{op} \to \mathcal{C}$, where Δ is the category of finite sequences $\mathbf{n} = (0, 1, \ldots, n)$ $(n \in \mathbb{N})$, with order-preserving maps as morphisms. The category of simplicial objects over $\mathcal{C}$ will be denoted by $s\mathcal{C}$.

Of course, $\mathcal{C} = \mathcal{DGL}_0$ (as well as $\mathcal{DGA}_1$) is itself a model category (as shown by Quillen in [**42**]), so in considering simplicial DGLs we can take advantage of the procedure described by Dwyer, Kan, and Stover for generating a model category structure on $s\mathcal{C}$. Essentially, this consists of:

- choosing a certain set $\{M_\alpha\}_{\alpha \in A}$ of "good" objects in $\mathcal{C}$ (corresponding to the projectives of homological algebra);
- creating simplicial objects $\Sigma^n M_\alpha$ for each $n \in \mathbb{N}$ and $\alpha \in A$ by placing M_α in simplicial dimension n; and using these to
- define weak equivalences (which are detected by maps out of $\Sigma^n M_\alpha$), and
- construct the cofibrant objects (corresponding to projective resolutions), and more generally all cofibrations.

See [**19**] for more details; as extended by Bousfield (who calls it a "resolution model category") in [**14**], this procedure works in great generality. In particular, it includes the model categories for universal algebras defined by Quillen in [**41**, II, §4], as well as the graded version (see, e.g., [**12**, §2.2]). The original example considered by Dwyer, Kan, and Stover was that of simplicial spaces; this can be readily modified to cover simplicial rational spaces (equivalently, simplicial DGLs or DGAs).

3.3. Simplicial DGLs. In applying the above procedure to $\mathcal{C} = \mathcal{DGL}_0$, we have a natural candidate for the "good objects" – namely, the DGL spheres $\{\mathcal{S}^n\}_{n=2}^{\infty}$ (which are just minimal models for $\{S^n_{\mathbb{Q}}\}_{n=2}^{\infty}$ – i.e., $\mathcal{S}^n = \mathcal{S}^n_{(x)}$ is a free graded Lie algebra on a single generator x in dimension n).

This determines what we mean by a (simplicial) *resolution* of a given DGL (L, ∂) – namely, a cofibrant simplicial DGL V, with each V_n homotopy equivalent to a coproduct $\coprod_{i \in I_n} \mathcal{S}^{n_i}$, such that the augmented simplicial DGL $V \to L$ is weakly contractible. A map $\varphi : V \to X$ of simplicial DGLs is a weak equivalence if and only if after taking (internal) homology in each simplicial dimension, the resulting map of graded simplicial abelian groups $\varphi_* : H_*V \to H_*X$ induces an isomorphism $H_n(\varphi_*) : H_n(H_*(V)) \xrightarrow{\cong} H_n(H_*(X))$ for each $n \geq 0$.

By [**42**, Thm. 2.1], we know that the model category $\mathcal{DGL}_0$ is equivalent to the category $s\mathcal{L}ie_1$ of connected simplicial Lie algebras; similarly for bigraded differential Lie algebras and simplicial graded Lie algebras by [**7**, Prop. 2.9]. In fact, it turns out that there is also a one-to-one correspondence between filtered DGLs and (cofibrant) simplicial DGL resolutions (see [**7**, Prop. 5.13]), so the DGL deformation theory mentioned in §3.1 can be translated into the language of simplicial resolutions.

3.4. Moduli and realizations. One can analyze $\mathcal{M}_P$ in terms of a *realization problem*: which distinct rational homotopy types of 1-connected topological spaces have the given graded Lie algebra P as their rational homotopy groups? In view of the equivalences of categories noted above, we can replace the DGL bigraded model (B, ∂) for $(P, 0)$ by a free simplicial resolution $G \to P$ of graded Lie algebras, and observe that if P itself can be realized by a rational space X (and so by a DGL (L^X, ∂)), then G can be realized by a *simplicial* space (or DGL) V. Moreover, any weak equivalence $\varphi : V \to W$ of simplicial spaces (or DGLs) induces a weak equivalence $\varphi_\# : \pi_*V \otimes \mathbb{Q} \to \pi_*W \otimes \mathbb{Q}$, and since $\pi_*V \otimes \mathbb{Q} = G$ is a resolution of P, this is just a choice of an isomorphism between two isomorphic copies of P.

As we shall see below (§6), there is a "moduli space" associated to any model category, and the various levels of classification which are a feature of moduli spaces in general (§1.1(i)) can be understood in terms of "structure-reducing" functors between the corresponding model categories – here exemplified by $\pi_*(\ ) \otimes \mathbb{Q} : \mathcal{T}_1 \to \operatorname{gr}\mathcal{L}ie$ (extended to simplicial objects).

4. Moduli and higher homotopy operations

Ideally, we want not only an algebraic description of the components of $\mathcal{M}_H$ in terms of the cohomology of the graded algebra H, say, as in §2.10, but also a geometric interpretation of this description – with the potential for lifting to the integral case.

Note that there is an obvious candidate for geometric invariants to distinguish between inequivalent rational spaces with the same cohomology – namely, Massey products, and their higher order analogues (see [**38, 33**]). In fact, Halperin and Stasheff's original motivation for their obstruction theory was to make precise a folk theorem stating that a space is formal if and only if all higher order Massey products – or more generally, all higher-order cohomology operations – vanish.

One difficulty is in the very definition of higher cohomology operations (see, e.g., [**37**]). This can be bypassed to some extent by re-interpreting them in terms of differentials in a spectral sequence (cf. [**27**, §7]) – at the price of losing both their geometric content, and the ability to identify the same higher operation in different contexts.

Because their Eckmann-Hilton dual – namely, higher homotopy operations – is somewhat more intuitive than the cohomology version, we shall concentrate on these. A precise definition requires some care; in the following sections, we present the version of [**10**]:

4.1. DEFINITION. A *lattice* is a finite directed non-unital category Γ (that is, we omit the identity maps), equipped with two objects $v_{\text{init}} = v_{\text{init}}(\Gamma)$ and $v_{\text{fin}} = v_{\text{fin}}(\Gamma)$ with a unique map $\phi_{\max} : v_{\text{init}} \to v_{\text{fin}}$, and for every $w \in V := \text{Obj}\,\Gamma$, there is at least one map from v_{init} to w, and at least one from w to v_{fin}. A composable sequence of k arrows in Γ will be called a *k-chain.*

4.2. The *W*-construction. Given a lattice Γ, one can define a new category $W\Gamma$ enriched over cubical sets, with the same set of objects V (cf. [**13**, III, §1]):

For $u, v \in V$, let $\Gamma_{n+1}(u,v)$ be the set of $(n+1)$-chains from u to v in Γ, and

$$W\Gamma(u,v) := \coprod_{n\geq 0} \Gamma_{n+1}(u,v) \times I^n / \sim,$$

where I is the unit interval. Write

$$f_1 \circ_{t_1} f_2 \cdots f_n \circ_{t_n} f_{n+1}$$

for

$$\langle u \xrightarrow{f_{n+1}} v_n \cdots \to v_1 \xrightarrow{f_1} v\rangle \times (t_n, \dots, t_1)$$

in $\Gamma_{n+1}(u,v) \times I^n$; then the relation $\sim$ is generated by

$$f_1 \circ_{t_1} f_2 \cdots f_n \circ_{t_n} f_{n+1} \;\sim\; f_1 \circ_{t_1} \cdots \circ_{t_{i-1}} (f_i f_{i+1}) \circ_{t_{i+1}} \cdots \circ_{t_n} \circ f_{n+1},$$

if $t_i = 0$ for $1 \leq i \leq n$, where $(f_i f_{i+1})$ denotes f_i composed with f_{i+1}.

The categorial composition in $W\Gamma$ is given by the concatenation:

$$(f_1 \circ_{t_1} \cdots \circ_{t_l} f_{l+1}) \circ (g_1 \circ_{u_1} \cdots \circ_{u_k} g_{k+1}) := (f_1 \circ_{t_1} \cdots \circ_{t_l} f_{l+1} \circ_1 g_1 \circ_{u_1} \cdots \circ_{u_k} g_{k+1}).$$

We write $\mathcal{P} := W\Gamma(v_{\text{init}}, vf)$.

4.3. DEFINITION. The *basis* category $bW\Gamma$ for a lattice Γ is defined to be the cubical subcategory of $W\Gamma$ with the same objects, and with morphisms given by $bW\Gamma(u,v) := W\Gamma(u,v)$ if $(u,v) \neq (v_{\text{init}}, v_{\text{fin}})$, while

$$bW\Gamma(v_{\text{init}}, v_{\text{fin}}) := \bigcup \{\alpha \circ \beta \mid \beta \in W\Gamma(v_{\text{init}}, w), \alpha \in W\Gamma(w, v_{\text{fin}}), v_{\text{init}} \neq w \neq v_{\text{fin}}\},$$

so that $b\mathcal{P} := bW\Gamma(v_{\text{init}}, v_{\text{fin}})$ consists of all decomposable morphisms.

4.4. FACT ([**10**, Proposition 2.15]). *For any lattice Γ, $W\Gamma(v_{\text{init}}, v_{\text{fin}})$ is isomorphic to the cone on its basis, with vertex corresponding to the unique maximal 1-chain.*

We can use the W-construction to define a higher homotopy operations, as follows:

4.5. Definition. Given $\mathcal{A} : \Gamma \to \mathrm{ho}\,\mathcal{T}_*$ for a lattice Γ as above, the corresponding *higher order homotopy operation* is the subset

$$\langle\langle \mathcal{A} \rangle\rangle \subset [\mathcal{A}(v_{\mathrm{init}}) \rtimes b\mathcal{P}, \mathcal{A}(v_{\mathrm{fin}})]_{\mathrm{ho}\,\mathcal{T}_*}$$

of the homotopy equivalence classes of maps

$$C\mathcal{A}|_{bW\Gamma(v_{\mathrm{init}}, v_{\mathrm{fin}})} : bW\Gamma(v_{\mathrm{init}}, v_{\mathrm{fin}}) = b\mathcal{P} \longrightarrow \mathcal{T}_*(\mathcal{A}(v_{\mathrm{init}}), \mathcal{A}(v_{\mathrm{fin}}))$$

induced by all possible continuous functors $C\mathcal{A} : bW\Gamma \to \mathcal{T}_*$ such that

$$\pi \circ C\mathcal{A} = \mathcal{A} \circ (\varepsilon|_{bW\Gamma}),$$

where the half-smash is defined

$$X \rtimes K := (X \times K)/(\{*\} \times K) = X \wedge K_+.$$

$\langle\langle \mathcal{A} \rangle\rangle$ is said to *vanish* if it contains the homotopy class of a constant map $b\mathcal{P} \longrightarrow \mathcal{T}_*(\mathcal{A}(v_{\mathrm{init}}), \mathcal{A}(v_{\mathrm{fin}}))$.

Given a lattice Γ, a *rectification* of a functor $\mathcal{A} : \Gamma \to \mathrm{ho}\,\mathcal{T}_*$ is a functor $F : \Gamma \to \mathcal{T}_*$ with $\pi \circ F$ naturally isomorphic to $\mathcal{A}$ (where $\pi : \mathcal{T}_* \to \mathrm{ho}\,\mathcal{T}_*$ is the obvious projection functor).

4.6. Proposition ([**10**, Thm. 3.8]). *$\mathcal{A}$ has a rectification if and only if the homotopy operation $\langle\langle \mathcal{A} \rangle\rangle$ vanishes (and in particular, is defined)*

With this at hand, we can now reformulate the deformation theory for the bigraded DGL model for $(P, 0)$ in the following form:

4.7. Theorem ([**7**, Thm. 7.14]). *For each connected graded Lie algebra P of finite type, there is a tree T_P, with each node α indexed by a cofibrant DGL L^α such that $H_* L^\alpha \cong P$, starting with $L^0 \simeq (P, 0)$ at the root 0. For each node α there is an integer $n_\alpha > 0$ such that L^α agrees with its successors in degrees $\leq n_\alpha$, with $n_0 = 1$, so that the sequential colimit $L^{(\infty)} = \mathrm{colim}_k L^{(k)}$ along any branch is well defined, and for any rational homotopy type in $\mathcal{M}_P$ there exists such a tree. Furthermore, for each two immediate successors β, γ of a node α, there is a higher homotopy operation $\langle\langle \beta, \gamma \rangle\rangle \subseteq [\mathbb{S}^n, L]$ which vanishes for $L = L^\beta$, but not for $L = L^\gamma$ (or conversely).*

Thus we have reformulated the deformation of the bigraded model as an obstruction theory for distinguishing between the various realizations of a given rational graded Lie algebra P in terms of higher homotopy operations. This can also be done integrally (cf. [**4**]), and one could identify these operations with appropriate cohomology classes in the dual version of the Halperin-Stasheff obstructions (see [**5**]).

4.8. Example of a DGL deformation. We shall not explain how the higher operations $\langle\langle \beta, \gamma \rangle\rangle$ are defined, since the construction is rather technical; see [**7**, §4] for the details. Instead, the reader might find the following example instructive:

Consider the graded Lie algebra $P = \mathbb{L}\langle a_1, b_1, c_2 \rangle / I$, where $\mathbb{L}\langle T_* \rangle$ is the free graded Lie algebra on a graded set of generators T_*, (with the subscripts indicating the dimension), and I is here the Lie ideal generated by $[a, a]$ and $[[c, a], [b, a]]$.

Step I: The minimal model for the coformal DGL $(P,0) \in \mathcal{D}GL_0$ is (B,∂_B), where B in dimensions ≤ 7 is $\mathbb{L}\langle a_1, b_1, c_2, x_3, y_5, w_6, z_7\rangle$, with

$$\begin{aligned} \partial_B(x) &= [a,a], \\ \partial_B(y) &= 3[x,a], \\ \partial_B(w) &= [[c,a],[b,a]], \\ \partial_B(z) &= 4[y,a]+3[x,x]. \end{aligned}$$

The bigraded model A_{**} is obtained from B by introducing an additional (homological) grading: $a,b \in A_{0,1}$, $c \in A_{0,2}$, $x \in A_{1,3}$, $w \in A_{1,6}$, $y \in A_{2,5}$, $z \in A_{3,7}$, and so on.

Step II: The free simplicial DGL resolution $C_\bullet \to L = (\mathcal{L},0)$ may be described in homological dimensions ≤ 3 as follows:

(1) $C_0^{(0)}$ is the DGL coproduct of

$$\begin{aligned} \mathcal{S}^1_{(a^{(0)})} &= \mathcal{S}^1_{(\langle a\rangle)}, \\ \mathcal{S}^1_{(b^{(0)})} &= \mathcal{S}^1_{(\langle b\rangle)}, \\ \mathcal{S}^2_{(c^{(0)})} &= \mathcal{S}^2_{(\langle c\rangle)}. \end{aligned}$$

(2) $C_1^{(0)} = \mathcal{S}^2_{(x^{(0)})} \amalg \mathcal{S}^6_{(w^{(0)})}$, where

$$\begin{aligned} x^{(0)} &= \langle[\langle a\rangle,\langle a\rangle]\rangle, \\ w^{(0)} &= \langle[[\langle c\rangle,\langle a\rangle],[\langle b\rangle,\langle a\rangle]]\rangle. \end{aligned}$$

(3) $C_2^{(0)}$ consists of $\mathcal{S}^5_{(y^{(0)})}$, where

$$y^{(0)} = \langle 3[\langle[\langle a\rangle,\langle a\rangle]\rangle,\langle\langle a\rangle\rangle]\rangle.$$

(4) $C_3^{(0)}$ consists of $\mathcal{S}^6_{(z^{(0)})}$, where

$$z^{(0)} = \langle 4[\langle 3[\langle[\langle a\rangle,\langle a\rangle]\rangle,\langle\langle a\rangle\rangle]\rangle,\langle\langle\langle a\rangle\rangle\rangle] + 6[\langle[\langle\langle a\rangle\rangle,\langle\langle a\rangle\rangle]\rangle,\langle\langle[\langle a\rangle,\langle a\rangle]\rangle\rangle]\rangle.$$

In analogy with the DGL sphere of §3.3, the DGL n-disk $\mathcal{D}^n_{(x)}$ is the free graded Lie algebra on two generators: x in dimension n and $\partial(x)$ in dimension $n-1$. With this notation, for $C_\bullet^{(1)}$ we need in addition:

(1) $\mathcal{D}^3_{(x^{(1)})} \hookrightarrow C_0^{(1)}$ with

$$\partial_W(x^{(1)}) = d_1(x^{(0)}) = \langle[a,a]\rangle.$$

(2) $\mathcal{D}^6_{(y^{(1)})} \hookrightarrow C_1^{(1)}$ with

$$\begin{aligned} y^{(1)} &= \langle 3[\langle x\rangle,\langle a\rangle]\rangle \\ \partial_W(y^{(1)}) = d_2(y^{(0)}) &= \langle 3[\langle[a,a]\rangle,\langle a\rangle]\rangle. \end{aligned}$$

(3) $\mathcal{D}^7_{(z^{(1)})} \hookrightarrow C_2^{(1)}$ with

$$\begin{aligned} z^{(1)} &= \langle 4[\langle 3[\langle x\rangle,\langle a\rangle]\rangle,\langle\langle a\rangle\rangle] + 6[\langle[\langle a\rangle,\langle a\rangle]\rangle,\langle\langle x\rangle\rangle]\rangle \\ \partial_W(z^{(1)}) = d_3(z^{(0)}) &= \langle 4[\langle 3[\langle[a,a]\rangle,\langle a\rangle]\rangle,\langle\langle a\rangle\rangle] + 6[\langle[\langle a\rangle,\langle a\rangle]\rangle,\langle\langle[a,a]\rangle\rangle]\rangle. \end{aligned}$$

For $C_\bullet^{(2)}$ we need in addition

(1) $\mathcal{D}^7_{(y^{(2)})} \hookrightarrow C_0^{(2)}$ with

$$\partial_W(y^{(2)}) = d_1(y^{(1)}) = \langle 3[x,a]\rangle.$$

(2) $\mathcal{D}^8_{(z^{(2)})} \hookrightarrow C_1^{(2)}$ with

$$\begin{aligned} z^{(2)} &= \langle 4[\langle y\rangle, \langle a\rangle] + 3[\langle x\rangle, \langle x\rangle]\rangle \\ \partial_W(z^{(2)}) = d_2(z^{(1)}) &= \langle 4[\langle 3[x,a]\rangle, \langle a\rangle] + 6[\langle [a,a]\rangle, \langle x\rangle]\rangle. \end{aligned}$$

For $C_\bullet^{(3)}$ we must add $\mathcal{D}^9_{(z^{(3)})} \hookrightarrow C_0^{(3)}$ with

$$\partial_W(z^{(1)}) = d_1(z^{(2)}) = \langle 4[y,a] + 3[x,x]\rangle.$$

Step III: So far, we have constructed the DGL simplicial resolution of the coformal DGL $(B,\partial_B) \simeq (P,0)$. Now we consider a different DGL having the same homology (i.e., a non-weakly equivalent topological space with the same rational homotopy groups):

Consider the DGL $(B',\partial_{B'})$ where $B' := \mathbb{L}\langle a_1, b_1, c_2, x_3, y_5, z_7, \ldots\rangle$, is a free Lie algebra with $\partial_{B'}(x) = [a,a]$, $\partial_{B'}(y) = 3[x,a] - [[b,a],c]$, $\partial_{B'}(z) = 4[y,a] + 3[x,x]$, and so on.

Here $P \cong H_*(B) = \mathbb{L}\langle a_1, b_1, c_2\rangle / \langle [a,a], [[c,a],[b,a]]\rangle$, so the bigraded model for $(P,0)$ is (A_{**}, ∂_B) of Step II above, and the filtered model is obtained from it by setting $D(y) = 3[x,a] - [[b,a],c]$ and $D(z) = 4[y,a] + 3[x,x] - 4w + 2[[x,b],c]$.

The corresponding free simplicial DGL resolution is obtained from $C_\bullet$ of Step II by making the following changes:

(1) Set

$$y^{(1)} := \langle 3[\langle x\rangle, \langle a\rangle] - [[\langle b\rangle, \langle a\rangle], \langle c\rangle]\rangle,$$

with

$$\partial_W(y^{(1)}) = d_2(y^{(0)}) = \langle 3[\langle [a,a]\rangle, \langle a\rangle]\rangle$$

as before (but now $\partial_W(y^{(2)}) = \langle 3[x,a] - [[b,a],c]\rangle$, of course).

(2) Set

$$\begin{aligned} z^{(1)} := {} & \langle 4[\langle 3[\langle x\rangle, \langle a\rangle]\rangle, \langle\langle a\rangle\rangle] + 6[\langle [\langle a\rangle, \langle a\rangle]\rangle, \langle\langle x\rangle\rangle] \\ & - 4\langle [[\langle c\rangle, \langle a\rangle], [\langle b\rangle, \langle a\rangle]]\rangle + 2[[\langle [\langle a\rangle, \langle a\rangle]\rangle, \langle\langle b\rangle\rangle], \langle\langle c\rangle\rangle]\rangle \end{aligned}$$

(with $\partial_W(z^{(1)})$ unchanged).

(3) Set

$$z^{(2)} := \langle 4[\langle y\rangle, \langle a\rangle] + 6[\langle x\rangle, \langle x\rangle] - 4\langle w\rangle + 2[[\langle x\rangle, \langle b\rangle], \langle c\rangle]\rangle,$$

with

$$\begin{aligned} \partial_W(z^{(2)}) = d_3(z^{(1)}) = {} & \langle 4[\langle 3[x,a] - [[b,a],c]\rangle, \langle a\rangle] \\ & + 6[\langle [a,a]\rangle, \langle x\rangle] - 4\langle [[c,a],[b,a]]\rangle + 2[[\langle [a,a]\rangle, \langle b\rangle], \langle c\rangle]\rangle \end{aligned}$$

(4) Finally,

$$\partial_W(z^{(3)}) = \langle 4[y,a] + 3[x,x] - 4w + 2[[x,b],c]\rangle.$$

Step IV: In order to define the higher homotopy operations which distinguish B' from B, observe that the required half-smashed are defined as follows:

$$\mathcal{S}^2_{(x)} \rtimes D[1] = (\mathbb{L}\langle X_* \rangle, \partial'),$$

where

$$\begin{aligned} X_2 &= \{(x,(d_0)),(x,(d_1))\} \\ X_3 &= \{(x,(\mathrm{Id}))\}, \end{aligned}$$

and

$$\partial'(x,(\mathrm{Id})) = (x,(d_0)) - (x,(d_0)).$$

Similarly,

$$\mathcal{S}^3_{(y)} \rtimes D[2] = (\mathbb{L}\langle Y_* \rangle, \partial'),$$

where

$$\begin{aligned} Y_3 &= \{(y,(d_0d_1)),(y,(d_0d_2)),(y,(d_1d_2))\}, \\ Y_4 &= \{(y,(d_0)),(y,(d_1)),(y,(d_2))\}, \\ Y_5 &= \{(y,(\mathrm{Id}))\}, \end{aligned}$$

with

$$\begin{aligned} \partial'(y,(\mathrm{Id})) &= -(y,(d_0)) + (y,(d_1)) - (y,(d_2)), \\ \partial'(y,(d_0)) &= -(y,(d_0d_1)) + (y,(d_0d_2)), \\ \partial'(y,(d_1)) &= -(y,(d_0d_1)) + (y,(d_1d_2)), \\ \partial'(y,(d_2)) &= -(y,(d_0d_2)) + (y,(d_1d_2)). \end{aligned}$$

Step V: For the DGL B' of Step III, with $C_\bullet$ as in Step II, we define $h_0 = f_0 : C_{0,*} \to B'$ by setting $f_0(\langle a \rangle) = a$, $f_0(\langle b \rangle) = b$, $f_0(\langle c \rangle) = c$, and $f_0 = 0$ for all other disks in $C_{0,*}$ (so for example $f_0(\langle [a,a] \rangle) = 0$).

Thus on $\mathcal{S}^2_{(x^{(0)})} \rtimes D[1]$ we have

$$\begin{aligned} h_1(x^{(0)},(d_0)) &= [a,a], \\ h_1(x^{(0)},(d_1)) &= 0, \end{aligned}$$

(see [**7**, Def. 4.7]), so we must choose $h_1(x^{(0)},(\mathrm{Id})) = x \in B'_3$.

Now on $\mathcal{S}^3_{(y^{(0)})} \rtimes D[2]$ we have

$$\begin{aligned} h_2(y^{(0)},(d_0d_1)) = h_2(y^{(0)},(d_0d_2)) = h_2(y^{(0)},(d_1d_2)) &= 0, \\ h_2(y^{(0)},(d_0)) = (h_1 \circ d_0)(y^{(0)},(d_0)) &= 3[x,a], \\ h_2(y^{(0)},(d_1)) = h_2(y^{(0)},(d_2)) &= 0. \end{aligned}$$

Thus in Definition 4.5 we shall be interested in a lattice Γ corresponding to the face maps of a simplicial object, with $bW\Gamma \cong \partial D[2]$. We have just defined a continuous functor $C\mathcal{A} : bW\Gamma \to \mathcal{T}_*$ for $C_\bullet$, as required there, and the resulting secondary operation is $\langle\langle 2, y^{(0)} \rangle\rangle = \{\langle 3[x,a] \rangle]\} \subseteq H_4B'$ (remember that the homology of a DGL corresponds to the rational homotopy groups of the associated space). Since $3[x,a]$ does not bound in B', $\langle\langle 2 \rangle\rangle$ does not vanish, and we have found an obstruction to the coformality of B'.

Unfortunately, DGL higher homotopy operations such as $\langle\langle 2, y^{(0)} \rangle\rangle$ are unsatisfactory, in as much as one cannot translate them canonically into integral homotopy operations. One way to avoid this difficulty is to use more general non-associative algebras, rather than Lie algebras, as our basic models. Thus, consider the category $\mathcal{DGN}$ of non-associative differential graded algebras. A DGN whose homology happens to be a graded Lie algebra will be called a *Jacobi* algebra. Every DGL has a free Jacobi model, and these can be used to resolve any DGL in $s\mathcal{DGN}$ (see [**7**, §7]).

Step VI: Consider a Jacobi model G for the coformal DGL P of Step I. We construct the minimal Jacobi resolution $J_\bullet \to G$ corresponding to the bigraded model for L (embedded in the canonical Jacobi resolution $U_\bullet \to G$) by modifying the free resolution $C_\bullet \to L$ of Step II, as follows:

For each $n \geq 0$ define the Jacobi algebras $J_n^{(0)}$ $(n = 0, 1, \dots)$ to be the coproducts

$$\begin{aligned} J_0^{(0)} &= \mathbb{S}^1_{(\langle a\rangle)} \amalg \mathbb{S}^1_{(\langle b\rangle)} \amalg \mathbb{S}^2_{(\langle c\rangle)} \\ J_1^{(0)} &= \mathbb{S}^2_{(x^{(0)})} \amalg \mathbb{S}^6_{(w^{(0)})} \\ J_2^{(0)} &= \mathbb{S}^5_{(y^{(0)})} \\ J_3^{(0)} &= \mathbb{S}^6_{(z^{(0)})}, \end{aligned}$$

and so on.

The face maps d_i $(i = 0, 1, \dots, n)$ are defined as follows: write $\langle x\rangle \in F(B)$ for the generator corresponding to an element $x \in B'_*$, then recursively a typical DGL generator for $W_n = W_{n,*}$ (in the canonical Stover resolution $W_\bullet(B)$, defined in [**7**, §5.3]) is $\langle \alpha\rangle$, for $\alpha \in W_{n-1}$, so an element of W_n is a sum of iterated Lie products of elements of B'_*, arranged within $n+1$ nested pairs of brackets $\langle\langle \cdots \rangle\rangle$. With this notation, the i-th face map of $W_\bullet$ is "omit i-th pair of brackets", and the j-th degeneracy map is "repeat j-th pair of brackets". We assume the bracket operation $\langle - \rangle$ is linear – i.e., that $\langle \alpha x + \beta y\rangle = \alpha\langle x\rangle + \beta\langle y\rangle$ for $\alpha, \beta \in \mathbb{Q}$ and $x, y \in B$.

1. For $y^{(0)} = \langle 3[\langle[\langle a\rangle, \langle a\rangle]\rangle, \langle\langle a\rangle\rangle]\rangle \in J_2^{(0)}3$, we have

$$\begin{aligned} d_0(y^{(0)}) &= 3[\langle[\langle a\rangle, \langle a\rangle]\rangle, \langle\langle a\rangle\rangle] = 3[x^{(0)}, a^{(0)}], \\ d_2(y^{(0)}) &= \langle 3[\langle[a, a]\rangle, \langle a\rangle]\rangle, \\ d_1(y^{(0)}) &= \langle 3[[\langle a\rangle, \langle a\rangle], \langle a\rangle]\rangle. \end{aligned}$$

This no longer vanishes as in Step II, since the Jacobi identity does not hold in $\mathcal{DGN}$, but we have an element $y^{(1;1)} := \langle \lambda_3(\langle a\rangle \otimes \langle a\rangle \otimes \langle a\rangle)\rangle \in J_1^{(1)}4$ (in the notation of [**7**, Example 7.8], with $\partial(y^{(1;1)}) = d_1(y^{(0)})$.

On the other hand, we also have an element $y^{(1;2)} := \langle 3[\langle x\rangle, \langle a\rangle]\rangle \in J_1^{(1)}4$ (which we denoted simply by $y^{(1)}$ in Step II, with $\partial(y^{(1;2)}) = d_2(y^{(0)})$. The simplicial identity $d_1 d_2 = d_1 d_1$ implies that

$$d_1(y^{(1;2)}) - d_1(y^{(1;1)}) = \langle 3[x, a] - \lambda_3(a \otimes a \otimes a)\rangle$$

is a ∂_J-cycle, so we have $y^{(2;1,2)} \in J_0^{(2)}5$ with

$$\partial_J(y^{(2;1,2)}) = \langle 3[x, a] - \lambda_3(a \otimes a \otimes a)\rangle.$$

2. For $z^{(0)} = \langle 4[\langle 3[\langle[\langle a\rangle, \langle a\rangle]\rangle, \langle\langle a\rangle\rangle]\rangle, \langle\langle\langle a\rangle\rangle\rangle] + 6[\langle[\langle\langle a\rangle\rangle, \langle\langle a\rangle\rangle]\rangle, \langle\langle[\langle a\rangle, \langle a\rangle]\rangle\rangle]\rangle$ in $J_3^{(0)}4$:

(a) $z^{(1;1)} := \langle \lambda_3(x^{(0)} \otimes \langle\langle a\rangle\rangle \otimes \langle\langle a\rangle\rangle)\rangle$, with

$$d_1(z^{(0)}) = \langle 6(2[[x^{(0)}, \langle\langle a\rangle\rangle], \langle\langle a\rangle\rangle], +[[\langle\langle a\rangle\rangle, \langle\langle a\rangle\rangle], x^{(0)}])\rangle = \partial_J(z^{(1;1)}),$$

(b) $z^{(1;2)} := \langle 4[\lambda_3(\langle a\rangle \otimes \langle a\rangle \otimes \langle a\rangle), \langle\langle a\rangle\rangle]\rangle$, with $d_2(z^{(0)}) = \langle 4[[[a^{(0)}, a^{(0)}], a^{(0)}], \langle\langle a\rangle\rangle]\rangle = \partial_J(z^{(1;2)})$ since $[x^{(0)}, x^{(0)}] = 0)$.

(c) $z^{(1;3)} = \langle 4[\langle 3[\langle [a,a]\rangle, \langle a\rangle]\rangle, \langle\langle a\rangle\rangle] + 6[\langle[\langle a\rangle, \langle a\rangle]\rangle, \langle\langle [a,a]\rangle\rangle]\rangle \in J_2^{(1)}5$, with

$$d_3(z^{(0)}) = \partial_J(z^{(1;3)}).$$

3. Next, in $J_\bullet^{(2)}$:

(a) The simplicial identity $d_1 d_1 = d_1 d_2$ implies that

$$\begin{aligned} d_1(z^{(1;1)}) - d_1(z^{(1;2)}) = \langle & 4[\lambda_3(\langle a\rangle \otimes \langle a\rangle \otimes \langle a\rangle), \langle a\rangle] \\ & - 6\lambda_3([\langle a\rangle, \langle a\rangle] \otimes \langle a\rangle \otimes \langle a\rangle)\rangle \end{aligned}$$

is a ∂_J-cycle – and indeed we have

$$z^{(2;1,2)} := \langle \lambda_4(\langle a\rangle \otimes \langle a\rangle \otimes \langle a\rangle \otimes \langle a\rangle)\rangle \in J_2^{(2)}6$$

with

$$\partial(z^{(2;1,2)}) = d_1(z^{(1;1)}) - d_1(z^{(1;2)}).$$

(b) Similarly,

$$\begin{aligned} d_1(z^{(1;3)}) - d_2(z^{(1;1)}) = \langle & 12[[\langle x\rangle, \langle a\rangle], \langle a\rangle] + 6[[\langle a\rangle, \langle a\rangle], \langle x\rangle] \\ & - 6\lambda_3(\langle [a,a]\rangle \otimes \langle a\rangle \otimes \langle a\rangle)\rangle \end{aligned}$$

is a ∂_J-cycle, so we have

$$z^{(2;1,3)} := \langle 6\lambda_3(\langle x\rangle \otimes \langle a\rangle \otimes \langle a\rangle)\rangle \in J_2^{(2)}6$$

with

$$\partial(z^{(2;1,3)}) = d_1(z^{(1;3)}) - d_2(z^{(1;1)}).$$

(c)
$$\begin{aligned} d_2(z^{(1;3)}) - d_2(z^{(1;2)}) = \; & \langle 4[\langle 3[x,a]\rangle, \langle a\rangle] + 6[\langle [a,a]\rangle, \langle x\rangle] \\ & - 4[\langle \lambda_3(a \otimes a \otimes a)\rangle, \langle a\rangle]\rangle \end{aligned}$$

is a ∂_J-cycle, hit by

$$z^{(2;2,3)} := \langle 4[\langle y\rangle, \langle a\rangle] + 3[\langle x\rangle, \langle x\rangle]\rangle \in J_2^{(2)}6.$$

4. Finally, we have

(a) For $d_1 z^{(2;1,2)} = \langle \lambda_4(a \otimes a \otimes a \otimes a)\rangle$ we have

$$\begin{aligned} \partial_J(d_1 z^{(2;1,2)}) &= d_1 d_2(z^{(1;2)}) - d_1 d_2(z^{(1;1)}) \\ &= \langle 4[\lambda_3(a \otimes a \otimes a), a] - 6\lambda_3([a,a] \otimes a \otimes a)\rangle \end{aligned}$$

(b) For $d_1 z^{(2;1,3)} = \langle 6\lambda_3(x \otimes a \otimes a)\rangle$ we have

$$\begin{aligned} \partial_J(d_1 z^{(2;1,3)}) &= d_1 d_2(z^{(1;1)}) - d_1 d_1(z^{(1;3)}) \\ &= \langle 6\lambda_3([a,a] \otimes a \otimes a) - 12[[x,a],a] - 6[[a,a],x]\rangle \end{aligned}$$

(c) For $d_1 z^{(2;2,3)} = \langle 4[y,a] + 3[x,a]\rangle$ we have

$$\begin{aligned} \partial_J(d_1 z^{(2;2,3)}) &= d_1 d_2(z^{(1;2)}) - d_1 d_1(z^{(1;3)}) \\ &= \langle 4[\lambda_3(a \otimes a \otimes a), a] - 12[[x,a],a] - 6[[a,a],x]\rangle, \end{aligned}$$

So there is an element $z^{(3;1,2,3)} \in J_0^{(3)}7$ with

$$\partial_J(z^{(3;1,2,3)}) = \langle \lambda_4(a \otimes a \otimes a \otimes a) - 6\lambda_3(x \otimes a \otimes a) + 4[y,a] + 3[x,a] \rangle.$$

This defines the second-order homotopy operation we are interested in (which also has an integral version).

5. Refined moduli spaces

Since neither the cohomology algebra nor the homotopy Lie algebra of a rational space determine the other, we can refine the coarse partition of the rational moduli space $\mathcal{M}$ by specifying both:

5.1. DEFINITION. For each 1-connected graded-commutative algebra H and graded Lie algebra P over $\mathbb{Q}$, let $\mathcal{M}_{H,P}$ denote the collection of all simply-connected rational homotopy types of spaces X with $H^*(X;\mathbb{Q}) \cong H$ and $\pi_*(X) \otimes \mathbb{Q} \cong P$.

Note that we cannot simply identify this as a quotient variety with $\mathcal{M}_H \cap \mathcal{M}_P$ in any natural way, since points in $\mathcal{M}_H$ are represented by filtered DGA models, while those in $\mathcal{M}_P$ are represented by filtered DGA models.

The set $\mathcal{M}_{H,P}$ may, of course, be empty; but Lemaire and Sigrist have shown that it can also be infinite (see [**34**, §3]). In order to analyze this refined moduli space, we shall need one additional ingredient.

5.2. The Quillen equivalences. The fact that $\operatorname{ho}\mathcal{DGL}_0$ and $\operatorname{ho}\mathcal{DGA}_1$ are equivalent to each other, leads us to expect a direct algebraic relationship between the corresponding model categories, which in fact exists, and may be described as follows:

For simplicity, restrict attention to simply-connected spaces of finite type. We may therefore assume that A_X^i is finite dimensional for each $i \geq 0$ (and of course $A_X^0 = \mathbb{Q}$, $A_X^1 = 0$). Taking the vector space dual of A_X^*, we obtain a 1-connected differential graded-cocommutative coalgebra (C_*^X, δ), whose homology is $H_*(X;\mathbb{Q})$, with the usual coalgebra structure (dual to the ring structure in cohomology). Let $\mathcal{DGC}$ denote the category of such coalgebras.

Quillen defined a pair of adjoint functors

$$\mathcal{DGL}_0 \underset{\mathcal{L}}{\overset{\mathcal{C}}{\rightleftarrows}} \mathcal{DGC}$$

as follows:

I: Given a DGC (C_*, d), let $\mathcal{L}(C_*) := (\operatorname{Prim}(\Omega C_*), \partial)$ denote the graded Lie algebra of primitives in the cobar construction of C_*, constructed as follows:

If $C_* \cong \mathbb{Q} \oplus \bar{C}_*$ (where $\bar{C}_* = C_{\geq 2}$, in our case), and $\Sigma^{-1}V$ is the graded vector space V shifted downwards (so that $(\Sigma^{-1}V)_i := V_{i+1}$, with $\sigma^{-1}v \leftrightarrow v$), let $\Omega C_* := T(\Sigma^{-1}\bar{C}_*)$ denote the tensor algebra on $\Sigma^{-1}\bar{C}_*$ with

$$\partial(\sigma^{-1}x) := -\sigma^{-1}(dx) + \frac{1}{2}\sum_i (-1)^{|c_i'|}[\sigma^{-1}c_i', \sigma^{-1}c_i''] ,$$

where $\bar{\Delta}c := \sum_i c_i' \otimes c_i''$ is the (reduced) comultiplication in $\bar{C}_*$, and $[\ ,\ ]$ is the commutator in $T(\Sigma^{-1}\bar{C}_*)$.

II: Given a DGL (L_*, ∂), let $\mathcal{C}(L_*, d)$ be the DGC $(\Lambda(\Sigma L_*), d)$, where ΣL_* is L_* shifted upwards, ΛV denotes the cofree graded coalgebra cogenerated by the graded vector space V, and the coderivation ∂ is defined by $d = d' - d''$, where

(5.1)

$$d'(\sigma x_1 \wedge \ldots \wedge \sigma x_n) := \sum_{i=1}^{n} (-1)^{i+\sum_{j<i}|x_j|} \sigma x_1 \wedge \ldots \wedge \sigma\partial x_i \wedge \ldots \wedge \sigma x_n),$$

$$d''(\sigma x_1 \wedge \ldots \wedge \sigma x_n) := \sum_{1\leq i<j\leq n} (-1)^{|x_i|}(\pm)\sigma[x_i, x_j] \wedge \sigma x_1 \ldots \widehat{\sigma x_i} \ldots \widehat{\sigma x_j} \ldots \sigma x_n) \ ,$$

and the sign $(\pm)$ is determined by

$$\sigma x_1 \wedge \ldots \wedge \sigma x_n = (\pm)\sigma x_i \wedge \sigma x_j \wedge \sigma x_1 \wedge \ldots \widehat{\sigma x_i} \ldots \widehat{\sigma x_j} \ldots \wedge \sigma x_n) \ .$$

See [**42**, App. B] or [**25**, §22] for the details.

5.3. Joint deformations. This suggests the following approach to the refined moduli space for a graded Lie algebra P and graded cocommutative coalgebra H, where we assume H and P are both of finite type, H is n-connected and P is $(n-1)$-connected $(n \geq 1)$, and $H_{n+1} \cong P_n$:

Let (B, ∂) be the bigraded model for $(P, 0)$, and $(C, d) = (C_*B, d' - d'')$ the corresponding coalgebra (§5.2 II). Note that (C, d) is cofibrant, but not necessarily minimal. As we deform (B, ∂) to produce all of $\mathcal{M}_P$, we change only d' in the coalgebra model, obtaining $(\hat{B}, \hat{\partial})$, say. At the same time, we have the deformations $(\tilde{A}, \tilde{d})$ say, of the formal DGC $(H, 0)$; and for each pair $\langle \hat{B}, \tilde{A} \rangle$, we have the variety $V_{\langle \hat{B}, \tilde{A} \rangle}$ of all DGC maps $\phi : (\hat{B}, \hat{\partial}) \to (\tilde{A}, \tilde{d})$ between them. Since both are cofree, these are determined by maps of graded vector spaces. The condition that ϕ be a quasi-isomorphism translates into a series of rank conditions, so that we get a semi-algebraic set parametrizing all such pairs which are equipped with a quasi-isomorphism between them – and in particular, the requisite homology H and homotopy groups P.

6. Nerves and moduli spaces

Since ordinary (integral) homotopy types do not have any known differential graded models, there is no hope of generalizing the deformation approach of sections 2-5 to cover them, too. However, in [**18**], Dwyer and Kan suggested an approach to such "moduli problems" based on the concept of nerves, which has proved useful conceptually in a number of contexts.

6.1. DEFINITION. The *nerve* $\mathcal{N}C$ of a small category C is a simplicial set whose k-simplices are the sequences of k composable arrows in C, with d_i="delete i-th object and compose", s_j="insert identity arrow after j-th object". The geometric realization of the nerve is called the *classifying space* of $\mathcal{C}$, written $BC := |\mathcal{N}C|$.

The nerve was originally defined by Segal in [**48**], based on ideas of Grothendieck; Quillen, in [**43**] helped clarify the close connection between nerves and homotopy theory, as evinced in the following properties:

(1) The functor $\mathcal{N} : \mathcal{C}at \to \mathcal{S}$ takes natural transformations to homotopies (cf. [**43**, §2, Prop. 2]), so that

(2) if a functor F has a left or right adjoint, then $\mathcal{N}F$ is a homotopy equivalence; more general conditions are provided by [**43**, §2, Thm. A]). If C has an initial or final object, then $\mathcal{N}C \simeq *$.
(3) The nerve of a functor is not often a fibration; conditions when $\mathcal{N}F$ fits into a quasi-fibration sequence are provided by [**43**, §2, Thm. B].
(4) Similarly, $\mathcal{N}C$ is not usually a Kan complex, unless C is a groupoid (cf. [**26**, I, Lemma 3.5]).

6.2. Definition. A *classification complex* in a model category $\mathcal{C}$ (see [**18**, §2.1]) is the nerve of some subcategory $\mathcal{D}$ of $\mathcal{C}$, all of whose maps are weak equivalences, and which includes all weak equivalences whose source or target is in $\mathcal{D}$. The category $\mathcal{D}$ is only required to be *homotopically* small (cf. [**16**, §2.2]) – that is, $\mathcal{N}\mathcal{D}$ has a set of components, and its homotopy groups (at each vertex) are small.

This construction has two main properties:

(a) The components of $\mathcal{N}\mathcal{D}$ are in one-to-one correspondence with the weak homotopy types of $\mathcal{D}$ in $\mathcal{C}$.
(b) The component of $\mathcal{N}\mathcal{D}$ corresponding to an object $X \in \mathcal{C}$ is weakly homotopy equivalent to the classifying space of the monoid $\operatorname{haut} X$ of self weak equivalences of X (cf. [**18**, Prop. 2.3]).

6.3. Remark. Taking the model category $\mathcal{C} := \mathcal{T}_0$ of connected pointed spaces, with $\mathcal{D} = \mathcal{W}_\mathcal{C}$ the subcategory of *all* weak homotopy equivalences in $\mathcal{C}$, $\mathcal{M}^\mathcal{C} := \mathcal{N}\mathcal{W}_\mathcal{C}$ is the candidate suggested by the Dwyer-Kan approach for the moduli space of pointed homotopy types. There is also an unpointed version, of course; on the other hand, allowing for non-connected spaces merely complicates the combinatorics of $\mathcal{M}$, without adding any new information.

One might ask in what sense this qualifies as a moduli space (aside from having the right components). Even though the analogy with the classical examples mentioned in the introduction is not clear-cut, note that $\mathcal{M}^{\mathcal{T}_0}$ as defined using the nerve (which is natural choice for a space to associate to a category) is related to $\mathcal{M}^{\mathcal{T}_1^{\mathbb{Q}}}$ of §2.5, and the latter does exhibit many of the attributes listed in §1.1. Moreover, as we shall see in the next section, this construction can be used to interpret the obstruction theory of §2.10, as well as its integral analogue, and also to describe $\mathcal{M}$ as the limit of a tower of fibrations, which give increasingly accurate approximations to $\mathcal{M}$.

Finally, the monoid $\operatorname{haut} X$ of self equivalences corresponds to the mapping class group of self diffeomorphisms of a surface (Sec. 1), whose classifying space is closely related (or even homotopy equivalent) to the moduli space.

6.4. The model categories. In order to provide a uniform treatment in different model categories, and in particular to allow for the comparisons mentioned in §3.4, it is useful to consider a resolution model category structure (§3.2) on a category $\mathcal{C} = s\mathcal{E}$ of simplicial objects over another category $\mathcal{E}$. In fact, one can do this at two different levels, and in some sense the comparison between these is the heart of this approach:

I. The *topological* level – where $\mathcal{E}$ can take several forms:

(a) The category $\mathcal{T}_0$ of connected pointed spaces.

(b) The subcategory $\mathcal{T}_0^{\mathbb{Q}}$ of pointed connected spaces having rational universal covers (but arbitrary fundamental group).

Note that one can approximate any pointed connected space $X \in \mathcal{T}_0$ by its fibrewise rationalization $X'_{\mathbb{Q}}$ (cf. [**15**, I, 8.2]), which lies in $\mathcal{T}_0^{\mathbb{Q}}$: if $\tilde{X} \to X \to B(\pi_1 X)$ is the universal covering space fibration for X, then $X'_{\mathbb{Q}}$ fits into a functorial fibration sequence $\widetilde{X_{\mathbb{Q}}} \to X'_{\mathbb{Q}} \to B(\pi_1 X)$, in which $\widetilde{X_{\mathbb{Q}}}$ is the usual rationalization of the universal cover. However, algebraic DGL, DGA, or DGC models do not generally extend to this case (unless $\pi_1 X$ is finite and acts nilpotently on the higher groups – see, e.g., [**54**]).

(c) The subcategory $\mathcal{T}_1^{\mathbb{Q}}$ of 1-connected pointed rational spaces (in $\mathcal{T}_0^{\mathbb{Q}}$). This can be replaced by the Quillen equivalent model categories $\mathcal{D}GL_0$ or $\mathcal{D}GA_1$.

(d) Other variants are possible – for instance, we could consider functor categories over $\mathcal{E}$ – that is, diagrams $D \to \mathcal{E}$ for a fixed small category D (see [**9**]).

II. The *algebraic* level – where $\mathcal{E}$ is correspondingly:

(a) The category Π-$\mathcal{A}lg$ of Π-*algebras* – that is, positively graded groups G_*, abelian in dimensions ≥ 2, equipped with an action of the primary homotopy operations (Whitehead products, compositions, and G_1-action) satisfying the usual identities (see [**2**] for a more explicit definition).

(b) The subcategory Π-$\mathcal{A}lg^{\mathbb{Q}}$ of *rational* Π-*algebras*, which are (positively) graded Lie algebras equipped with a "fundamental group action" of an arbitrary group π, satisfying the usual identities (see [**30**]).

(c) When $\pi = 0$, we obtain the subcategory Π-$\mathcal{A}lg_1^{\mathbb{Q}} \cong \operatorname{gr}\mathcal{L}ie$ of simply-connected rational Π-algebras, which are just graded Lie algebras over $\mathbb{Q}$.

(d) There is also a concept of Π-algebras for arbitrary diagrams (again, see [**9**]).

7. Approximating classification complexes

The classification complex of a model category $\mathcal{C}$, as defined in §6.2, may appear to be a somewhat artificial marriage of a traditional moduli space, whose set of components correspond to the (weak) homotopy types of $\mathcal{C}$, and the individual components, which are classifying spaces of the form $B\operatorname{haut} X$. It is not clear at first glance why such complexes might be useful. To understand this, we show how $\mathcal{M}^{\mathcal{C}}$ can be approximated by a tower of fibrations, in a way that elucidates the obstruction theory of §2.10:

7.1. Postnikov systems and Eilenberg-MacLane objects. Recall from [**17**, §1.2] that functorial Postnikov towers

$$X \to \ldots \to P_n X \to P_{n-1} X \to \ldots \to P_0 X$$

may be defined in categories of the form $\mathcal{C} = s\mathcal{E}$ using the matching space construction of [**15**, X,§4.5]. By considering the fibers of successive Postnikov sections, we can define the analogue of homotopy groups in each such category, and find that they are corepresented, as one might expect, by appropriate suspensions of the "good objects" in $\mathcal{E}$. For both $\mathcal{E} = \mathcal{T}_0$ and $\mathcal{E} = \Pi\text{-}\mathcal{A}lg$, we find that the natural "homotopy groups" $\hat{\pi}_n X$ of any $X \in s\mathcal{E}$ (the *bigraded groups* of [**20**]) take values in $\Pi\text{-}\mathcal{A}lg$, with the obvious modifications for the rational variants.

Note that for a simplicial space $X \in s\mathcal{T}_0$, applying π_* in each simplicial dimension yields a simplicial Π-algebra $\pi_* X$, and the two sequences of Π-algebras $(\hat{\pi}_n X)_{n=0}^{\infty}$ and $(\pi_n \pi_* X)_{n=0}^{\infty}$ fit into a "spiral long exact sequence":

(7.2)
$$\dots \pi_{n+1}\pi_* X \xrightarrow{\partial^\star_{n+1}} \Omega\hat{\pi}_{n-1} X \xrightarrow{s_n} \hat{\pi}_n X \xrightarrow{h_n} \pi_n\pi_* X \xrightarrow{\partial^\star_n} \cdots \hat{\pi}_0 X \xrightarrow{h_0} \pi_0\pi_* X \to 0$$

(see [**20**, 8.1]), in which each term is not only a Π-algebra, but a module over $\hat{\pi}_0 X \cong \pi_0\pi_* X$ under a "fundamental group action", for $n \geq 1$. Here $\Omega\Lambda$ denotes the abelian Π-algebra obtained from a Π-algebra Λ by re-indexing and suspending the operations (so that in particular $\Omega\pi_* X \cong \pi_*\Omega X$ for any space X).

Furthermore, one can construct *classifying objects* $\mathbf{B}\Lambda \in s\mathcal{T}_0$ for any Π-algebra Λ (with $\hat{\pi}_0\mathbf{B}\Lambda = \Lambda$ and $\hat{\pi}_i\mathbf{B}\Lambda = 0$ otherwise); *Eilenberg-MacLane* objects $\mathbf{B}(\Lambda, n)$; and *twisted* Eilenberg-MacLane objects $\mathbf{B}_\Lambda(M, n)$ for any Π-algebra Λ, Λ-module M, and $n \geq 1$, with

$$\hat{\pi}_i\mathbf{B}_\Lambda(M,n) \cong \begin{cases} \Lambda & \text{if } i = 0 \\ M & (\text{as a } \Lambda\text{-module}) \text{ if } i = n \\ 0 & \text{otherwise.} \end{cases}$$

This is true more generally for resolution model categories, under reasonable assumptions. To simplify the notation we denote by K_Λ the analogous simplicial Π-algebra (with $\pi_n K_\Lambda \cong \Lambda$ for $n = 0$, and 0 otherwise), and similarly the Eilenberg-MacLane objects $K(\Lambda, n)$, and twisted Eilenberg-MacLane objects $K_\Lambda(M, n)$ in $s\Pi\text{-}\mathcal{A}lg$. We shall use boldface in general to indicate constructions in $s\mathcal{T}_0$, as opposed to $s\Pi\text{-}\mathcal{A}lg$.

Finally, one can define natural k-invariants for the Postnikov system of a simplicial object $X \in s\mathcal{E}$, as in [**8**, Prop. 6.4], and these take values in appropriate André-Quillen cohomology groups (represented by the twisted Eilenberg-MacLane objects). These groups are denoted respectively by $H^n(X/\Lambda; M) := [X, \mathbf{B}_\Lambda(M,n)]_{\mathbf{B}\Lambda}$, for a simplicial space X equipped with a map to $\mathbf{B}\Lambda$, and $H^n(G/\Lambda; M) := [G, K_\Lambda(M,n)]_{K_\Lambda}$, for a simplicial Π-algebra G equipped with a map to K_Λ.

It turns out that for any simplicial space X there is a natural isomorphism

$$H^n(X/\Lambda; M) \xrightarrow{\cong} H^n(\pi_* X/\Lambda; M) \tag{7.3}$$

for every $n \geq 1$ (cf. [**8**, Prop. 8.7]).

7.4. Relating classification complexes. We shall now show how when $\mathcal{E} = \mathcal{T}_0$ – and more generally – the classification complex $\mathcal{M}^{\mathcal{T}_0}$ can be exhibited as the homotopy limit of a tower of fibrations, where the successive fibers have a cohomological description showing the relationship between $\pi_0\mathcal{M}^{\mathcal{T}_0}$ and the higher homotopy groups. The tower in question is constructed essentially by taking successive Postnikov sections. The idea is an old one, and is useful even in analyzing the self-equivalences of a single space X (see, for example, [**55**]).

7.5. DEFINITION. For a given Π-algebra Λ, we denote by $\mathcal{D}(\Lambda) = \mathcal{D}^{s\mathcal{T}_0}(\Lambda)$ the category of simplicial spaces $X \in s\mathcal{T}_0$ such that $\pi_* X \simeq \mathbf{B}\Lambda$ (in $s\Pi\text{-}\mathcal{A}lg$) (that is, $\pi_n \pi_* X \cong \Lambda$ for $n = 0$, and $\pi_n \pi_* X - 0$ otherwise). The nerve of $\mathcal{D}^{s\mathcal{T}_0}(\Lambda)$ will be denoted by $\mathcal{M}_\Lambda$.

The "pointed" version is the nerve of the category $\mathcal{R}(\Lambda)$ of pairs (X, ρ), where $X \in s\mathcal{T}_0$ and $\rho : \mathbf{B}\Lambda \to \pi_* X$ is a specified weak equivalence in $s\Pi$-$\mathcal{A}lg$ (again with weak equivalences as morphisms).

Although $\mathcal{M}_\Lambda$ is the more natural object of interest in our context, we actually study $\mathcal{R}(\Lambda)$. As noted in [**8**, §1.1], there is a fibration sequence

$$\mathcal{N}\mathcal{R}(\Lambda) \to \mathcal{M}_\Lambda \to B\,\mathrm{Aut}(\Lambda)\,, \tag{7.6}$$

where $\mathrm{Aut}(\Lambda)$ is the group of automorphisms of the Π-algebra Λ.

7.7. Definition. For each $n \geq 1$, let $\mathcal{R}_n(\Lambda)$ denote the category of n-Postnikov sections under K_Λ – that is, the objects of $\mathcal{R}_n(\Lambda)$ are pairs $(X\langle n\rangle, \rho)$, where $X\langle n\rangle \in s\mathcal{T}_0$ is a simplicial space such that $\mathbf{P}_n X\langle n\rangle \simeq X\langle n\rangle$ (Postnikov sections in $s\mathcal{T}_0$), and $\rho : P_n K_\Lambda \to P_n \pi_* X\langle n\rangle$ is a weak equivalence. The morphisms of $\mathcal{R}_n(\Lambda)$ are weak equivalences of simplicial spaces compatible with the maps ρ up to weak equivalence of simplicial Π-algebras.

7.8. A tower of realization spaces. Given Λ, the Postnikov section functors $\mathbf{P}_n : s\mathcal{T}_0 \to s\mathcal{T}_0$ of §7.1 induce compatible functors $\Phi_n : \mathcal{R}(\Lambda) \to \mathcal{R}_n(\Lambda)$ and $F_n : \mathcal{R}_{n+1}(\Lambda) \to \mathcal{R}_n(\Lambda)$; and as in [**8**, Thm. 9.4] and [**18**, Thm. 3.4], these in turn induce a weak equivalence

$$\mathcal{N}\mathcal{R}(\Lambda) \to \mathrm{holim}_n\, \mathcal{N}\mathcal{R}_n(\Lambda). \tag{7.9}$$

Combining (7.6) and (7.9), we may try to obtain information about the space of realizations $\mathcal{M}_\Lambda$ by studying the successive stages in the tower

$$\ldots \mathcal{N}\mathcal{R}_{n+1}(\Lambda) \xrightarrow{\mathcal{N}F_n} \mathcal{N}\mathcal{R}_n(\Lambda) \xrightarrow{\mathcal{N}F_{n-1}} \ldots \to \mathcal{N}\mathcal{R}_1(\Lambda). \tag{7.10}$$

8. Analyzing the tower

The first step in analyzing the tower (7.10) is to understand when the successive fibers are non-empty, and if so, to count their components. In the rational case, too, our main task was identifying the components of the space of rational homotopy types, and a partial ordering on the components was induced by the successive deformations. The problem of empty fibers did not arise there, since all DGLs (or DGAs) are realizable by rational spaces. However, the fact that we have ordered the successive choices in a tower, rather than a tree as in Theorem 4.7, suggests that we can describe them by means of an obstruction theory, as follows:

Assume given a point $(X\langle n\rangle, \rho)$ in $\mathcal{N}\mathcal{R}_n(\Lambda)$, so that $X\langle n\rangle \in s\mathcal{T}_0$ is a simplicial space which is an n-Postnikov stage (for some putative simplicial space Y realizing the given Π-algebra Λ), and $\rho : \mathbf{K}_\Lambda \to P_n \pi_* X\langle n\rangle$ is a choice of a weak equivalence. We can then reinterpret [**8**, Prop. 9.11] as saying:

8.1. Proposition. *$X\langle n\rangle \in \mathcal{R}_n(\Lambda)$ extends to an $(n+1)$-Postnikov stage in $\mathcal{R}_{n+1}(\Lambda)$ if an only if the $n+1$-st k-invariant for the simplicial Π-algebra $\pi_* X$ vanishes in $H^{n+3}(\Lambda; \Omega^{n+1}\Lambda)$.*

The spiral long exact sequence (7.2) actually determines the homotopy "groups" of the simplicial Π-algebra $\pi_* X\langle n\rangle$ completely:

$$\pi_k\pi_* X\langle n\rangle \cong \begin{cases} \Lambda & \text{for } k=0, \\ \Omega^{n+1}\Lambda & \text{for } k=n+2 \\ 0 & \text{otherwise.} \end{cases} \tag{8.2}$$

However, (8.2) in itself does not imply that $\pi_* X\langle n\rangle$ is a twisted Eilenberg-MacLane object $K_\Lambda(\Omega^{n+1}\Lambda, n+2)$ in $s\Pi\text{-}\mathcal{A}lg$ – for that to happen, the map $\pi_* X\langle n\rangle \to K_\Lambda$ must have a section (whose existence is equivalent to the vanishing of the k-invariant in Proposition 8.1).

8.3. Remark. It may help to understand why if one considers a simplicial group $K \in \mathcal{G}$ (as a model for a connected topological space). The fundamental group then appears as $\Gamma = \pi_0 K$, the set of path components of K (all homotopy equivalent to each other) – which happens to have a group structure. When $K = K^\Gamma(M,n) = K(M,n) \ltimes \Gamma$ is a twisted Eilenberg-MacLane object, the choice of the section $s : B\Gamma \to K$ is what distinguishes K from a disjoint union (of cardinality $|\Gamma|$) of copies of ordinary Eilenberg-MacLane objects $K(M,n)$ – that is, $L := \coprod_{|\Gamma|} K(M,n)$. Even though we can put a group structure on $\pi_0 L$ so as to make it abstractly isomorphic to Γ, we will not get the right action of Γ on M (which, in the case of K, appears in the usual way by conjugation with any representative of $\gamma \in \Gamma = \pi_0 K$).

8.4. Distinguishing between liftings. Since the weak homotopy types of the realizations of a given Π-algebra Λ (that is, components of $\mathcal{M}_\Lambda^{\mathcal{T}_0}$) are in one-to-one correspondence with the weak homotopy types of the simplicial spaces Y realizing K_Λ in $s\mathcal{T}_0$ (if any), we should be able to describe them inductively in terms of the simplicial-space k-invariants of Y, and thus of the successive Postnikov approximations $X\langle n\rangle$.

So, one naturally expects that there will be a correspondence between the sections s_n and the k-invariants $\mathbf{k}_n$ in $s\mathcal{T}_0$. However, by [**8**, Prop 9.11] (using the identification of (7.3)) we know that the map $\phi : K_\Lambda(\Omega^{n+1}\Lambda, n+2) \to K_\Lambda(\Omega^{n+1}\Lambda, n+2)$ corresponding to $\mathbf{k}_n$ must be a weak equivalence, in our case – that is, it is within the indeterminacy of the **k**-invariants, which is certainly contained in $\mathrm{Aut}_\Lambda(\Omega^{n+1}\Lambda)$.

Thus, the relevant information for constructing successive stages in a Postnikov tower for the simplicial space Y with $\pi_* Y \simeq K_\Lambda$ (if it exists), aside from the Π-algebra Λ, is *not* its **k**-invariants, nor the k-invariants for the simplicial Π-algebras $\pi_*\mathbf{P}_n\mathbf{B}\Lambda$ $(n = 1, 2, \dots)$. Instead, it is the seemingly innocuous choice of the section $s_n : K_\Lambda \to K_\Lambda(\Omega^{n+1}\Lambda, n+2)$, with which any twisted Eilenberg-MacLane object (of simplicial spaces or Π-algebras) is equipped:

8.5. Proposition. *If $X\langle n\rangle \in \mathcal{R}_n(\Lambda)$ can be lifted to $\mathcal{R}_n(\Lambda)$, the possible choices for the $(n+1)$-st Postnikov stage $X\langle n+1\rangle$ are determined by the choices of sections $s_n : K_\Lambda \to \pi_* X\langle n\rangle$, which correspond to elements in $H^{n+2}(\Lambda; \Omega^{n+1}\Lambda)$.*

8.6. The components. As noted above, the Dwyer-Kan concept of classification complexes has one advantage over the traditional moduli spaces, in that the topology of each component encodes further information about the corresponding homotopy type: namely, the topological group of self-equivalences.

Such groups are generally hard to analyze, as one might expect from the topological analogue (see, e.g., [**44**]). One advantage of the tower (7.10) is that the successive fibers are generalized Eilenberg-MacLane spaces:

8.7. PROPOSITION ([**8**, Prop. 9.6*ff.*]). *For any* Π*-algebra* Λ *and* $n \geq 0$, *the fiber of* $\mathcal{R}_{n+1}(\Lambda) \to \mathcal{R}_n(\Lambda)$ *is either empty, or has all components weakly equivalent to* $\prod_{i=1}^{n+1} K(H_\Lambda^{n+3-i}(\Lambda, \Omega^{n+1}\Lambda), i)$.

9. Comparing the rational and integral versions

The constructions of the rational moduli space as an infinite-dimensional algebraic variety over $\mathbb{Q}$, and of the integral moduli space as a classification complex, are not related in any evident way. However, given a Π-algebra $\Lambda \in \Pi\text{-}\mathcal{A}lg$ and its rationalization $\Lambda^{\mathbb{Q}} \in \Pi\text{-}\mathcal{A}lg^{\mathbb{Q}}$ (defined $\Lambda_1^{\mathbb{Q}} := \Lambda_1$ and $\Lambda_i^{\mathbb{Q}} := \Lambda_i \otimes \mathbb{Q}$ for $i \geq 2$), there is a connection between the set of components of $\mathcal{M}_{\Lambda^{\mathbb{Q}}}$ and that of $\mathcal{M}_\Lambda$, and also between the respective towers (7.10).

9.1. The preimage of rationalization. Clearly, Λ is realizable only if $\Lambda^{\mathbb{Q}}$ is (which is automatic in the simply-connected case), but the converse need not hold (as the example of a non-realizable torsion Π-algebra in [**3**, Prop. 4.3.6] shows). Therefore, the number of components of $\mathcal{M}_{\Lambda^{\mathbb{Q}}}$ need have no relation to that of $\mathcal{M}_\Lambda$, since there seems to be no known method for determining all the (integral) homotopy types X whose fibrewise rationalizations $X'_{\mathbb{Q}}$ (§6.4I(b)) are homotopy equivalent. However, the obstruction theories of §2.10 and Propositions 8.1-8.5, respectively, do provide a framework for studying the preimage of the (fibrewise) rationalization:

Given a rationalized Π-algebra A, we have

(i) an algebraic question: which Π-algebras Λ have $\Lambda^{\mathbb{Q}} \cong A$?
(ii) a homotopy-theoretic question: what are the components of $\mathcal{M}_\Lambda$ (though different components can map to the same component of $\mathcal{M}_A$ under the fibrewise rationalization functor)?

For the first question, restrict attention to the simply-connected case, so that A is just a (connected) graded Lie algebra over $\mathbb{Q}$. In this case we must first consider which connected graded Lie algebras over $\mathbb{Z}$ have $L \otimes \mathbb{Q} \cong A$. Given such an L, we would like to classify all possible Π-algebra structures on L; one possible approach to this problem is given by the obstruction theory of [**11**], in terms of certain relative cohomology groups (where "relative" involves comparing the same object in different categories via a forgetful functor).

Theoretically, the obstruction theory of Section 8 provides an answer to the second question, although it does not tell us how to identify all integral homotopy types having weakly equivalent rationalizations.

9.2. Comparing cohomology theories. In view of the obstruction theories of Sections 2 and 8, a first step towards understanding the relation between the integral and rational classifications is to compare the cohomology theories that house the respective obstructions. There are two main cases to consider:

I. We can compare the cohomology of a Π-algebra Λ, with coefficients in some Λ-module M (say, $M = \Omega^n\Lambda$) with the cohomology of the rationalization $\Lambda^{\mathbb{Q}}$, with rationalized coefficients in $M^{\mathbb{Q}}$ (which is a Λ-module, so in particular a $\Lambda^{\mathbb{Q}}$-module).

In this case the exact sequence of Λ-modules $0 \to K \to M \to M^{Q} \to C \to 0$ (for $K := \mathrm{Ker}\,(q)$ and $C = \mathrm{Coker}(q)$) yields two long exact sequences in cohomology, in the usual way. Moreover, the functors

$$\Pi\text{-}\mathcal{A}lg \xrightarrow{T} \Pi\text{-}\mathcal{A}lg^{\mathbb{Q}} \xrightarrow{S} \mathcal{A}b\mathcal{G}p,$$

where $T(-) := (-)^{\mathbb{Q}}$ and $S(-) := \mathrm{Hom}_{\Lambda}(-, M^{\mathbb{Q}})$, satisfy the conditions of [**12**, Thm. 4.4], since a rationalized free Π-algebra is free (and so H-acyclic) in $\Pi\text{-}\mathcal{A}lg^{\mathbb{Q}}$. Thus we have a generalized Grothendieck spectral sequence with

$$E^2_{s,t} = (L_s\bar{S}_t)(L_*T)\Lambda \;\Rightarrow\; H^{s+t}(\Lambda^{\mathbb{Q}}; M^{\mathbb{Q}}) \tag{9.3}$$

converging to the cohomology in the category $\Pi\text{-}\mathcal{A}lg^{\mathbb{Q}}$). Here L_s denotes the s-th left derived functor, and (for simply-connected Λ) $\bar{S} : \mathrm{bg}\,\mathfrak{L}ie \to \mathrm{gr}\,\mathcal{A}b\mathcal{G}p$ is the functor induced by S, which exists because the homotopy groups of any simplicial graded Lie algebra over $\mathbb{Q}$ actually take value in the category $\mathrm{bg}\,\mathfrak{L}ie$ of bigraded Lie algebras.

For general $\Lambda \in \Pi\text{-}\mathcal{A}lg$, by [**12**, Prop. 3.2.3], we have instead a functor $\bar{S} : (\Pi\text{-}\mathcal{A}lg^{\mathbb{Q}})\text{-}\Pi\text{-}\mathcal{A}lg \to \mathrm{gr}\,\mathcal{A}b\mathcal{G}p$ whose domain is the analogue for $\Pi\text{-}\mathcal{A}lg^{\mathbb{Q}}$ of the category of Π-algebras for spaces. Its objects are bigraded groups, endowed with an action of all primary homotopy operations which exist for the homotopy groups of a simplicial rational Π-algebra. As in the simply-connected case, these include the bigraded Lie bracket mentioned above, and presumably others.

II. Alternatively, one could start with a rationalized Π-algebra A – for simplicity, a graded Lie algebra – and try to compare its cohomology (with coefficients in an A-module M) taken in the category of Lie algebras with that obtained by thinking of it as an ordinary Π-algebra. This means taking the derived functors of the composite of

$$\Pi\text{-}\mathcal{A}lg^{\mathbb{Q}} \xrightarrow{I} \Pi\text{-}\mathcal{A}lg \xrightarrow{S} \mathcal{A}b\mathcal{G}p,$$

where I is the inclusion, and $S(-) := \mathrm{Hom}_{\Pi\text{-}\mathcal{A}lg/A}(-, M) = \mathrm{Hom}_{\Pi\text{-}\mathcal{A}lg/I(A)}(-, M)$.

These do not satisfy the conditions of [**12**, Thm. 4.4], since a rationalized free Π-algebra is not a free Π-algebra. However, we can take a simplicial resolution $V \to A$ in the category $s\Pi\text{-}\mathcal{A}lg^{\mathbb{Q}}$ of simplicial graded Lie algebras, and then resolve each V_n functorially in $s\Pi\text{-}\mathcal{A}lg$ to get a bisimplicial free Π-algebra $W_{\bullet\bullet}$, with $Xd := \mathrm{diag}\, W_{\bullet\bullet}$ a free Π-algebra resolution of $I(A)$. If $\mathcal{A}b(-)$ denotes the abelianization in the over category $\Pi\text{-}\mathcal{A}lg/A$, then $\mathcal{A}bW_{\bullet\bullet}$ is a bisimplicial abelian object, or equivalently, a double complex in $(\Pi\text{-}\mathcal{A}lg/A)_{\mathrm{ab}}$, and the bicosimplicial abelian group $SW_{\bullet\bullet}$ is $\mathrm{Hom}_{\Pi\text{-}\mathcal{A}lg/A}(\mathcal{A}bW_{\bullet\bullet}, M)$. Since

$$\begin{aligned}
\mathrm{Hom}_{\Pi\text{-}\mathcal{A}lg/A}(X, M) &= \mathrm{Hom}_{(\Pi\text{-}\mathcal{A}lg/A)_{\mathrm{ab}}}(\mathrm{diag}\,\mathcal{A}bW_{\bullet\bullet}, M)\\
&\simeq \mathrm{Tot}\,\mathrm{Hom}_{(\Pi\text{-}\mathcal{A}lg/A)_{\mathrm{ab}}}(\mathcal{A}bW_{\bullet\bullet}, M)\\
&= \mathrm{Tot}\,\mathrm{Hom}_{\Pi\text{-}\mathcal{A}lg/A}(W_{\bullet\bullet}, M)
\end{aligned}$$

by the generalized Eilenberg-Zilber Theorem, we see that both the spectral sequences for the bicomplex $SW_{\bullet\bullet}$ converge to the cohomology of $I(A)$. Since by definition

$$\pi^t_v SW_{n\bullet} \cong H^t(IQ_n, M) = H^t(L_nI(-), M)(A),$$

we obtain a cohomological spectral sequence with

$$E_2^{s,t} = (\pi^s H^t(-,M))(L_* I)A \Rightarrow H^{s+t}(I(A),M). \tag{9.4}$$

This is less useful than (9.3), since we cannot identify the E_2-term explicitly as a derived functor.

9.5. REMARK. As was pointed out in Section 4, another way to relate the integral and rational moduli spaces is geometrically, using higher homotopy operations. This was the motivations behind [**7**] (in conjunction with [**4**]). The main difficulty in using homotopy operations in any systematic way for such a purpose is the lack of an appropriate taxonomy. The most promising way to overcome this would be by establishing a clear correspondence between higher operations and suitable cohomology groups.

References

1. H.J. Baues and J.-M. Lemaire, *Minimal models in homotopy theory*, Math. Ann. **225** (1977), 219-242.
2. D. Blanc, *Abelian Π-algebras and their projective dimension*, Algebraic Topology: Oaxtepec 1991 (M.C. Tangora, ed.), Contemp. Math. **146**, AMS, Providence, RI, 1993, pp. 39-48.
3. ———, *Operations on resolutions and the reverse Adams spectral sequence*, Trans. Amer. Math. Soc. **342** (1994), 197-213.
4. ———, *Higher homotopy operations and the realizability of homotopy groups*, Proc. Lond. Math. Soc. (3) **70** (1995), 214-240.
5. ———, *Algebraic invariants for homotopy types*, Math. Proc. Camb. Phil. Soc. **27** (1999), 497-523.
6. ———, *Realizing coalgebras over the Steenrod algebra*, Topology **40** (2001), 993-1016.
7. ———, *Homotopy operations and rational homotopy type*, Algebraic Topology: Categorical decomposition techniques, Prog. in Math. **215**, Birkhäuser, Boston-Basel, 2003, pp. 47-75.
8. D. Blanc, W.G. Dywer, and P.G. Goerss, *The realization space of a Π-algebra: a moduli problem in algebraic topology*, Topology **43** (2004), 857-892.
9. D. Blanc, M.W. Johnson, and J.M. Turner, *On realizing diagrams of Π-algebras*, preprint, 2004.
10. D. Blanc and M. Markl, *Higher homotopy operations*, Math. Zeit. **345** (2003), 1-29.
11. D. Blanc and G. Peschke, *The fiber of functors between categories of algebras*, preprint, 2004.
12. D. Blanc and C.S. Stover, *A generalized Grothendieck spectral sequence*, Adams Memorial Symposium on Algebraic Topology, Vol. 1 (N. Ray and G. Walker, eds.), Lond. Math. Soc. Lec. Notes Ser. **175**, Cambridge U. Press, Cambridge, 1992, pp. 145-161.
13. J.M. Boardman and R.M. Vogt, *Homotopy Invariant Algebraic Structures on Topological Spaces*, Springer-Verlag Lec. Notes Math. **347**, Berlin-New York, 1973.
14. A.K. Bousfield, *Cosimplicial resolutions and homotopy spectral sequences in model categories*, Geom. & Topology **7** (2003), 1001-1053.
15. A.K. Bousfield and D.M. Kan, *Homotopy Limits, Completions, and Localizations*, Lec. Notes Math. **304**, Springer-Verlag, Berlin-New York, 1972.
16. W.G. Dwyer and D.M. Kan, *Function complexes in homotopical algebra*, Topology **19** (1980), 427-440.
17. ———, *An obstruction theory for diagrams of simplicial sets*, Proc. Kon. Ned. Akad. Wet. - Ind. Math. **46** (1984), 139-146.
18. ———, *A classification theorem for diagrams of simplicial sets*, Topology **23** (1984), 139-155.
19. W.G. Dwyer, D.M. Kan, and C.R. Stover, *An E^2 model category structure for pointed simplicial spaces*, J. Pure Appl. Alg. **90** (1993), 137-152.
20. ———, *The bigraded homotopy groups $\pi_{i,j}X$ of a pointed simplicial space*, J. Pure Appl. Alg. **103** (1995), 167-188.
21. E.R. Fadell and S.Y. Husseini, *Geometry and Topology of Configuration Spaces*, Springer Monog. in Math., Springer-Verlag, Berlin-New York, 2001.

22. Y. Félix, *Dèformation d'espaces rationnels de même cohomologie*, C. R. Acad. Sci., Paris, Sér. A-B **290** (1980), A41-A43.
23. ______, *Dènombrement des Types de k-Homotopie: Théorie de la Déformation*, Mem. Soc. Math. France **3**, Paris, 1980/81.
24. ______, *Modèles bifiltrés: une plaque tournant en homotopie rationelle*, Can. J. Math. **33** (1981), 1448-1458.
25. Y. Félix, S. Halperin, and J.-C. Thomas, *Rational Homotopy Theory*, *Grad. Texts Math.* **203**, Springer-Verlag, Berlin-New York, 2001.
26. P.G. Goerss and J.F. Jardine, *Simplicial Homotopy Theory*, Birkhaüser, Boston-Basel, 1999.
27. S. Halperin and J.D. Stasheff, *Obstructions to homotopy equivalences*, Adv. in Math. **32** (1979), 233-279.
28. Y. Haralambous, *Perturbation d'algèbres de Lie différentielles et coformalité modérée'*, Bull. Soc. Math. Belg. Sér. B **43** (1991), 59-67.
29. J.L. Harer, *Stability of the homology of the mapping class groups of orientable surfaces*, Ann. Math. (2) **121** (1985), 215-249.
30. P.J. Hilton, *Note on the Jacobi identity for Whitehead products*, Proc. Camb. Phil. Soc. **57** (1961), 180-182.
31. Y. Imayoshi and M. Taniguchi, *An Introduction to Teichmüller Spaces*, Springer-Verlag, Berlin-New York, 1992.
32. T. Jozefiak, *Tate resolutions for commutative graded algebras over a local ring*, Fund. Math. **74** (1972), 209-231.
33. D.P. Kraines, *Massey higher products*, Trans. Amer. Math. Soc. **124** (1966), 431-449.
34. J.-M. Lemaire and F. Sigrist, *Dénombrement de types d'homotopie rationelle*, C. R. Acad. Sci., Paris, I **287** (1978), 109-112.
35. I.H. Madsen and M.S. Weiss, *The stable moduli space of Riemann surfaces: Mumford's conjecture*, preprint, 2002 (`math.AT/0212321`).
36. M.A. Mandell, *E_∞-algebras and p-adic homotopy theory*, Topology **40** (2001), 43-94.
37. C.R.F. Maunder, *Cohomology operations of the N-th kind*, Proc. Lond. Math. Soc. Ser. (2) **13** (1963), 125-154.
38. W.S. Massey, *A new cohomology invariant of topological spaces*, Bull. Amer. Math. Soc. **57** (1951), p. 74.
39. T. Nishimoto, H. Shiga, and T. Yamaguchi, *Elliptic rational spaces whose cohomologies and homotopies are isomorphic*, Topology **42** (2003), 1397-1401.
40. A. Oukili, *Sur l'homologie d'une algébre différentielle de Lie*, Ph.D. thesis, Univ. de Nice, 1978.
41. D.G. Quillen, *Homotopical Algebra*, Lec. Notes Math. **20**, Springer-Verlag, Berlin-New York, 1963.
42. ______, *Rational homotopy theory*, Ann. Math. **90** (1969), 205-295.
43. ______, *Higher K-theory, I*, Algebraic K-Theory, I: Higher K-Theories (Battelle Mem. Inst., 1972) (H. Bass, ed.), Lec. Notes Math. **341**, Springer-Verlag, Berlin-New York, 1973, pp. 85-147.
44. J.W. Rutter, *Spaces of Homotopy Self-equivalences: A Aurvey*, Lec. Notes Math. **1662**, Springer-Verlag, Berlin-New York, 1997.
45. J.W. Schlesinger, *The semi-simplicial free Lie ring*, Trans. Amer. Math. Soc. **122** (1966), 436-442.
46. M. Schlessinger and J.D. Stasheff, *The Lie algebra structure of tangent cohomology and deformation theory*, J. Pure Appl. Alg. **38** (1985), 313-322.
47. ______, *Deformation theory and rational homotopy type*, preprint, 1999.
48. G.B. Segal, *Classifying spaces and spectral sequences*, Pub. Math. Inst. Hautes Et. Scient. **34** (1968), 105-112.
49. J.D. Stasheff, *Rational homotopy obstruction and perturbation theory*, Algebraic Topology - Vancouver 1977 (P. Hoffman, R.A. Piccinini, and D. Sjerve, eds.), Lec. Notes Math. **673**, Springer-Verlag, Berlin-New York, 1978, pp. 7-31.
50. D.P. Sullivan, *Infinitesimal computations in topology*, Pub. Math. IHES **47** (1978), 269-332.
51. D. Sundararaman, *Moduli, Deformations and Classifications of Compact Complex Manifolds*, Pitman Research Notes in Math. **45**, Boston-London, 1980.
52. J. Tate, *Homology of noetherian rings and local rings*, Ill. J. Math. **1** (1957), 14-27.

53. U. Tillmann, *Strings and the stable cohomology of mapping class groups*, Proceedings of the International Congress of Mathematicians (Beijing, 2002), Vol. 2, Higher Ed. Press, Beijing, 2002, pp. 447-456.
54. G.V. Triantafillou, *Equivariant minimal models*, Trans. Amer. Math. Soc. **274** (1982), 509-532.
55. C.W. Wilkerson, *Classification of spaces of the same n-type for all n*, Proc. Amer. Math. Soc. **60** (1976), 279-285.

Department of Mathematics, University of Haifa, Haifa 31905, Israel
E-mail address: blanc@math.haifa.ac.il

Contemporary Mathematics
Volume **387**, 2005

K-regular Graphs and Hecke Surfaces

R. Brooks and M. Monastyrsky

Abstract. We discuss some relations between Hecke surfaces and k-regular graphs. Hecke surfaces connect with Hecke groups and are a natural extension of the modular group $PSL(2, Z)$. Hecke surfaces are dense in the moduli space of Riemann surfaces. We study spectral characteristics of Hecke surfaces and show that most results which were proved for modular surface remain valid to Hecke surfaces. One of the main result is to exhibit some universal properties of Hecke surfaces with cusps of length $\gg L$ for some large L.

Preface

With great sorrow I submit my joint paper with Robert Brooks to the volume dedicated to his memory. His premature and sudden death was a great shock and loss for all his friends. Bob had the very rare ability to gain new insights into problems discussed by prominent mathematicians.

We started to work together a couple years ago when we found some common themes between monodromial representations of braid groups (my field of interest) and his very original investigations of spectral characteristics of Laplacians on "typical" Riemann surfaces. We had found very intriguing coincidences between these problems and proposed to study these relations in depth. – Unfortunately, we had time to prepare only a small part of our preliminary results. However, I believe that even this portion of our work may be of interest and hence worthy of publication.

M. Monastyrsky

1. The Basic Construction

Let Γ be a finite k-regular graph.

Definition 1.1. An orientation $\mathcal{O}$ on Γ is an assignment for each vertex $v \in \Gamma$, of a cyclic ordering of the edges emanating from v.

A graph $(\Gamma, \mathcal{O})$ with orientation is often referred to in the literature as a *fat graph.*

Generalizing the construction of [**5**] for the case $k = 3$, we will associate to the oriented graph, $(\Gamma, \mathcal{O})$ a pair of Riemann surfaces, and $S^O(\Gamma, \mathcal{O})$ and $S^C(\Gamma, \mathcal{O})$.

2000 *Mathematics Subject Classification.* Primary 20H10, 57M15; Secondary 22E40, 05C10.

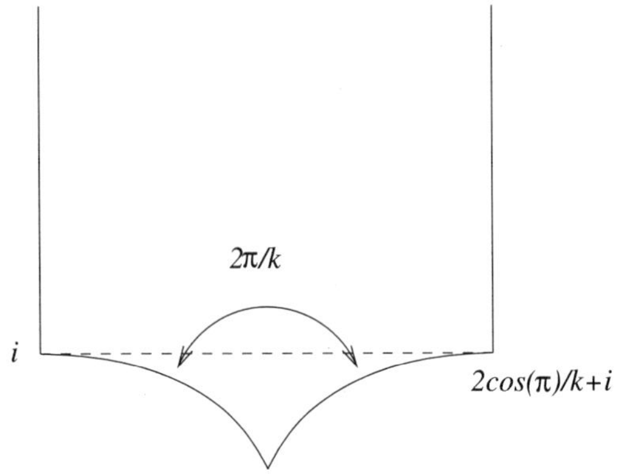

FIGURE 1. The fundamental domain $\mathbf{H}_k$

$S^O(\Gamma, \mathcal{O})$ will be a finite-area Riemann surface, and $S^C(\Gamma, \mathcal{O})$ will denote its conformal compactification. As in [**5**], the idea is that the spectral geometry of the non-compact surface $S^O(\Gamma, \mathcal{O})$ is controlled (up to geometric constants) by the spectral geometry of the oriented graph $(\Gamma, \mathcal{O})$, which may then be studied combinatorially. The spectral geometry of the closed surface $S^C(\Gamma, \mathcal{O})$ will be close to the spectral geometry of the open surface $S^O(\Gamma, \mathcal{O})$, provided that $S^O(\Gamma, \mathcal{O})$ satisfies *a large cusps condition*, which will be explained below.

A central part of the construction is the following.

DEFINITION 1.2. For given k, the Hecke group $\mathbf{H}_k$ is the discrete subgroup of $PSL(2, \mathbb{R})$ generated by the matrices

$$A_k = \begin{pmatrix} 1 & 2\cos(\pi/k) \\ 0 & 1 \end{pmatrix} \qquad B_k = \begin{pmatrix} 0 & -1 \\ 1 & 0 \end{pmatrix}.$$

A fundamental domain F_k for H_k is given by the region shown in Figure 1, where ρ_0 is the intersection in the upper half plane of the circles of radius 1 centered at 0 and $2\cos(\pi/k)$. Noting that i is the fixed point of B_k, we see that ρ_0 is the fixed point of

$$A_k B_k = \begin{pmatrix} 2\cos(\pi/k) & -1 \\ 1 & 0 \end{pmatrix},$$

and hence

$$\rho_0 = \cos(\pi/k) + i \sin(\pi/k).$$

The corresponding circles meet at ρ_0 with angle $2\pi/k$.

The fact that A_k and B_k generate a discrete group can be read off from the Poincaré Polygon Theorem, the fact that A_k preserves the horocycle $y = 1$, and the fact that $A_k B_k$ is a rotation through angle $2\pi/k$ about ro_0 and sends i to $2\cos(\pi/k) + i$.

In the particular case $k = 3$, we have $2\cos(\pi/k) = 1$, and we have the well-known generators and the fundamental domain for $PSL(2, \mathbb{Z})$.

2. Hecke Surfaces

Let $\mathcal{H}$ denote the collection of surfaces

$$\mathcal{H}_k = \{S : S = S^C(\Gamma, \mathcal{O}) \text{ for some k-regular } (\Gamma, \mathcal{O})\}.$$

Note that $\mathcal{H}_k$ is precisely the set $\mathcal{B}$ of *Belyi surfaces*, for which several characterizations are known, see [**1, 6**].

We prove

THEOREM 2.1. *For each k,*

$$\mathcal{H}_k = \mathcal{B}.$$

It follows, for instance, that for any Riemann surface S and for any ε, there is a k-regular $(\Gamma, \mathcal{O})$ such that $S^C(\Gamma, \mathcal{O})$ is ε-close to S (for any reasonable metric on the moduli space of surfaces).

The point here is that the description of S as $S = S^C(\Gamma_k, \mathcal{O}_k)$ for some k may be very complicated, while for another k', the graph $\Gamma_{k'}, \mathcal{O}_{k'}$ might be quite simple.

PROOF. By definition, a surface S lies in $\mathcal{B}$ if there is a holomorphic mapping

$$S \to S^2$$

with precisely three branch values.

Now suppose that $S = S^C(\Gamma, \mathcal{O})$ for a k-regular Γ. Then there are finitely many points $\{p_i\}$ on S such that

$$S^O = S - \cup\{p_i\} = \mathbb{H}/H,$$

where H is a finite-index subgroup of the Hecke group $\mathbf{H_k}$. Thus, S^O covers $\mathbb{H}/\mathbf{H_k}$, which is S^2 minus three singular points. This covering map extends over the points $\{p_i\}$ to give a holomorphic map $S \to S^2$ branched over three points. This shows that

$$\mathcal{H}_k \subset \mathcal{B}.$$

To show the other direction, we begin by considering the k-regular oriented graph $(\Gamma_{2,k}, \mathcal{O}_{2,k})$ on two vertices v_1 and v_2. The cyclic ordering at v_1 is the cyclic ordering $(0, 1, \ldots, k-1)$, while the cyclic ordering at v_2 is the inverse ordering $(0, k-1, \ldots, 1)$. Each *LHT* path is of length 2.

It is then easy to calculate that $S^0(\Gamma_{2,k}, \mathcal{O}_{2,k})$ is S^2 with punctures at the k points $e^{2\pi i l/k}$.

Thus, S will lie in $\mathcal{H}_k$ if there are finitely many points $\{p_i\}$ such that $S_{\bigcup}\{p_i\}$ covers $S^O(\Gamma_{2,k}, \mathcal{O}_{2,k})$. But if S is a Belyi surface, there is a holomorphic map $\phi : S \to S^2$ branched over three points, which we may take to be $1, e^{2\pi i/k}, e^{4\pi i/k}$. Deleting from S the inverse image under ϕ of these points and also the inverse images of the other points $e^{2\pi l/k}$ exhibits S minus a finite number of points as a cover of $S^O(\Gamma_{2,k}, \mathcal{O}_{2,k})$.

This shows that

$$\mathcal{B} \subset \mathcal{H}_k,$$

completing the proof of the theorem. □

We would like to mention that the result of the theorem is not new. For a similar description of Belyi surfaces, see [**10, 8**]. We give only the new and simple proof, related to graph theory.

3. Orientations

In this section, we review some standard constructions of k-regular graphs, and show that they admit orientations in a more-or-less standard way. In §5 below we will provide constructions of orientations which have more unusual properties and are less canonical.

EXAMPLE 3.1 (Cayley graphs). Let G be a finite group, with a given set of generators $\mathcal{S} = \{g_1, \ldots, g_s, h_1, [\ldots, h_t]\}$, where the h_j's satisfy $h_j^2 = id$, and the brackets indicate that we allow the possibility that the set of h_j's may be empty. We do not assume that the set of generators §5 is symmetric.

The Cayley graph $\Gamma(G, \mathcal{S})$ of G with respect to the set of generators $\mathcal{S}$ is the $2s+t$-regular graph whose vertices are the elements of G, and where g is joined to g' by an edge if

$$g' = g_i^{\pm 1} g \quad \text{or} \quad g' = h_j g \quad \text{for some } i \text{ or } j.$$

Each edge is therefore labeled by an element of the generating set, and is a directed edge if the label is a g_i and is undirected if the label is an h_j.

We make pick an orientation $\mathcal{O}$ of $\Gamma(G, \mathcal{S})$ in a left-invariant manner by picking a cyclic ordering of the edges at the identity vertex, and transporting this cyclic ordering to all vertices by left translation. We will call such an orientation a homogeneous orientation on $\Gamma(G, \mathcal{S})$.

It is not difficult to describe the LHT paths for a homogeneous orientation. Beginning at the origin, pick out a new generator or the inverse of a generator. The orientation picks out a new generator or its inverse. Continuing this process, we arrive at a cyclic word W. If we denote by l the length of this cyclic word, then the length of the LHT path corresponding to this edge is $l|W|$, where by $|W|$ we denote the order of the word W in G. It may happen that this LHT path passes through the identity a number of times before closing up, if an initial segment of W^l is already the identity element. The structure of LHT paths at every point is the same as the structure of LHT paths at the identity.

We illustrate this construction in the case where $s = t = 1$. In this case, we have a 3-regular graph, and the choice of cyclic ordering at the identity does not make any difference. In this case, there will be one LHT path through each point of length the order of g. There will be either one or two LHT paths of length twice the order of gh, depending on whether there is a subword of gh^l which represents the identity.

As a particular case, we consider the group $PSL(2, \mathbb{Z}/5)$ with generators

$$h = \begin{pmatrix} 0 & 1 \\ -1 & 0 \end{pmatrix}, \qquad g = \begin{pmatrix} 1 & a \\ 0 & 1 \end{pmatrix}.$$

Then

$$gh = \begin{pmatrix} -a & 1 \\ -1 & 0 \end{pmatrix},$$

so choosing $a = \pm 1$, we get the familiar soccerball graph, while for a $a = \pm 2$, we find that each vertex has one LHT path of length 5 and two of length 10, so that $S^C(\Gamma(PSL(2, Z/5), \mathcal{S}), \mathcal{O}))$ is a surface of genus 4.

When there are no h_i's, one convenient class of cyclic orderings at the identity is shown in Figure 2. We call such an ordering *symmetric*.

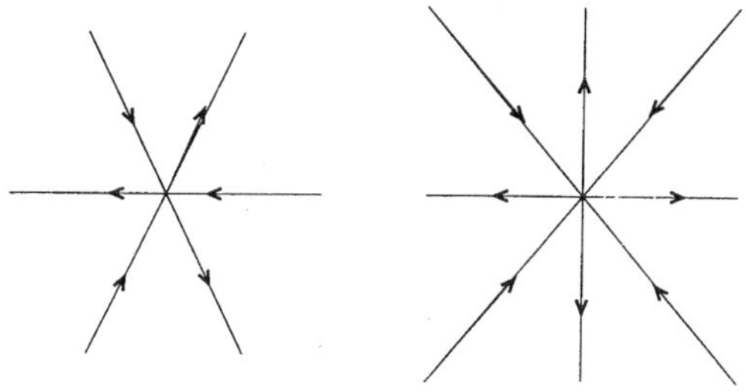

FIGURE 2. Symmetric cyclic orderings

In this case, when k is odd, we have the two words

$$W_1 = \Pi_{j=0}^{k-1} g_{k-j} \qquad \text{and} \qquad W_2 = \Pi_{j=0}^{k-1} g_{k-j}^{-1}$$

of length k. When k is even there is just one word of length $2k$, given by

$$W_1 = \Pi_{j=0}^{k-1} g_{k-j}^{-1} \Pi_{j=0}^{k-1} g_{k-j}.$$

REMARK 3.2. In the recent work [**7**], R. Brooks and A. Zuk studied Cheeger constants of some Hecke surfaces, including the last example.

EXAMPLE 3.3. Let π_k be the k-th Platonic graph of [**3**]. It is the k-regular graph defined by

$$\{(a,b) \in \mathbb{Z}/k \times \mathbb{Z}/k, a, b \quad \text{relatively prime to } k\}/(a,b) \sim (-a,-b).$$

Two vertices (a,b) and (c,d) are joined by an edge provided that

$$\det \begin{pmatrix} a & b \\ c & d \end{pmatrix} \equiv \pm 1 (\text{mod } k).$$

An orientation $\mathcal{O}$ on π_k may be defined as follows: at the vertex (a,b), let $<(a,b),(c,d)>$ be an edge. We choose the sign of (c,d) so that

$$\det \begin{pmatrix} a & b \\ c & d \end{pmatrix} \equiv 1 (\text{mod } k).$$

Then the next edge in the cyclic order at (a,b) is $<(a,b),(c-a),(d-b)>$.

With this orientation, all *LHT* paths are of length 3 by virtue of the sequence

$$<(a,b),(c,d)> \rightarrow <(c-a,d-b),(-a,-b)> \rightarrow <(c,d),(a-c,b-d)>.$$

The surface $S^C(\pi_k, \mathcal{O})$ is the *Platonic surface* P_k, which is the compactification of the modular surface

$$S_k = \mathbb{H}^2/\Gamma_k, \qquad \Gamma_k = \left\{ \begin{pmatrix} a & b \\ c & d \end{pmatrix} \equiv \pm \begin{pmatrix} 1 & 0 \\ 0 & 1 \end{pmatrix} (\text{mod } k) \right\}.$$

EXAMPLE 3.4. Let $(\Gamma, \mathcal{O})$ be an oriented k-regular graph, all of whose *LHT* paths have length l. Then we may define the dual oriented graph $(\Gamma^*, \mathcal{O}^*)$ as an l-regular graph, all of whose *LHT* paths are of length k, as follows: the vertices of Γ^* are the *LHT* paths of Γ. The edges $\{e\}$ correspond to the edges e of Γ, and $\{e\}$ joins the two (not necessarily distinct) *LHT* paths to which e belongs.

The orientation $\mathcal{O}^*$ on Γ is given as follows: given a LHT path γ and an edge e on Γ, the next element in the cyclic ordering at $\{\gamma\}$ is $\{e'\}$, where e' is the edge following e along the path γ.

THEOREM 3.5.

$$S^C(\Gamma,\mathcal{O}) = S^C(\Gamma^*,\mathcal{O}^*).$$

This duality concerns two types of compactification of surface by horocycles and to add the points of absolute by geodesics going to cusps.

The point here is that it may be that $(\Gamma,\mathcal{O})$ is difficult to analyze, but $(\Gamma^*,\mathcal{O}^*)$ may be relatively easy to understand. For instance, bounding the Cheeger constant and first eigenvalue of the dual Platonic graphs π_k^* uniformly from below is equivalent to Selberg's Theorem [**13**] up to constant, but the Cheeger constant and first eigenvalue of π_k^* may be calculated in an elementary manner [**3**].

4. Large Cusps

In this section, we study the relationship between the spectral geometry of $(\Gamma,\mathcal{O})$, $S^O(\Gamma,\mathcal{O})$ and $S^C(\Gamma,\mathcal{O})$. As this is in large measure a repetition of the material of [**5**], we will be brief.

As the oriented graph $(\Gamma,\mathcal{O})$ describes $S^O(\Gamma,\mathcal{O})$ as an orbifold covering of $\mathbb{H}^2/\mathbf{H}_k$, we have

THEOREM 4.1. *There are constants $C_1(k)$, $C_1'(k)$, $C_2(k)$, $C_2'(k)$, $C_3(k)$, $C_3'(k)$ such that*

(i) $C_1'(k)h(\Gamma) \le h(S^O(\Gamma,\mathcal{O})) \le C_1(k)h(\Gamma)$,

(ii) $C_2'(k)\gamma_1(\Gamma) \le \gamma_1(S^O(\Gamma,\mathcal{O})) \le C_2(k)\gamma_1(\Gamma)$,

(iii) $C_3'(k)\,\mathrm{diam}(\Gamma,\mathcal{O}) \le \widetilde{\mathrm{diam}}(S^O(\Gamma,\mathcal{O})) \le C_3(k)\,\mathrm{diam}\,(\Gamma,\mathcal{O})$,

where "$\widetilde{\mathrm{diam}}$" denotes the diameter of the "thick part", suitably defined, of $S^O(\Gamma,\mathcal{O})$.

We are interested in the extent to which the spectral geometry of $S^O(\Gamma,\mathcal{O})$ carries over to that of $S^C(\Gamma,\mathcal{O})$. We may pose the more general question of when the conformal compactification S^C of a finite-area Riemann surface S^O has spectral geometry close to that of S^O. The answer is contained in [**4**]. It requires the following

DEFINITION 4.2. Let S^O be a finite-area Riemann surface. S^O has cusps of length $\ge L$ if there is a system C_i of closed horocycles such that

(i) Each horocycle has length at least L.

(ii) Each cusp is contained in the interior of one of the C_i's.

(iii) The interiors of the C_i's are disjoint.

We then have

THEOREM 4.3. *For all ε, there exists L, such that if S^O is a finite-area Riemann surface with cusps of length $\ge L$, then there exist neighborhoods of the cusps in S^O and S^C, of size depending only on L, such that outside the cusp neighborhoods, the hyperbolic metrics ds_O^2 and ds_C^2 on S^O and S^C satisfy*

$$\frac{1}{1+\varepsilon}ds_O^2 \le ds_C^2 \le (1+\varepsilon)ds_O^2.$$

As is proved in [**4**], it is sufficient to show

COROLLARY 4.1. *Given ε, there exists L such that if S^O has cusps of length $\geq L$, then*

(i) $\frac{1}{1+\varepsilon}h(S^O) \leq h(S^C) \leq (1+\varepsilon)h(S^O)$.

(ii) $\frac{1}{1+\varepsilon}\gamma_1(S^O) \leq \gamma_1(S^C) \leq (1+\varepsilon)\gamma_1(S^O)$.

(iii) $\frac{1}{1+\varepsilon}\widetilde{\operatorname{diam}}(S^O) \leq \operatorname{diam}(S^C) \leq (1+\varepsilon)\widetilde{\operatorname{diam}}(S^O)$.

In short, when S^O satisfies the large cusps condition, the spectral geometries of S^O and S^C are close.

Since the fundamental domain $\mathbf{F}_k$ has a horocycle of length $2\cos(\pi/k)$, if the *LHT* path on the graph $(\Gamma, \mathcal{O})$ corresponding to a cusp has length l, then the l horocycle pieces of the corresponding fundamental domains come together to give a closed horocycle of length $2l\cos(\pi/k)$ about the cusp, and these will be disjoint for different cusps. So if the graph $(\Gamma, \mathcal{O})$ has *LHT* paths all of length l, then $S^O(\Gamma, \mathcal{O})$ will have cusps of length $\geq 2l\cos(\pi/k)$. We will call the horocycles constructed this way the *canonical horocycles* of $S^O(\Gamma, \mathcal{O})$.

As in [**5**], it may happen that the canonical horocycles of $S^O(\Gamma, \mathcal{O})$ may not have length $\geq L$, but that by making some horocycles larger and other horocycles smaller, the resulting system of horocycles will have length $\geq L$. Thus, the precise condition for the surface $S^O(\Gamma, \mathcal{O})$to have cusps of length $\geq L$ can be expressed purely in terms of the graph $(\Gamma, \mathcal{O})$, but the precise graph-theoretic statement is a bit complicated. It amounts to saying that the small cusps are sufficiently far apart, and sufficiently far away from short closed geodesics.

The theory as described in [**4**] is qualitative, but was made quantitative in [**11**]. We give an outline of the proof. If each cusp has a horocycle of length at least 2π, than you can close off the cusp with a metric of negative curvature by changing the metric conformally inside the cusp. The number 2π arises as necessary condition for this by Gauss-Bonnet theorem. D. Mangoubi [**11**] shows it is sufficient. The corresponding number for k-regular graphs would be the first integer m such that $2m\cos(\pi/k) > 2\pi$. So the limiting behavior as k is going to infinity is $m = 4$ and for $k = 3$ (modular group), $m = 7$. In particular, Mangoubi calculates how long the cusps must be to guarantee that S^C carries a metric of negative curvature. He shows that this will be the case provided that the cusps have length $\geq 2\pi$.

We remark that the large cusps condition *does not imply* that all the closed paths on the graph Γ are short. It is a condition only on the *LHT* paths. Thus, the oriented graph $(\Gamma, \mathcal{O})$ may have plenty of short geodesics, while still having cusps of length $\geq L$ for some large L.

Of course, it is not always convenient to change the metric within closed horocycles. For instance, the Platonic graphs π_k have *LHT* paths all of length 3, and so do not have large cusps. In [**11**] it is shown by example that one cannot weaken the large cusps condition by, for instance, replacing horocycles with a general condition such as large geodesic curvature and convexity. However, in special cases, we may still modify the metric on $S^O(\Gamma, \mathcal{O})$ in a canonical way to obtain the desired results. Here is an example geared to handle the Platonic graphs.

THEOREM 4.4. *There exists a k_0 and a number d_0 with the following property: let $(\Gamma, \mathcal{O})$ be a k-regular graph, for $k \geq k_0$, such that all the LHT paths have length equal to* 3. *Then there exist neighborhoods of the cusps and of the vertices* *of* $S^O(\Gamma, \mathcal{O})$ *and* $S^C(\Gamma, \mathcal{O})$*, depending only on* k_0*, such that outside of these neighborhoods, the metrics* ds_C^2 *and* ds_O^2 *satisfy*

$$\frac{1}{d_0} ds_O^2 \leq ds_C^2 \leq d_0 ds_O^2.$$

The notation is meant to emphasize that we do not have $d_0 \to 0$ as $k \to \infty$, in contrast to the situation in Theorems 4.2 and 4.1.

5. Prime Surfaces

In this section, we will study a class of surfaces which we call *prime surfaces.* They are surfaces of the form $S^C(\Gamma, \mathcal{O})$, where $(\Gamma, \mathcal{O})$ has precisely one left-hand-turn path. Geometrically, they are surfaces whose geometry is dominated by one large embedded geodesic ball.

One expects that if $S^C(\Gamma, \mathcal{O})$ is prime, and the Cheeger constant of Γ is large, then the Cheeger constant of $S^C(\Gamma, \mathcal{O})$ is exceptionally large. The intuitive reason for this is that a curve cutting Γ efficiently into two pieces can't have a component which is exceptionally short, because then it will bound a disc, and so have Cheeger constant larger tan 1. It cannot also have a component which is exceptionally long, since then it will have to cross the geodesic ball so that two strands are very close for a very long stretch, and this too is inefficient.

An orientation $\mathcal{O}$ on Γ is said to be *a prime orientation* if $(\Gamma, \mathcal{O})$ has precisely one left-hand turn path.

The terminology comes from a theorem of Billingsley [**2**], which says that prime numbers occur as frequently as cyclic permutations in the symmetric group, and an analogy between *LHT* paths on graphs and the cycle decomposition of an element of the symmetric group. This last analogy is not quite precise: a k-regular graph Γ on n vertices will not admit a prime orientation if k is odd and $n \equiv 0 \pmod 4$; if $k \equiv 0 \pmod 4$ and n even, or if $k \equiv 2 \pmod 4$. Somewhat curiously, most known constructions of graphs with large Cheeger constant or large first eigenvalue fall into one of these categories.

On the other hand, almost all graphs with k odd and $n \equiv 2 \pmod 4$ or $k \equiv 0 \pmod 4$ and n odd carry such orientations, as was shown in [**6**]:

THEOREM 5.1. *([**6**], see [**14**]) Suppose that* k *is odd and* $n \equiv 2 \pmod 4$ *or that* $k \equiv 0 \pmod 4$ *and* n *is odd.*

(a) *Let* Γ *be a* k*-regular graph, and* T *a spanning tree of* Γ *such that* $\Gamma - T$ *is connected (or more generally, so that each component of* $\Gamma - T$ *has an even number of edges). Then* Γ *admits a prime orientation.*
(b) *This last condition holds with asymptotic probability* $\to 1$ *as* $n \to \infty$.

If k is even and $(k-2)n \equiv 2 \pmod 4$, and Γ has a spanning tree T such that $\Gamma - T$ is connected, then Γ admits an orientation with precisely two *LHT* paths.

We remark that for graphs satisfying the condition (a), the construction of a prime orientation can be carried out algorithmically, once the spanning tree is given.

We now provide a collection of examples of graphs carrying prime orientations, with particular emphasis on graphs having large Cheeger constants.

EXAMPLE 5.2. If Γ is a k-regular graph, denote by $l\Gamma$ the kl-regular graph obtained by replacing each edge by l edges joining the same vertices. Let n be the number of vertices of Γ (and hence also $l\Gamma$).

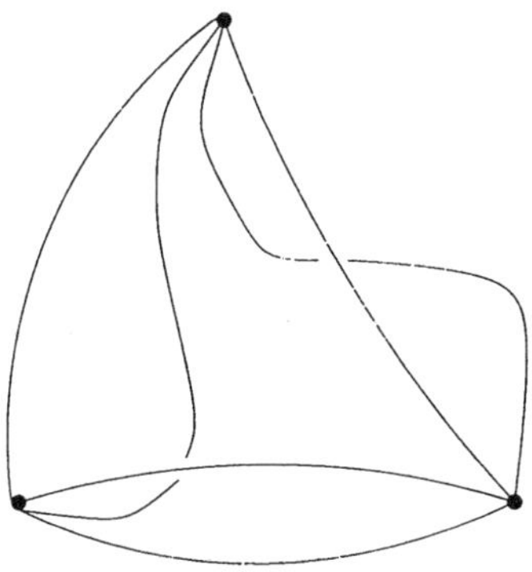

FIGURE 3. A prime orientation on $2\Gamma_3$

(i) If k is odd and $n = 2 \pmod 4$, then $l\Gamma$ admits a prime orientation for all odd l.
(ii) If k is even, then $l\Gamma$ admits a prime orientation provided that $kl \equiv 0 \pmod 4$.

Taking a specific case, let Γ_3 be the two-regular graph on 3 vertices. Then $l\Gamma_3$ admits a prime orientation for any even l. Note that the Cheeger constant of $l\Gamma$ is $2l$, which is also the regularity of $l\Gamma_3$.

A prime orientation on $2\Gamma_3$ is shown in Figure 3. Note that the genus of $S^C(l\Gamma_3, \mathcal{O})$ is $3(l/2) - 1 = 2, 5 \ldots$.

EXAMPLE 5.3.

(a) Let K^m be the complete graph on m vertices. Then K^m carries a prime orientation provided that $l \equiv 1$ or $2 \pmod 4$.
(b) Let $K^{2,m}$ be the complete graph on $2m$ vertices. Then $K^{2,m}$ carries a prime orientation provided that m is odd.

Conclusion

The main result of this paper concerns the relation of special class of Fuchsian groups and k-regular graphs. Hecke surfaces appear in several subjects. It follows from Theorem 2.1 that such surfaces make up an everywhere dense subset in the space of all Riemann surfaces. On the other hand, Hecke surfaces arise in the investigation of monodromial representations of the braid groups [**12**]. In the special case of the braid groups B_3 and C_4 (the "colored braid group"), these representations determine correlation functions of certain two-dimensional conformal models [**12**].

The technique of fat graphs allows one to obtain explicit coordinates on the Fricke (Teichmüller) spaces [**9**].

All these topics will be discussed in forthcoming papers.

Acknowledgements

I would like to thank the Department of Mathematics at the Technion for the invitation to give a talk at the conference on Geometry and Spectral Theory dedicated to the memory of Robert Brooks. The discussion with participants of the conference, especially A. Gamburd and A. Zuk, was very valuable.

I would like also to thank the Department of Mathematics and the Institute of Theoretical Physics at the Technion for financial support. The research of M. Monastyrsky was supported in part by the Russian Foundation for Basic Research (Grant No. 02-01-00734) and Scientific Schools (Grant No. 99-2185-1).

References

1. G. Belyi, *On Galois extentions of a maximal cyclotomic field*, Math. USSR, Izv. **14** (1980), 247-256.
2. P. Bilingsley, *Probability and Measure*, John Willey, NY, 3ed, 1995.
3. R. Brooks, *Spectral geometry and the Cheeger constant*, Expanding Graphs, Proc. Dimacs Workshop (J. Friedman, ed.), Amer. Math. Soc. (1993), pp. 5-19.
4. ______, *Platonic Surfaces*, Comm. Math. Helv. **74** (1999), 156-170.
5. R. Brooks and E. Makover, *Random construction of Riemann surfaces*, preprint, Technion, Haifa, 2000.
6. ______, *Belyi surfaces*, Israel Math. Conf. Proc. **15** (2001), 37-46.
7. R. Brooks, A. Zuk, *On the asymptotic isoperimetric constants for graphs and Riemann surfaces*, J. of Diff. Geometry **62** (2002), 49-78.
8. P. B. Cohen, C. Itzykson, and J. Wolfart, *Fuchsian triangle groups and Grothendieck dessins. Variations on a theme of Belyi*, Commun. Math. Physics. **163** (1994), 605-627.
9. V. V. Fock, *Conbinatorial description of the moduli space of projective structures*, preprint hep-th/9312193.
10. G. Jones and D. Singermann, *Belyi functions, hypermaps and Galois groups*, Bull. London Math. Soc. **28** (1996), 561-590.
11. D. Mangoubi, *Riemann surfaces and 3-regular graphs*, J. Anal. Math. 91 (2003), 193–209.
12. M. Monastyrsky and S. Nechaev, *Correlation function for some conformal theories on Riemann surfaces*, Mod. Phys. Lett. A. **12** (1997) 589-596.
13. A. Selberg, *On the estimation of Fourier coefficients of modular forms*, Proc. Sympos. Pure Math. **8** (1965), 1-15.
14. N. H. Xuong, *Sur quelques classes des graphes possedant des propriétés remarquables*, C. R. Acad. Sci. Paris **283** (1976), 813-816.

Dept. of Mathematics, Technion-Israel Institute of Technology, Haifa, 32000 Israel

Institute for Theoretical and Experimental Physics (ITEP), 117259 Moscow, Russia

E-mail address: monastyrsky@itep.ru

Contemporary Mathematics
Volume **387**, 2005

Isospectrality and Spectral Rigidity of Surfaces with Small Topology

Peter Buser and Klaus-Dieter Semmler

Dedicated to the memory of Robert Brooks

Abstract. Brooks and Tse have constructed pairs of isospectral non-isometric surfaces of constant negative curvature for any genus $g \geq 4$. They also found pairs with variable curvature for $g = 3$. We describe this construction in a slightly different way and show where the example fails when the curvature is constant. In the second part of this article we outline that for sufficiently small topology, here signature $(0; 2, 2, 3, 3; 0)$, no such pairs with constant negative curvature exist.

1. Introduction

In 1980, Marie-France Vignéras [**15**] constructed for the first time pairs of compact Riemann surfaces which are non-isometric but have the same spectrum (either Laplace spectrum or length spectrum). Her construction used quaternion algebras over number fields, and the genus of her first example was rather large. A few years later, Sunada [**14**] found a general construction allowing one to build examples using finite groups. Soon the question came up whether pairs of isospectral non-isometric Riemann surfaces (with the hyperbolic metric of constant negative curvature) exist for any genus $g \geq 2$.

In [**4**], after having scanned numerous possibilities, Robert Brooks and Richard Tse published examples for any genus $g \geq 4$, but were unable to settle the question for $g = 3$ and $g = 2$. This question is still open.

In [**4**] Brooks and Tse also gave an example for $g = 3$ with variable curvature. In the case of constant curvature however, the non-isometry breaks down. We give a detailed description of this example in Section 2 and show that one may take the curvature as close to -1 as one wishes.

In the second half of this article we consider the wider class of hyperbolic orbifolds, that is, quotient $\mathbb{H}/\Gamma$ of the hyperbolic plane $\mathbb{H}$ by a Fuchsian group Γ, where Γ may have elliptic and parabolic elements. In this wider class there are

2000 *Mathematics Subject Classification.* Primary 58J53; Secondary 30F35, 05C25.
Key words and phrases. Isospectrality, Fuchsian groups, moduli.
The authors were supported by the Swiss National Science Foundation, FNS-grants 20-68181, 20-101639, 21-65270.

cases, where the topology is, in a way, smaller than 2 (the area is smaller than that of a surface of genus 2), and for which the moduli space has *spectral rigidity*, i.e., distinct members always have distinct spectra. The first example of such a moduli space was given by Haas [**9**] who considered one-holed tori with fixed given boundary length, then later by Buser-Semmler in [**6**] for signature $(1,1)$ in general, and by Dianu [**8**] for signature $(1;p;0)$. All these cases correspond to 2-generator Fuchsian groups. In Section 3 we will describe a moduli space with spectral rigidity coming from a 3-generator group.

2. The Example of Genus 3 by Brooks and Tse

The constructions in [**4**] use Sunada's technique from [**14**] which we recall here in the following form. Let G be a finite group, H_1, H_2 two subgroups of G with the property that for any $g \in G$ the conjugacy class $[g]$ has as many representatives in H_1, as in H_2:

$$\text{card}([g] \cap H_1) = \text{card}([g] \cap H_2), \quad \forall g \in G. \tag{2.1}$$

Sunada's theorem (of which there are numerous variants, as e.g., in [**2**]) now says that if M is a Riemannian manifold and G acts as a subgroup of the isometry group of M such that H_1 and H_2 act freely, then the quotients

$$M_1 = M/H_1, \quad M_2 = M/H_2 \tag{2.2}$$

(with the induced metric) are isospectral.

One can also show that if H_1 and H_2 are non-conjugate in G and if M/G (which may be an orbifold) has trivial isometry group, then M_1 and M_2 are non-isometric.

Brooks and Tse [**4**] consider the following example. Let from now on G denote the group

$$G = \text{PSL}(3,2),$$

the group of all 3×3 matrices over the field $\mathbb{F}_2$. The following

$$A = \begin{pmatrix} 1 & 0 & 1 \\ 1 & 1 & 0 \\ 0 & 1 & 0 \end{pmatrix}, \quad B = \begin{pmatrix} 0 & 0 & 1 \\ 1 & 0 & 0 \\ 1 & 1 & 0 \end{pmatrix} \tag{2.3}$$

are generators of G satisfying

$$A^7 = B^7 = 1. \tag{2.4}$$

It is shown, e.g., in [**4**] or [**12**] that the two subgroups of index 7

$$H_1 = \left\{ \begin{pmatrix} 1 & * & * \\ 0 & * & * \\ 0 & * & * \end{pmatrix} \in \text{PSL}(3,2) \right\}, \quad H_2 = \left\{ \begin{pmatrix} 1 & 0 & 0 \\ * & * & * \\ * & * & * \end{pmatrix} \in \text{PSL}(3,2) \right\} \tag{2.5}$$

satisfy condition (2.1) and are non-conjugate in $\text{PSL}(3,2)$.

In order to visualize the examples obtained by Sunada's method one may work with *Schreier graphs*. Let $\mathcal{G}$ be any finite group with generators $\mathcal{A}_1, \dots, \mathcal{A}_m$, and let $\mathcal{H}$ be a subgroup of $\mathcal{G}$ of index l, say. The Schreier graph for this setting is defined as follows. The vertices are the cosets, $g_0\mathcal{H}, \dots, g_{l-1}\mathcal{H}$, and for any g_i and any generator $\mathcal{A}_j$, an edge is drawn from vertex $g_i\mathcal{H}$ to vertex $g_{i'}\mathcal{H}$, where the latter is defined by the condition $\mathcal{A}_j g_i \mathcal{H} = g_{i'}\mathcal{H}$.

For G and the two subgroups H_1, H_2 as above, we take $g_i = A^i$, $i = 0, \dots, 6$, so that the vertices of the two Schreier graphs S_1 and S_2 become $H_\delta, AH_\delta, \dots, A^6 H_\delta$,

($\delta = 1, 2$). This enumeration of the cosets has the advantage that edges corresponding to generator A are simply the ones going from $A^i H_\delta$ to $A^{i+1} H_\delta$.

Now let us take a convex geodesic quadrangle Q in the hyperbolic plane $\mathbb{H}$ with consecutive sides $\alpha, \beta', \beta, \alpha'$, where α and α' have the same length, and β and β' have the same length. Let us also fix an orientation of the boundary of Q. For $\delta = 1, 2$ an orientable surface M_δ is obtained by replacing each vertex $A^i H_\delta$ of S_δ by a copy Q_δ^i of Q, and by pasting these copies together along their sides according to the following rules, where all exponents are to be understood modulo 7: Side α of Q_δ^i is pasted to side α' of Q_δ^{i+1}, and side β of Q_δ^i is pasted to side β' of $Q_\delta^{i'}$, where i' is the exponent given by $BA^i H_\delta = A^{i'} H_\delta$, $i = 0, \ldots, 6$. In order to get orientable surfaces the pastings have to be such that matching sides meet with opposite orientations.

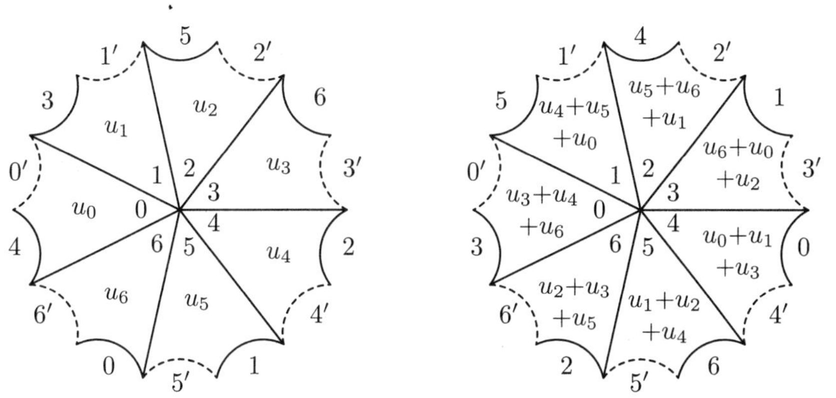

FIGURE 1. Either polygon yields a surface of genus 3

Figure 1 shows the copies of Q (labelled $0, \ldots, 6$) drawn in a way such that side α of Q_δ^i coincides with side α' of Q_δ^{i+1}, i.e., the pastings corresponding to generator A are given by the adjacencies in the figure. The remaining pastings are indicated by the labels of the sides: Side β of Q_δ^i labelled k is pasted to side β' of Q_δ^k labelled k'. This coincides with the figure given by Brooks and Tse [**4**, p. 15. Fig. 2], where in the first example Q_1^i is marked with the first column of A^i, and in the second example Q_2^i is marked with the first row of A^{-i}.

Looking at the Euler characteristic we see that the surfaces M_1, M_2 obtained in this way have genus 3.

Let us now take Q to be the geodesic quadrangle obtained by pasting together two equilateral triangles with angles $\pi/7$. It is an exercise to see that this is the unique quadrangle for which M_1 and M_2 (with the metric induced by the hyperbolic metrics of the quadrangles) have smooth hyperbolic metrics. With the quadrangle thus set we get an isospectral pair.

Unfortunately, the two examples are isometric: The orientation reversing isometry $\phi : \mathbb{H} \to \mathbb{H}$ that sends the Q_1^i on the left to the Q_2^{5-i} on the right is compatible with the pastings and thus induces an isometry from M_1 to M_2.

Here is a way to break the symmetry, however by introducing variable curvature. Take any open domain $\tilde{Q}$ in Q that has no symmetry (e.g., having the shape of the letter "Q") lying in a small neighbourhood of the vertex between sides α and

α'. Then change the metric on Q in such a way that it coincides with the initial metric on $Q \smallsetminus \tilde{Q}$ and such that on $\tilde{Q}$ it has non-constant curvature varying between $-1-\epsilon$ and $-1+\epsilon$, ϵ some small positive constant.

Now repeat the above construction for this modified Q. Considering how the copies of $\tilde{Q}$ are distributed on M_1 and M_2 we see that a possible isometry from M_1 to M_2 would have to come from an orientation preserving isometry between the two 14-gons. However, none of these isometries matches with the pasting schemes. Hence, M_1 and M_2 are non-isometric.

We finish this paragraph by proving the isospectrality of the Brooks-Tse example in a direct way using transplantation as in [**5**].

Let u be an eigenfunction of the Laplace operator on M_1 for some eigenvalue λ and denote by u_i the restriction of u to Q_1^i, $i = 0, \dots, 6$. For any i and j there exists a unique isometry φ_{ij} mapping Q_1^i to Q_2^j, and we may push forward the function u_i to Q_2^j, the push forward being the function $u_i \circ \varphi_{ij}^{-1}$ on Q_2^j. For simplicity, we denote this function u_i again.

In the interior of the Q_2^j the function thus obtained satisfy the eigenvalue equation for the Laplacian (because the φ_{ij} are isometries), but they do not match properly across the edges, except perhaps in some special cases. Now take on each Q_2^j the selection of three pushed forward functions as given by Figure 1 (u_3, u_4, u_6 on Q_2^0, etc.) and take their sums as indicated ($u_3 + u_4 + u_6$ on Q_2^0, etc). A diagram chasing reveals that these linear combinations match smoothly across all edges. Let us check just one case: in the example on the right hand side of Figure 1 we have the pair $u_3 + u_4 + u_6$, $u_6 + u_0 + u_2$ of functions meeting each other along the β-side of Q_2^0 (labelled "3"), and on the left hand side we have u_3 continuing smoothly over the β-side (labelled "6") into u_6, then in the same way u_4 over the β-side into u_2, and u_6 into u_0. Hence, $u_3 + u_4 + u_6$ continues smoothly over the β-side of Q_2^0 into $u_6 + u_0 + u_2$, and so forth.

We thus have a linear mapping from the eigenspace of eigenvalue λ on M_1 into the corresponding eigenspace on M_2 given symbolically by

$$\begin{bmatrix} v_0 \\ v_1 \\ v_2 \\ v_3 \\ v_4 \\ v_5 \\ v_6 \end{bmatrix} = \begin{bmatrix} 0 & 0 & 0 & 1 & 1 & 0 & 1 \\ 1 & 0 & 0 & 0 & 1 & 1 & 0 \\ 0 & 1 & 0 & 0 & 0 & 1 & 1 \\ 1 & 0 & 1 & 0 & 0 & 0 & 1 \\ 1 & 1 & 0 & 1 & 0 & 0 & 0 \\ 0 & 1 & 1 & 0 & 1 & 0 & 0 \\ 0 & 0 & 1 & 1 & 0 & 1 & 0 \end{bmatrix} \begin{bmatrix} u_0 \\ u_1 \\ u_2 \\ u_3 \\ u_4 \\ u_5 \\ u_6 \end{bmatrix} \tag{2.6}$$

The matrix (this one over the field $\mathbb{R}$) has determinant $\neq 0$, and so the mapping is one-to-one. Since a similar mapping exists in the inverse direction, it follows that the eigenspaces with eigenvalue λ on M_1 and M_2 have the same dimensions, i.e., M_1 and M_2 are isospectral.

3. Spectral Rigidity for Signature $(0; 2, 2, 3, 3; 0)$

While the preceding section dealt with the question of how far down in the topology we can go with the isospectral examples (that are known so far), we now want to look at how far we can come up with the spectrally rigid cases.

The goal is to give an outline of the spectral rigidity of the Riemann surfaces of signature $(0; 2, 2, 3, 3; 0)$, that is, (see Singerman [**13**] for the general notation) surfaces of genus 0 carrying a hyperbolic metric with four cone-like singularities of cone angles $2\pi/p$, where p has the respective values 2,2,3,3. The proof is rather involved and will appear in [**7**]. What we want to do here is to show how the approach goes in principle.

The general scheme of the approach to the spectrum in small topology is as follows.

(1) Select suitable geometric parameters that define the Riemann surface (of the given topology) uniquely. Take spectral data as parameters. This sets up a parameter space $\mathcal{T}$.
(2) A point (sometimes called "moduli") in $\mathcal{T}$ determines the corresponding Riemann surface uniquely, but different points may determine the same surface. Therefore, find a fundamental domain $\mathcal{F} \subset \mathcal{T}$ such that any surface of the given signature is represented by a unique point in $\mathcal{F}$. Describe $\mathcal{F}$ in terms of algebraic relations and inequalities between the parameters.
(3) Show that for a suitable finite part of the spectrum the algebraic relations and inequalities given to describe $\mathcal{F}$ are sufficient to tell us which among the "spectral lines" are the parameters. This will show that the spectrum – in fact a finite part of it – determines the parameters and hence the surface.

Let us now follow this scheme. The surfaces – which are orbifolds – will be given as quotients $\mathbb{H}/\Gamma$, where Γ is a Fuchsian group, and we will say that the signature of $\mathbb{H}/\Gamma$ is also the signature of Γ.

The spectrum of $\mathbb{H}/\Gamma$ here is the length spectrum or, what amounts to the same, the *trace spectrum* of Γ, that is, the increasing sequence of traces of the extended conjugacy classes of elements in Γ ("extended" means that the class also includes the conjugates of the inverse). The precise statement is this.

THEOREM 3.1. *Riemann surfaces of signature* $(0; 2, 2, 3, 3; 0)$ *are isospectral if and only if they are isometric.*

3.1. Generators. A Fuchsian group of signature $(0; 2, 2, 3, 3; 0)$ will be generated by four Möbius transformations $\alpha, \beta, \gamma, \delta$ given by the following matrices $A, B, C, D \in \mathrm{SL}(2, \mathbb{R})$ which are normalized such that the traces of A, B, C are ≥ 0, and the traces of AB, BC, AC are ≤ 0.

$$(3.1)\qquad \begin{aligned} A &= \begin{bmatrix} 0 & \sigma \\ -\sigma & 0 \end{bmatrix}, & D &= \frac{-1}{2s}\begin{bmatrix} (s+r\rho\sigma)\lambda & \frac{-(r^2+s^2)\rho\sigma}{\lambda} \\ \lambda\rho\sigma & \frac{(s-r\rho\sigma)}{\lambda} \end{bmatrix} \\ B &= \begin{bmatrix} 0 & \sigma\lambda \\ -\frac{\sigma}{\lambda} & 0 \end{bmatrix}, & C &= \frac{1}{2s}\begin{bmatrix} s-r\rho\sigma & (r^2+s^2)\rho\sigma \\ -\rho\sigma & s+r\rho\sigma \end{bmatrix} \end{aligned}$$

The numbers satisfy the following conditions

$$r \leq 0, \quad s > 0, \quad \lambda > 1, \quad \rho = \sqrt{3}, \quad \sigma = 1.$$

A straight forward calculation shows that

$$A^2 = B^2 = C^3 = -\mathbf{1}, \quad ABCD = \mathbf{1}.$$

We will have to determine λ such that the group generated by A, B, C is discrete and has signature $(0; 2, 2, 3, 3; 0)$. To this end we introduce the following parameters which will later parametrize our surfaces.

$$x = -\operatorname{tr} AB, \quad y = -\operatorname{tr} BC, \quad z = -\operatorname{tr} AC, \tag{3.2}$$

tr being the <u>*half trace*</u> which is more convenient for computations,

$$\operatorname{tr}(X) := \frac{1}{2}\operatorname{trace}(X). \tag{3.3}$$

One may check that

$$x = \frac{1}{2}\left(\frac{1}{\lambda} + \lambda\right), \quad y = \frac{\rho(r^2 + s^2 + \lambda^2)}{4\lambda s}, \quad z = \frac{\rho(r^2 + s^2 + 1)}{4s}, \tag{3.4}$$

and $x > 1$, $y \geq \rho/2$, $z \geq \rho/2$.

Note that in return,

$$\lambda = x + \sqrt{x^2 - 1}, \quad s = \frac{\rho(r^2 + s^2 + 1)}{4z}, \quad \text{with} \quad r^2 + s^2 = \frac{\lambda - \frac{y}{z}}{\frac{y}{z} - \frac{1}{\lambda}}. \tag{3.5}$$

The trace of D satisfies the remarkable equation

$$\operatorname{tr}^2(D) + \operatorname{tr}(D) + x^2 + y^2 + z^2 - 2xyz - \tfrac{3}{4} = 0. \tag{3.6}$$

We thus see that D is of order 3 (i.e., $|\operatorname{tr} D| = \frac{1}{2}$) if and only if the triple x, y, z satisfies the equation $P(x, y, z) = 0$, where P is the polynomial

$$P(x, y, z) = x^2 + y^2 + z^2 - 2xyz + \frac{x - 1}{2}. \tag{3.7}$$

It is not difficult to show that for $x > 1$ any y, z satisfying $P(x, y, z) = 0$ also satisfy $y > 1$ and $z > 1$. Furthermore, if $x, y, z > 1$ are given satisfying $P(x, y, z) = 0$, then (3.4) can be solved for r, s and λ with

$$\lambda = \frac{-s + r\rho\sigma}{s + r\rho\sigma}. \tag{3.8}$$

The fixed points a, b, c, d of A, B, C, D are now

$$a = i, \quad b = \lambda i, \quad c = r + is, \quad d = (r + is)/\lambda.$$

These fixed points are the vertices of a convex geodesic quadrilateral in $\mathbb{H}$ which has the following properties. The angle sum at vertices a and b is π, and the angle sum at vertices c and d is $2\pi/3$. Poincaré's polygon theorem (see, e.g., in Beardon's book [**1**]) tells us that the group $\tilde{\Gamma}$ generated by the reflections across the sides of the polygon is discrete with the polygon as fundamental domain, and it follows that the group Γ generated by the Möbius transformations $\alpha, \beta, \gamma, \delta$ (represented by A, B, C, D) has index 2 in $\tilde{\Gamma}$ and its signature is $(0; 2, 2, 3, 3; 0)$.

3.2. The parameter space. The preceding considerations lead us to the following parameter space, where any point represents a Fuchsian group of this signature,

$$\mathcal{T} = \{(x, y, z) \in \mathbb{R}^3 \mid x, y, z > 1, \ P(x, y, z) = 0\}. \tag{3.9}$$

For the next proposition we want to give two proofs in order to illustrate the different ways one may want to proceed with the spectrum. The first proof uses arguments in $\mathrm{SL}(2, \mathbb{R})$ and is given in [**7**] so that we just mention the steps. The second proof uses arguments from surface geometry.

PROPOSITION 3.2. *Any Riemann surface of signature* $(0; 2, 2, 3, 3; 0)$ *arises as* $\mathbb{H}/\Gamma$ *with* Γ *as constructed as above.*

PROOF in $\mathrm{SL}(2, \mathbb{R})$. The idea is to consider *any* subgroup of $\mathrm{PSL}(2, \mathbb{R})$ generated by elliptic elements $\alpha, \beta, \gamma, \delta$ of respective orders 2, 2, 3, 3, satisfying $\alpha\beta\gamma\delta = \mathbf{1}$, and represent the generators by matrices $A, B, C, D \in \mathrm{SL}(2, \mathbb{R})$ with the sign convention $\operatorname{tr} C = \operatorname{tr} D = \frac{1}{2}$, $\operatorname{tr} AB, \operatorname{tr} BC, \operatorname{tr} AC \leq 0$. One then has $ABCD = \varepsilon \mathbf{1}$ with $\varepsilon = \pm 1$.

The groups generated in this way cover all those of signature $(0; 2, 2, 3, 3; 0)$, but also many others, most of them non-discrete.

It then comes out as a surprising fact that the polygon $\mathcal{P}$ formed by the fixed points a, b, c, d of A, B, C, D is convex for all groups with $\varepsilon = 1$ and non-convex or degenerate for all groups with $\varepsilon = -1$. Poincaré's polygon theorem then allows one to complete the proof.

PROOF using surface geometry. Let S be a given surface of signature $(0; 2, 2, 3, 3; 0)$ and denote by a, b the cone points of order 2, and by c, d, the cone points of order 3. Draw the shortest geodesic arc ω from a to b on S. Since the cone angles at c and d are smaller than 2π, these points do not lie on the minimal arc ω.

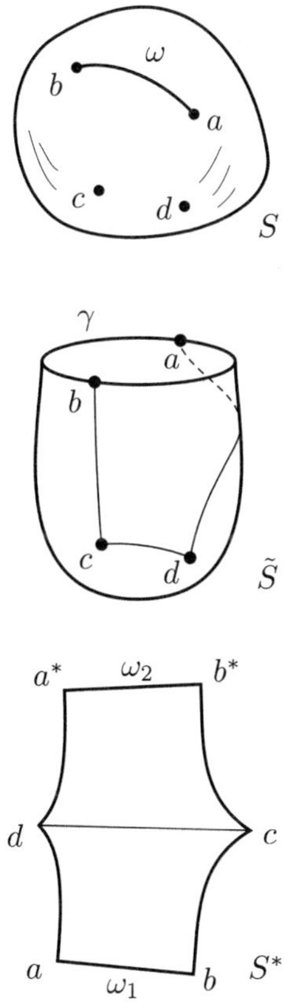

FIGURE 2

Now cut S open along ω to get a hyperbolic surface $\tilde{S}$ with a smooth boundary geodesic γ and the two cone points c and d in its interior.

Connecting c and d on $\tilde{S}$ with a geodesic arc and then drawing the orthogonals from c and d to γ (not shown in the figure) we get a decomposition of $\tilde{S}$ into two geodesic quadrilaterals with two right angles. Such a quadrilateral is determined up to isometry by the lengths of three of its sides. So the two are symmetric across their sides. Moreover, the angles at c and d are both $2\pi/6$. We therefore also have an orientation preserving isometry $\varphi : \tilde{S} \to \tilde{S}$ interchanging a with b and c with d.

φ also interchanges the shortest arc on $\tilde{S}$ from c to b with the shortest arc from d to a. Cutting $\tilde{S}$ open along these arcs we thus get a convex hyperbolic geodesic polygon S^* as shown in the figure, with consecutive vertices a, b, c, b^*, a^*, d, where sides bc, cb^*, a^*d and da have the same lengths, sides b^*a^* and ab have the same lengths, the angles at c and d are $2\pi/3$ and the angles at a and b, respectively a^* and b^* add up to π. This is, up to its position the fundamental domain of a Fuchsian group Γ as constructed above.

3.3. Some spectral lines. Let us consider a few traces (or "spectral lines") from the spectrum of $S = \mathbb{H}/\Gamma$. They correspond to the closed geodesics shown in Figure 3.

Such figures are drawn as follows. Take the fundamental domain $\mathcal{P}$ of Γ as shown in Figure 2, paint part $abcd$ white and part a^*dcb^* black. The images under Γ tessellate $\mathbb{H}$ with alternatively black and white quadrilaterals. The images of a and b are surrounded by four quadrilaterals, those of c and d by six.

If now a word such as $CD^{-1}AB$ is given representing the element $g = \gamma\delta^{-1}\alpha\beta \in \Gamma$, then we draw a curve in $\mathbb{H}$ starting from the white part of $\mathcal{P}$, go around vertex c in the positive sense to the next white quadrilateral, there around the image of d in the negative sense to the next white quadrilateral, and so on. The curve drawn in this way projects to a closed curve in S whose free homotopy class (in the sense of orbifolds) corresponds to g.

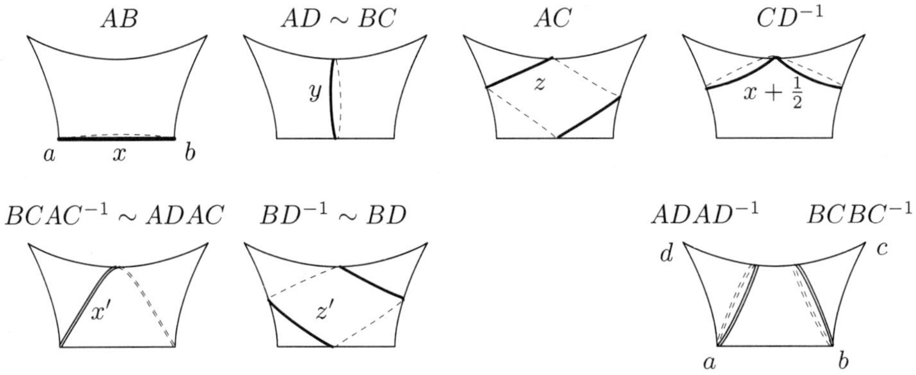

FIGURE 3. Closed geodesics on a surface of signature $(0; 2, 2, 3, 3; 0)$

Figure 3 shows copies of S which must be interpreted as surfaces of genus 0 with four cone points ("pillows").

The geodesic corresponding to AB goes from cone point a to cone point b and back to a again. This "bouncing back" at b is caused by the fact that the cone angle at b equals π. The geodesic also bounces back at a. The geodesic corresponding to $BCAC^{-1}$ bounces back at a and b in the same way. Since $ABCD = \mathbf{1}$ and $B^{-1} = -B$, we have $ADAC = -AC^{-1}BC$, where $AC^{-1}BC$ is conjugate to $BCAC^{-1}$, and so the words $ADAC$ and $BCAC^{-1}$ correspond to the same geodesic. This is indicated with the symbol $\sim$.

In order to compute the traces we use the following classical formula for $X, Y \in \mathrm{SL}(2, \mathbb{R})$, (e.g., [**11**]), where we recall that $\mathrm{tr} = \frac{1}{2}\mathrm{trace}$,

$$\mathrm{tr}\, XY + \mathrm{tr}\, XY^{-1} = 2\,\mathrm{tr}\, X\,\mathrm{tr}\, Y. \tag{3.10}$$

This formula and the relations $A^2 = B^2 = C^3 = D^3 = -\mathbf{1}$ and $ABCD = \mathbf{1}$ are all we need to compute the trace spectrum of S. Here is an initial list.

$$\mathrm{tr}\, AB = -x, \quad \mathrm{tr}\, BC = -y, \quad \mathrm{tr}\, AC = -z$$

$$\begin{aligned}
\mathrm{tr}\, CD^{-1} &= -\,\mathrm{tr}\, ABC^{-1} = -2\,\mathrm{tr}\, AB\,\mathrm{tr}\, C + \mathrm{tr}\, ABC &&= x + \tfrac{1}{2}\\
\mathrm{tr}\, BCAC^{-1} &= 2\,\mathrm{tr}\, BC\,\mathrm{tr}\, AC^{-1} - \mathrm{tr}\, BC^{-1}A &&= -\,2yz + x + \tfrac{1}{2}\\
\mathrm{tr}\, BD^{-1} &= \mathrm{tr}\, ABCB = 2\,\mathrm{tr}\, AB\,\mathrm{tr}\, BC - \mathrm{tr}\, AC^{-1} &&= 2xy - z.
\end{aligned}$$

In the same way we find $\mathrm{tr}\, ADAD^{-1} = \mathrm{tr}\, BCBC^{-1} = -2y^2 + \frac{1}{2}$ for the two geodesics in the last example of Figure 3.

3.4. The fundamental domain. Setting

$$A_1 = -A, \quad B_1 = -C^{-1}BC, \quad C_1 = C^{-1}, \quad D_1 = AD^{-1}A^{-1},$$

we get matrices of new generators of Γ that satisfy the same conditions as A, B, C, D, (e.g., $A_1B_1C_1D_1 = AC^{-1}BCC^{-1}AD^{-1}A^{-1} = AC^{-1}BAABCA^{-1} = \mathbf{1}$, etc). This gives us a new triple of parameters for Γ,

$$\begin{aligned} x_1 &= -\operatorname{tr} A_1B_1 = -\operatorname{tr} AC^{-1}BC = 2yz - x - \tfrac{1}{2} \\ y_1 &= -\operatorname{tr} B_1C_1 = \operatorname{tr} BC^{-1} = y \\ z_1 &= -\operatorname{tr} A_1B_1 = \operatorname{tr} AC^{-1} = z. \end{aligned}$$

Likewise, setting

$$A_2 = -B, \quad B_2 = -A, \quad C_2 = D^{-1}, \quad D_2 = C^{-1},$$

we get generators

$$\begin{aligned} x_2 &= -\operatorname{tr} A_2B_2 = -\operatorname{tr} AB = x \\ y_2 &= -\operatorname{tr} B_2C_2 = \operatorname{tr} AD^{-1} = -\operatorname{tr} BC = y \\ z_2 &= -\operatorname{tr} A_2B_2 = \operatorname{tr} BD^{-1} = 2xy - z. \end{aligned}$$

Finally, setting

$$A_3 = -B, \quad B_3 = -A, \quad C_3 = C^{-1}, \quad D_3 = CD^{-1}C^{-1},$$

we get generators and parameters

$$x_3 = x, \quad y_3 = z, \quad z_3 = y. \tag{3.11}$$

Hence, we have a group $\mathcal{M}$ acting on the parameter space $\mathcal{T}$ generated by the transformations

$$(x, y, z) \mapsto m_i(x, y, z) := (x_i, y_i, z_i), \quad i = 1, 2, 3.$$

The essential changes here are given by the functions

$$x' = 2yz - x - \tfrac{1}{2}, \quad z' = 2xy - z.$$

The corresponding replacements of the geodesics are shown in Figure 3. The algebraic interpretation is as follows. The polynomial P (in (3.7)) is symmetric in y and z, and m_3 is just the automorphism of $\mathcal{T}$ that takes care of this fact. If y and z are given, and if x is a root of the quadratic polynomial $P(\,.\,, y, z)$, then x' is the other root and vice-versa. Thus, m_1 exchanges these roots. Similarly, m_2 exchanges the roots of $P(x, y, \,.\,)$.

In order to set up a fundamental domain for the action of $\mathcal{M}$ we use the following

Algorithm.

(1) *If* $y > z$ *apply* m_3.
(2) *If* $z > xy$ *apply* m_2.
(3) *If* $x > yz - \frac{1}{4}$ *apply* m_1.

Note that xy is the middle between the two roots of $P(x, y, \,.\,)$ and $yz - \frac{1}{4}$ is the middle between the roots of $P(\,.\,, y, z)$. The three coordinates are therefore monotone decreasing under this algorithm, and since the trace spectrum of Γ is discrete, the algorithm will eventually stop.

The domain $\mathcal{F}$ is defined as the set of all triples that cannot be made smaller by the algorithm.

$$\mathcal{F} = \{(x, y, z) \in \mathcal{T} \mid y \leq z \leq xy, \quad x \leq yz - \tfrac{1}{4}\}. \tag{3.12}$$

From what has just been said we have the following.

PROPOSITION 3.3. *Any Riemann surface of signature* $(0; 2, 2, 3, 3; 0)$ *is represented by a triple in* $\mathcal{F}$.

By Theorem 3.1, distinct points in $\mathcal{F}$ never represent the same group. As a consequence, $\mathcal{M}$ is the full modular group on the parameter space $\mathcal{T}$, and $\mathcal{F}$ is a fundamental domain of $\mathcal{M}$.

Below we will need the following.

LEMMA 3.4. *Let* $(x, y, z) \in \mathcal{F}$. *Then*

(i) $y > x + \frac{1}{2} \Longrightarrow x < \frac{1+\sqrt{17}}{4}$,
(ii) $x = y + \frac{1}{2} \Longrightarrow y \geq \frac{1+\sqrt{17}}{4}$,
(iii) $z = y + \frac{1}{2} \Longrightarrow y \geq \frac{3+\sqrt{5}}{4}$.

PROOF. These are just consequences of the defining equations and inequalities of $\mathcal{F}$. As an illustration we prove (ii) which is the simplest.

Writing P in the following form

$$P(x, y, z) = (z - xy)^2 + \frac{x+1}{2} - (x^2 - 1)(y^2 - 1)$$

and recalling that $P = 0$ we see that $2(x-1)(y^2-1) \geq 1$ for any $(x, y, z) \in \mathcal{F}$. If now $x = y + \frac{1}{2}$ then $(2y-1)(y^2-1) \geq 1$, which is equivalent to $y(2y^2 - y - 2) \geq 0$, and the right hand side in (ii) is the root larger than 1 of $2y^2 - y - 2$. $\square$

3.5. The bottom of the spectrum. The hardest part in the proof of Theorem 3.1 is the estimate of the initial part of the spectrum and we will state the result without proofs.

For the following we denote by Λ_{CD} the subsequence in the trace spectrum of Γ formed by the traces that correspond to elements whose matrices can be written as words in C and D. In a similar manner we denote by Λ_{AD} and Λ_{BC} the subsequences corresponding to words in A and D, respectively B and C. Finally we denote by Λ the trace spectrum of Γ itself.

PROPOSITION 3.5. *For any* Γ *with parameters* $(x, y, z) \in \mathcal{F}$ *the following hold.*

(i) *The first line in* Λ *is either* x *or* y.
(ii) *The first two lines in* Λ_{CD} *are* x *and* $x + \frac{1}{2}$. *The third line in* Λ_{CD} *is larger than* $x + \frac{1}{2}$.
(iii) *The first two lines in* $\Lambda_{AD} \cup \Lambda_{BC}$ *are* y *and* $2y^2 - \frac{1}{2}$.

For the second proposition we introduce the following subsets of $\mathcal{F}$,

$$\mathcal{F}^a = \{(x, y, z) \in \mathcal{F} \mid y > x + \tfrac{1}{2}\}$$
$$\mathcal{F}^c = \{(x, y, z) \in \mathcal{F} \mid x, z > y + \tfrac{1}{2}\}$$
$$\mathcal{F}^b = \mathcal{F} \smallsetminus (\mathcal{F}^a \cup \mathcal{F}^c).$$

The statement then is

PROPOSITION 3.6. *For any* Γ *with parameters* $(x, y, z) \in \mathcal{F}$ *the following hold.*

(1) *If* $(x, y, z) \in \mathcal{F}^a$, *then the first line in* $\Lambda \setminus \Lambda_{CD}$ *is* y.
(2) *If* $(x, y, z) \in \mathcal{F}^b$, *then the first four lines in* Λ *are among* $x, y, z, x+\frac{1}{2}, x', z'$.
(3) *If* $(x, y, z) \in \mathcal{F}^c$, *then the first two lines in* $\Lambda \setminus (\Lambda_{AD} \cup \Lambda_{BC})$ *are* x *and* z.

3.6. Some spectral arguments. Let us now illustrate how one may get geometric information out of spectral data. In all the arguments that follow, Γ and $\tilde{\Gamma}$ are groups of signature $(0; 2, 2, 3, 3; 0)$ as above with parameters (x, y, z) and $(\tilde{x}, \tilde{y}, \tilde{z})$ in $\mathcal{F}$.

The first argument is the most simple one and shows rather nicely how algebraic relations come in.

LEMMA 3.7. *If* $\{x, y, z\} = \{\tilde{x}, \tilde{y}, \tilde{z}\}$, *then* $(x, y, z) = (\tilde{x}, \tilde{y}, \tilde{z})$.

PROOF. Both triples satisfy $P = 0$. Therefore, by the symmetry properties of P,

$$0 = P(x, y, z) - P(\tilde{x}, \tilde{y}, \tilde{z}) = \tfrac{1}{2}(x - \tilde{x}). \tag{3.13}$$

Thus, $x = \tilde{x}$, and since $y \leq z$ and $\tilde{y} \leq \tilde{z}$ (by the definition of $\mathcal{F}$), it follows that $y = \tilde{y}$ and $z = \tilde{z}$. □

The next lemma gives a spectral characterization of the sets $\mathcal{F}^a, \mathcal{F}^b, \mathcal{F}^c$.

LEMMA 3.8. *Let* $\ell_1 \leq \ell_2 \leq \ell_3$ *be the first three spectral lines of* Γ. *Then*

$$(x, y, z) \in \mathcal{F}^a \iff \ell_1 < \tfrac{1+\sqrt{17}}{4} \text{ and } \ell_1 + \tfrac{1}{2} = \ell_2 < \ell_3$$
$$(x, y, z) \in \mathcal{F}^c \iff \ell_1 + \tfrac{1}{2} < \ell_2.$$

PROOF. Assume $(x, y, z) \in \mathcal{F}^a$. By Proposition 3.5(ii), the first two lines in Λ_{CD} are x and $x + \frac{1}{2}$, while the third line in Λ_{CD} is larger than $x + \frac{1}{2}$. By Proposition 3.6 (a), the first line outside Λ_{CD} is y, where by hypothesis $y > x + \frac{1}{2}$. This shows that $\ell_3 > \ell_2 = \ell_1 + \frac{1}{2}$. The upper bound for ℓ_1 is given by Lemma 3.4(i).

Assume $(x, y, z) \in \mathcal{F}^b$. Then $y \leq x + \frac{1}{2}$. In the case where $x \leq y$ we have by Proposition 3.5(i) that $\ell_1 = x$, and it follows that $\ell_2 \leq y \leq x + \frac{1}{2} = \ell_1 + \frac{1}{2}$ with equality only if $y = x + \frac{1}{2}$, in which case we have $\ell_2 = \ell_3 = \ell_1 + \frac{1}{2}$. In the case where $x \geq y$, Proposition 3.5(i) tells us that $\ell_1 = y$. Since $(x, y, z) \notin \mathcal{F}^c$, either $x \leq y + \frac{1}{2}$ or $z \leq y + \frac{1}{2}$, and therefore $\ell_2 \leq \ell_1 + \frac{1}{2}$. Together with Lemma 3.4(ii)(iii) this shows that for $(x, y, z) \in \mathcal{F}^b$ we have $\ell_2 \leq \ell_1 + \frac{1}{2}$, and equality can only occur if either $\ell_2 = \ell_3$ or if $\ell_1 \geq \frac{1+\sqrt{17}}{4}$.

Finally assume $(x, y, z) \in \mathcal{F}^c$. Then $\ell_1 = y$. By Proposition 3.5(iii), the first two lines in $\Lambda_{AD} \cup \Lambda_{BC}$ are y and $2y^2 - \frac{1}{2} > y + \frac{1}{2}$. By Proposition 3.6(c), the first line outside $\Lambda_{AD} \cup \Lambda_{BC}$ is x or z, and both are larger than $y + \frac{1}{2}$. This completes the proof. □

As an immediate corollary we have

LEMMA 3.9. *If the first three lines in the spectra of* Γ *and* $\tilde{\Gamma}$ *coincide, then* (x, y, z) *and* $(\tilde{x}, \tilde{y}, \tilde{z})$ *lie in the same* $\mathcal{F}^i$, $(i = a, b, c)$.

This allows us to work on each $\mathcal{F}^i$ separately.

LEMMA 3.10. *If* $(x, y, z), (\tilde{x}, \tilde{y}, \tilde{z}) \in \mathfrak{F}^a$, *and if* Γ *and* $\tilde{\Gamma}$ *are isospectral, then* $(x, y, z) = (\tilde{x}, \tilde{y}, \tilde{z})$

PROOF. By Proposition 3.5(i), the first line for both Γ and $\tilde{\Gamma}$ is $\ell_1 = x = \tilde{x}$. For a matrix that is written as a word in C and D the trace depends solely on x (this may, e.g., be seen using formula (3.10)). Therefore, part Λ_{CD} of the spectrum *is the same* for Γ and $\tilde{\Gamma}$. Outside Λ_{CD} the first line that comes is y, respectively $\tilde{y}$ (Proposition 3.6(a)), so we also have $y = \tilde{y}$. Since on $\mathfrak{F}$, trace z (respectively $\tilde{z}$) is the smaller of the two solutions of $P(x, y, \,.\,) = 0$ we also have $z = \tilde{z}$. □

LEMMA 3.11. *If* $(x, y, z), (\tilde{x}, \tilde{y}, \tilde{z}) \in \mathfrak{F}^c$ *and* Γ, *and if* $\tilde{\Gamma}$ *are isospectral then* $(x, y, z) = (\tilde{x}, \tilde{y}, \tilde{z})$

PROOF. This time Proposition 3.5(i) tells us that $y = \tilde{y}$, and parts $\Lambda_{AD} \cup \Lambda_{BC}$ of the spectra of Γ and $\tilde{\Gamma}$ coincide. Outside these parts the first two lines (Proposition 3.6(c)) are x and z respectively, $\tilde{x}$ and $\tilde{z}$. So we have $\{x, z\} = \{\tilde{x}, \tilde{z}\}$, and Lemma 3.8 yields $(x, y, z) = (\tilde{x}, \tilde{y}, \tilde{z})$. □

What is now left is the proof of the spectral rigidity of surfaces with parameters in $\mathfrak{F}^b$. By Proposition 3.6(b), the first four spectral lines of any Γ with parameters in $\mathfrak{F}^b$ are among $x, y, z, x + \frac{1}{2}, x', z'$, and it is an interesting fact in itself that the "mini spectrum" formed by this sequence (ordered such that it is monotone increasing) allows us to determine the geometry. The statement is the following which holds in $\mathfrak{F}$ and not just $\mathfrak{F}^b$.

PROPOSITION 3.12. *If for* Γ *and* $\tilde{\Gamma}$ *with parameters in* $\mathfrak{F}$ *the smallest four lines among* $x, y, z, x + \frac{1}{2}, x', z'$ *respectively,* $\tilde{x}, \tilde{y}, \tilde{z}, \tilde{x} + \frac{1}{2}, \tilde{x}', \tilde{z}'$ *coincide, then* $(x, y, z) = (\tilde{x}, \tilde{y}, \tilde{z})$.

The proof reminds one of those puzzles, where six neighbors have to be found out, each living in a house with a different color, the person in the red house drinking tea, the neighbour of the coffee drinker not having a driver's license etc., but there is no room to give it here.

References

1. A. F. Beardon *The Geometry of Discrete Groups*, Graduate Texts in Mathematics, vol. 91, Springer-Verlag, New York, 1995.
2. P. Bérard and H. Pesce, *Construction de variétés isospectrales autour du théorème de T. Sunada*, Progress in Inverse Spectral Geometry, Birkhäuser, Basel, 1997, pp. 63–83.
3. A. Binotto, *Surfaces de Riemann de type* (0, 3) *et* (1, 1)*: Polynômes au service de l'étude spectrale*, Ph.D. Thesis, EPFL, Lausanne, 2000.
4. R. Brooks and R. Tse, *Isospectral surfaces of small genus*, Nagoya Math. J. **107** (1987), 13–24.
5. P. Buser, *Isospectral Riemann surfaces*, Ann. Inst. Fourier (Grenoble) **36** (1986), 167–192.
6. P. Buser and K.-D. Semmler, *The geometry and spectrum of the one holed torus*, Comment. Math. Helv. **63** (1988), 259–274.
7. P. Buser, N. Flach, and K.-D. Semmler *Spectral rigidity and discreteness of 2233 groups*, with an appendix by C. Maclachlan and G. Rosenberger, preprint, EPFL, Lausanne, 2004.
8. R. Dianu, *Sur le spectre des tores pointés*, Ph.D. Thesis, EPFL, Lausanne, 2000.
9. A. Haas, *Length spectra as moduli for hyperbolic surfaces*, Duke Math. J. **52** (1985), 923–934.
10. H. Helling, *Über den Raum der kompakten Riemannschen Flächen vom Geschlecht 2*, J. Reine Angew. Math. **268/269** (1974), 286–293.
11. R. D. Horowitz, *Characters of free groups represented in the two-dimensional special linear group*, Comm. Pure Appl. Math. **25** (1972), 635-649.
12. R. Perlis, *On the equation* $\zeta_K(s) = \zeta_{K'}(s)$, J. Number Theory **9** (1977), 342–360.

13. D. Singerman, *Finitely maximal Fuchsian groups*, J. London Math. Soc. **6** (1972), 29–38.
14. T. Sunada, *Riemannian coverings and isospectral manifolds*, Ann. of Math. **121** (1985), 169–186.
15. M.-F. Vignéras, *Variétés riemanniennes isospectrales et non isométriques*, Ann. of Math. **112** (1980), 21–32.

Section de Mathématiques, Ecole Polytechnique Fédérale de Lausanne, SB-IGAT-Geom Station 8, CH-1015 Lausanne, Switzerland
E-mail address: peter.buser@epfl.ch

Section de Mathématiques, Ecole Polytechnique Fédérale de Lausanne, SB-IGAT-Geom Station 8, CH-1015 Lausanne, Switzerland
E-mail address: klaus-dieter.semmler@epfl.ch

Contemporary Mathematics
Volume **387**, 2005

Topics in Isoperimetric Inequalities

Isaac Chavel

Dedicated to the memory of Robert Brooks

ABSTRACT. The first part of the paper gives four arguments for the isoperimetric inequality in Euclidean space (all arbitrary dimension), all of which use the arithmetic–geometric mean inequality — albeit in different fashions. The second part of the paper is a review of the Bol–Fiala inequality, with Osserman's refinement, in simply connected surfaces with Gauss curvature bounded from above.

1. Introduction: The isoperimetric problem

Let M be a Riemannian manifold. Whether or not it has a boundary, M will always refer to the interior. The *isoperimetric problem* is to find, for any given v less than the volume of M, the region of volume v with minimal area of its boundary. The analytic aspect of the problem is to write this minimal area as a function of v, namely,

DEFINITION. The *isoperimetric profile* $v \mapsto I_M(v)$ *of* M, defined for every $v \in (0, V(M))$ ($V(M)$ might be infinite), is given by

$$I_M(v) = \inf\{A(\partial D) : \ V(D) = v, \ D \subset\subset M\},$$

where D varies over relatively compact open submanifolds of M with smooth boundary in M. Such open subsets of M will simply be referred to as *smooth regions.* (In particular, they can have at most finite many components.)

One, naturally, seeks explicit knowledge of the function $I_M(v)$. Also, one wants to know if, given any v, a region Ω exists for which $V(\Omega) = v$ and $A(\partial\Omega) = I_M(v)$; such a region is called an *isoperimetric region* or *minimizer*, for short. Furthermore, one also wants know if the boundary of the isoperimetric region is smooth. Finally, if it exists, is it unique.

A weaker version of the problem is to offer an accessible lower bound $\phi(v)$ for $I_M(v)$, that is, an *isoperimetric inequality* of the form

$$A(\partial D) \geq \phi(V(D)) \tag{1.1}$$

2000 *Mathematics Subject Classification.* Primary 58E35; Secondary 49Q20.
Key words and phrases. Isoperimetric inequalities, Riemannian manifolds.

for all relatively compact smooth regions D.

The classical isoperimetric inequality in Euclidean space of all dimensions represents the best all worlds, namely, for any such region in $\mathbb{R}^n$, one has

$$\frac{A(\partial D)}{V(D)^{1-1/n}} \geq \frac{A(\mathbb{S}^{n-1})}{V(\mathbb{B}^n)^{1-1/n}}, \tag{1.2}$$

where $\mathbb{B}^n$ denotes the unit n-disk in $\mathbb{R}^n$ and $\mathbb{S}^{n-1}$ its boundary, the $(n-1)$-dimensional unit sphere in $\mathbb{R}^n$. One has equality in (1.2) if and only if D is a disk in $\mathbb{R}^n$. Since the quotient on the right hand side of (1.2) remains the same for disks of any radius r in $\mathbb{R}^n$, the inequality constitutes a complete solution of the isoperimetric problem. It says that for any given volume v, the disks of volume v are the only minimizers of the bounding area, and the isoperimetric profile $I_M(v)$ is known explicitly, namely,

$$\boldsymbol{\omega_n} := V(\mathbb{B}^n), \quad \mathbf{c_{n-1}} := A(\mathbb{S}^{n-1}) \quad \Rightarrow \quad \mathbf{c_{n-1}} = n\boldsymbol{\omega_n},$$

and

$$I_{\mathbb{R}^n}(v) = n{\boldsymbol{\omega_n}}^{1/n} v^{1-1/n}.$$

The classical isoperimetric inequality in $\mathbb{R}^n$ also has the advantage of possessing a veritable wealth of techniques (especially in the plane), each of which "represents" some subfield of analysis and/or geometry. Here we sample four different proofs of the inequality (1.2) in $\mathbb{R}^n$; all of which use the arithmetic–geometric mean inequality. We start with the Brunn–Minkowski inequality (even though it is well-known, one cannot resist including it), and then give a new argument of X. Cabré (2003) based on the solution to the Neumann problem in $\mathbb{R}^n$. We then present two uniqueness arguments — F. Almgren (1976) and M. Gromov (1986).

We note that the classical Euler–Lagrange equation for an extremal area subject to constant volume constraint is that the mean curvature of the boundary of the extremal domain have constant mean curvature. And A.D. Alexandrov (1962) proved that any domain in $\mathbb{R}^n$ with C^2 boundary of constant mean curvature must be isometric to a disk. So, in a sense, there is no longer any necessity for just uniqueness arguments. But our point, here, is that the two latter arguments are just first stages for more ambitious arguments. Almgren's argument is the heuristic for his theorem on minimal varieties of arbitrary codimension, and Gromov's argument is but a small piece — adapted to Euclidean space — of his relative isoperimetric inequalities on compact Riemannian manifolds of positive Ricci curvature. (See Chavel (1994, 2001); the suggestion to use part of Gromov's argument in Euclidean space appears in M. Berger (2003, pp. 63–66)).

The second part of the paper is a chapter in the Riemannian geometry of surfaces. It consists of a review (with quite a bit of detail) of the isoperimetric inequality of G. Bol (1941) and F. Fiala (1940) for simply connected domains on surfaces with Gauss curvature bounded from above, as sharpened by R. Osserman (1986). (Here the material requires some background from Riemannian geometry. The reader can find it in Chavel (1994, Chaps. II, III).) We conclude with remarks on B. Kleiner's (1992) generalization of their result to dimension 3. The goal here is to call attention to Aubin's conjecture (1976) that simply connected domains on manifolds of all dimensions ≥ 2 with nonpositive Riemannian sectional curvature satisfy (1.2), with equality if and only if the domain is a flat geodesic disk. Thus, even if one cannot solve the isoperimetric problem in such a manifold, one would have an informative "isoperimetric comparison theorem." The conjecture is still open for dimensions ≥ 5.

Even though we do not consider symmetrization arguments here, one must note the early papers of J. Steiner (1838), H.A. Schwarz (1884) and C. Caratheodory–E. Study (1909), who inaugurated the modern study of isoperimetric inequalities. These arguments were originally restricted to 2- and 3-dimensional Euclidean space. The isoperimetric problem for manifolds beyond the Euclidean plane and space was first solved for domains in the 2-sphere bounded by convex curves, by F. Bernstein (1905); and a unified proof for the simply connected spaces of constant sectional curvature of all dimensions (which might be viewed as a summary statement of all the previous work) using spherical symmetrization was given in E. Schmidt (1948). See also the treatment in Yu.D. Burago–V.A. Zalgaller (1988).

Yet another symmetrization argument of J. Steiner (1842) was, more recently, revived by W.Y. Hsiang (1991), H. Howard–M. Hutchings–F.M. Morgan (1999), and A. Ros (2001). Still another approach to symmetrization, two-point symmetrization, was initiated, it seems, by L. Ahlfors (1973), and used on the n-sphere by A. Baernstein–B.A. Taylor (1976).

For current presentations of symmetrization methods, see E.H. Lieb–M. Loss (1996) and A. Baernstein (2004). For the Brunn-Minkowski inequalities see the recent R.J. Gardner (2002).

While interest in isoperimetric inequalities in more general manifolds had been around for quite awhile, fresh insights and new directions of research were initiated by M. Gromov in (1980, 1986) and (1981, 1999).

For early surveys one has the books of W. Blaschke (1956), H. Hadwiger (1957), and, more recently, Yu.D. Burago–V.A. Zalgaller (1988) and I. Chavel (2001). Extended survey articles are R. Osserman (1978, 1979) and A. Ros (2001).

2. Isoperimetry and the arithmetic–geometric mean inequality

2.1. The Brunn–Minkowski theorem.

DEFINITION. Let K be a compact subset of $\mathbb{R}^n$. Define its *Minkowski area*, Mink (K) by

$$\text{Mink}\,(K) = \liminf_{h\downarrow 0} \frac{\mathbf{v}_n([K]_h) - \mathbf{v}_n(K)}{h},$$

where

$$[K]_h = \{x \in \mathbb{R}^n : \ d(x, K) \leq h\}.$$

Thus, $K \mapsto \text{Mink}\,(K)$ is a functional defined on collection compacts sets K — not on the boundary of K. Furthermore, the Minkowski area is a generalization of Riemannian $(n-1)$-dimensional area in that, when $K = \overline{\Omega}$ in $\mathbb{R}^n$ with $\partial\Omega \in C^1$, we have

$$\text{Mink}\,(K) = A(\partial\Omega),$$

where A denotes the standard Riemannian $(n-1)$-dimensional area. However, note that when $n = 1$, then $\text{Mink}\,\{x_o\} = 2$ at the same time that $\mathbf{v}_1(\{x_o\}) = 0$.

THEOREM 2.1. *Let X and Y be bounded measurable subsets of $\mathbb{R}^n$. Then*

$$V(X+Y)^{1/n} \geq V(X)^{1/n} + V(Y)^{1/n}. \tag{2.1}$$

PROOF. Assume X and Y consist of one "box" each, with no intersections (if they have an intersection then we may translate one of them — this will have no effect on the volume). Then

$$V(X) = \prod_{j=1}^{n} \alpha_j, \qquad V(Y) = \prod_{j=1}^{n} \beta_j,$$

and $X + Y$ is then a box with

$$V(X+Y) = \prod_{j=1}^{n} (\alpha_j + \beta_j).$$

The arithmetic–geometric mean inequality then implies

$$\left\{\frac{\prod \alpha_j}{\prod (\alpha_j + \beta_j)}\right\}^{1/n} + \left\{\frac{\prod \beta_j}{\prod (\alpha_j + \beta_j)}\right\}^{1/n} \le \frac{1}{n}\sum_{j=1}^{n} \frac{\alpha_j}{\alpha_j + \beta_j} + \frac{1}{n}\sum_{j=1}^{n} \frac{\beta_j}{\alpha_j + \beta_j} = 1,$$

which implies the inequality (2.1).

Now assume that we have the inequality when X and Y are each the union of non-overlapping boxes (except maybe at the boundary) such that the total number of boxes is $\le k-1$, $k > 2$. So consider the case when the union of X and Y consists of k boxes. Then either X or Y has at least 2 boxes. Assume it is X.

Then there exists a hyperplane in $\mathbb{R}^n$ which divides $\mathbb{R}^n$ into two closed half-spaces H^+ and H^-, such that any box in X is either in H^+ or H^-, and that each half-space has at least one box in X. Then one can translate the boxes of Y so that P divides Y into Y^+ and Y^- satisfying

$$\frac{V(X^+)}{V(X)} = \frac{V(Y^+)}{V(Y)}, \qquad \frac{V(X^-)}{V(X)} = \frac{V(Y^-)}{V(Y)}.$$

If X consists of ℓ boxes then Y consists of $k - \ell$ boxes, and both Y^+ and Y^- have at most $k - \ell$ boxes. But each of X^+ and X^- have strictly less than ℓ boxes. So both $X^+ \cup Y^+$ and $X^- \cup Y^-$ have strictly less than k boxes, each. By hypothesis, the theorem is valid for $X^+ \cup Y^+$ and $X^- \cup Y^-$. Therefore,

$$\begin{aligned}
V(X+Y) &\ge V(X^+ + Y^+) + V(X^- + Y^-) \\
&\ge \left\{V(X^+)^{1/n} + V(Y^+)^{1/n}\right\}^n + \left\{V(X^-)^{1/n} + V(Y^-)^{1/n}\right\}^n \\
&= V(X^+)\left\{1 + \left(\frac{V(Y^+)}{V(X^+)}\right)^{1/n}\right\}^n + V(X^-)\left\{1 + \left(\frac{V(Y^-)}{V(X^-)}\right)^{1/n}\right\}^n \\
&= V(X^+)\left\{1 + \left(\frac{V(Y)}{V(X)}\right)^{1/n}\right\}^n + V(X^-)\left\{1 + \left(\frac{V(Y)}{V(X)}\right)^{1/n}\right\}^n \\
&= \frac{V(X^+)}{V(X)}\left\{V(X)^{1/n} + V(Y)^{1/n}\right\}^n + \frac{V(X^-)}{V(X)}\left\{V(X)^{1/n} + V(Y)^{1/n}\right\}^n \\
&= \left\{V(X)^{1/n} + V(Y)^{1/n}\right\}^n.
\end{aligned}$$

which is (2.1) when X and Y consist of a finite number of boxes.

For arbitrary bounded measurable sets X and Y, one approximates them from within by finite unions of non-overlapping boxes. □

COROLLARY 2.1 (The isoperimetric inequality for Minkowski area). *If* $Y = \mathbb{B}^n$, *then* $X + \epsilon\mathbb{B}^n = [X]_\epsilon$, *which implies*

$$V([X]_\epsilon)^{1/n} \geq V(X)^{1/n} + \epsilon{\boldsymbol{\omega}_\mathbf{n}}^{1/n}, \tag{2.2}$$

which implies

$$\text{Mink}\,(X) \geq n{\boldsymbol{\omega}_\mathbf{n}}^{1/n}V(X)^{1-1/n} = \frac{A(\mathbb{S}^{n-1})}{V(\mathbb{B}^n)^{1-1/n}}V(X)^{1-1/n}. \tag{2.3}$$

Note that even if one has an easy characterization of equality in (2.2), it cannot guarantee that the argument itself will pass to the limiting result (2.3). However, D. Ohman (1955) proved that if A is non-convex, then there exists a constant c so that

$$V([X]_\epsilon) \geq \left\{V(X)^{1/n} + \epsilon{\boldsymbol{\omega}_\mathbf{n}}^{1/n}\right\}^n + c\epsilon.$$

See Burago–Zalgaller (1988, pp. 71–74). In particular, if A is non-convex one cannot have equality in the classical isoperimetric inequality. In the case where A is convex one can give the classical proof (based on calculus) to characterize equality in the isoperimetric inequality.

2.2. Solvability of a Neumann problem in $\mathbb{R}^n$. First note that for Euclidean space one may reduce the collection of competing regions to domains (that is, connected regions). Indeed, if $\Omega = \Omega_1 \cup \cdots \cup \Omega_k$ is a decomposition of Ω into domains then Minkowski's inequality (applied to the characteristic functions of the components of Ω) implies

$$\left\{\sum_{j=1}^{k} V(\Omega_j)\right\}^{(n-1)/n} \leq \sum_{j=1}^{k}\{V(\Omega_j)\}^{(n-1)/n},$$

with equality if and only if Ω is connected. Therefore knowledge of the isoperimetric inequality for domains implies its validity for relatively compact regions, with the same characterization of the case of equality.

Therefore, consider a domain $D \subset\subset \mathbb{R}^n$, with C^2 boundary. The argument of X. Cabré (2003) relies on the solvability of the Neumann problem

$$\Delta u = 1 \quad \text{on } D \qquad\qquad \frac{\partial u}{\partial \nu} = c \quad \text{on } \partial D.$$

A necessary condition is:

$$V(D) = \iint_D \Delta u\, dV = \int_{\partial D} \partial u/\partial\nu\, dA = cA(\partial D).$$

So c must given by $c = V/A$. The theory says that there is indeed a solution for such a c.

Now define

$$\Gamma^+ := \{y \in D :\ u(x) \geq u(y) + (\nabla u)(y)\cdot(x-y)\ \forall\, x \in D\}.$$

That is, the tangent hyperplane to the graph of u, at the point $(y, u(y))$, supports the graph in the upper half-space defined by the tangent hyperplane. At every point of Γ^+ the Hessian of u is positive semidefinite, so all of its eigenvalues are nonnegative.

We normally think of ∇u as a vector field on D, but we may also think of it as a mapping $\nabla u : D \to \mathbb{R}^n$. Then the Hessian of u is the Jacobian matrix of the mapping ∇u, which implies

$$\frac{V(D)}{n^n} \geq \frac{V(\Gamma^+)}{n^n} = \iint_{\Gamma^+} \left[\frac{\Delta u}{n}\right]^n dV = \iint_{\Gamma^+} \left[\frac{\text{tr Hess}\, u}{n}\right]^n dV = \cdots$$
$$\cdots \geq \iint_{\Gamma^+} \det \text{Hess}\, u \, dV \geq V\left[(\nabla u)(\Gamma^+)\right]$$

(the last inequality might come from many-to-one points of the mapping). It remains to show

$$(2.4) \qquad (\nabla u)(\Gamma^+) \supseteq \mathbb{B}^n(c).$$

If we do so, then we would have $V/n^n \geq \boldsymbol{\omega}_{\mathbf{n}} c^n = \boldsymbol{\omega}_{\mathbf{n}} V^n/A^n$, which is $A \geq n{\boldsymbol{\omega}_{\mathbf{n}}}^{1/n} V^{1-1/n}$, the isoperimetric inequality (1.2).

To prove (2.4) we argue as follows: For any $\xi \in \mathbb{R}^n$ consider a hyperplane in $\mathbb{R}^{n+1}$ moving up the x^{n+1}-axis from $-\infty$, with normal vector $\xi - \mathbf{e}_{n+1}$. Then there is a first point at which the hyperplane intersects — and therefore touches — the graph of $u : \overline{D} \to \mathbb{R}^n$. If that first point of contact is over ∂D then the slope of the hyperplane is $\geq c$, which implies the slope of the normal $\leq 1/c$, which implies $|\xi| \geq c$.

So for any ξ, satisfying $|\xi| < c$, the hyperplane must hit the graph at some $(y, u(y))$, $y \in D$, which implies $\xi = (\nabla u)(y)$; so $y \in \Gamma^+$, and this implies the (2.4).

If we have equality in the isoperimetric inequality, then (1) $(\nabla u)(\Gamma^+) = \mathbb{B}^n(c)$, the n-disk of radius c, (2) ∇u is a diffeomorphism, (3) $\Gamma^+ = D$, which implies u is convex, and (4) we have equality in the arithmetic-geometric mean inequality.

Therefore

$$\frac{\partial^2 u}{\partial {x^j}^2} = \frac{1}{n}, \qquad \frac{\partial^2 u}{\partial x^j \partial x^k} = 0 \quad \forall j \neq k,$$

which implies

$$\frac{\partial u}{\partial x^j} = \frac{x^j}{n} + \alpha_j,$$

that is,

$$(\nabla u)(x) = \frac{x}{n} + \boldsymbol{\alpha}.$$

Therefore, for all $w \in \partial D$ we have

$$|w + n\boldsymbol{\alpha}| = nc,$$

so ∂D is a sphere.

2.3. On the isoperimetric profile. For the sake of completeness, we quote a broad existence theorem.

MAIN THEOREM FOR M WITHOUT BOUNDARY. *If M^n is compact, or covers a compact, then, for any v, $0 < v < V(M)$, there exists $\Omega_v \subset\subset M$ whose boundary in M, $\Sigma_v = \partial\Omega_v$ minimizes area among smooth regions of volume v, that is, $A(\Sigma_v) = I_M(v)$. Moreover, except for a singular set of Hausdorff dimension at most $n - 8$, the boundary Σ_v of any minimizer is a smooth imbedded hypersurface with constant mean curvature.*

Of course, the theorem applies to the standard simply connected spaces of constant sectional curvature — Euclidean space, the sphere, and hyperbolic space of all dimensions.

MAIN THEOREM FOR M WITH BOUNDARY. *Assume, in the above, M has boundary and $\overline{M^n}$ is compact (the boundary may be $C^{1,1}$). Then, again, the minimizer Ω_v exists, with boundary Σ_v. Σ_v might include some of the boundary ∂M, in which case $\Sigma_v \in C^{1,1}$ in a neighborhood of ∂M. The mean curvature $\mathfrak{h} = \mathfrak{h}_v$ is constant on the set of all smooth points of $\Sigma_v \cap M$. The mean curvature of $\Sigma_v \cap \partial M$ must satisfy $\mathfrak{h} \leq \mathfrak{h}_v$.*

Note that in this case $A(\overline{\Omega}_v \cap \partial M)$ is included in the area $A(\Sigma_v)$. Also, the mean curvature of Σ_v is always relative to the *interior* unit normal vector field along Σ_v. Finally, the mean curvature of $\Sigma_v \cap \partial M$ is defined in a weak sense. This formulation of the main theorems is from A. Ros (2001); see the references there for the proof of the theorem. The proof of $C^{1,1}$ regularity in a neighborhood of the boundary is in B. White (1991) and E. Stredulinsky–W.P. Ziemer (1997).

Parallel/Fermi coordinates. Let Ω be an arbitrary region with compact closure and smooth boundary in M, and assume a 1-parameter family of regions is given by perturbing the boundary $\partial\Omega$ by

$$\Phi(\epsilon; w) = \operatorname{Exp} \epsilon\nu_w, \qquad w \in \partial\Omega,$$

where ν denotes the *exterior* unit normal vector field along $\partial\Omega$. Let Ω_ϵ be the region with boundary $\partial\Omega_\epsilon = \Phi_\epsilon(\partial\Omega)$, where $\Phi_\epsilon(w) = \Phi(\epsilon; w)$. Let ν_ϵ denote the exterior unit normal vector field along $\partial\Omega_\epsilon$, for every ϵ. Set $V(\epsilon) = V(\Omega_\epsilon)$ and $A(\epsilon) = A(\partial\Omega_\epsilon)$. Then the standard variational formulae read as:

$$V'(\epsilon) = A(\epsilon), \qquad A'(\epsilon) = \int_{\partial\Omega_\epsilon} \mathfrak{h}_\epsilon \, dA,$$

where $\mathfrak{h}_\epsilon$ is the mean curvature relative to $-\nu_\epsilon$, and

$$A''(\epsilon) = \int_{\partial\Omega_\epsilon} \{{\mathfrak{h}_\epsilon}^2 - |\mathcal{B}_\epsilon|^2 - \langle \mathfrak{Ric}\, \nu_\epsilon, \nu_\epsilon \rangle\} \, dA,$$

where $\mathfrak{Ric}$ denotes the Ricci curvature of M, $|\mathcal{B}_\epsilon|^2$ denotes the Gram–Schmidt norm of the second fundamental form $\mathcal{B}_\epsilon$ of $\partial\Omega_\epsilon$ relative to ν_ϵ, and $\langle\ ,\ \rangle$ denotes the Riemannian inner product in M.

THEOREM 2.2 (C. Bavard–P. Pansu (1986)). *Let Ω be an isoperimetric region with smooth boundary, and constant mean curvature $\mathfrak{h}$ with respect to the unit interior normal along the boundary. Then the isoperimetric profile I_M has weak (in the sense of Calabi) left- and right- first derivatives and second derivative satisfying*

$$\frac{D^+ I_M}{dv} \leq \mathfrak{h} \leq \frac{D^- I_M}{dv}, \qquad v = V(\Omega),$$

and

$$\frac{D^2 I_M}{dv^2} \leq -\frac{1}{{I_M}^2(v)} \int_{\partial\Omega} \{|\mathcal{B}|^2 + \langle \mathfrak{Ric}\, \nu, \nu \rangle\} \, dA, \qquad v = V(\Omega).$$

If the Ricci curvature of M is bounded from below, then

$$\frac{D^2}{dv^2} \left\{ {I_M}^2(v) + 2 \left(\inf_M \mathfrak{Ric} \right) v^2 \right\} \leq 0 \qquad \text{at } v = V(\Omega).$$

In particular, the function

$$v \mapsto {I_M}^2(v) + 2 \left(\inf_M \mathfrak{Ric} \right) v^2$$

is concave, which implies $I_M(v)$ is locally Lipschitz.

PROOF. Given an arbitrary function $f(x)$, to say that it satisfies

$$\frac{D^+ f}{dx}(x_o) \le C_1 \le \frac{D^- f}{dx}(x_o), \quad \text{and} \quad \frac{D^2 f}{dx^2}(x_o) \le C_2$$

at x_o in the sense of Calabi is to say that there exists a smooth function $\phi(x)$ defined on some neighborhood of x_o such that

$$\begin{array}{ll} f(x) \le \phi(x) \quad \forall\, x \quad , & \phi(x_o) = f(x_o), \\ \phi'(x_o) = C_1 & \phi''(x_o) = C_2. \end{array}$$

Let $\epsilon(v)$ denote the inverse function of $V = V(\epsilon)$, then

$$\frac{dA}{dv} = \frac{A'}{A} \qquad \Rightarrow \qquad \frac{d^2 A}{dv^2} = \frac{1}{A}\left\{\frac{A''}{A} - \left(\frac{A'}{A}\right)^2\right\}.$$

Therefore, if $\partial\Omega$ has constant mean curvature, then

$$\left.\frac{d^2 A}{dv^2}\right|_{\epsilon=0} = -\frac{1}{A^2}\int_{\partial\Omega} \{|\mathcal{B}|^2 + \langle \mathfrak{Ric}\,\nu, \nu\rangle\}\, dA. \tag{2.5}$$

Then for ϵ in a neighborhood of 0 we have $v(\epsilon)$ in a neighborhood of $V(\Omega)$, and

$$\begin{aligned} I_M(v) &\le A(v) \qquad \text{for } v \text{ in a neighborhood of } V(\Omega), \\ I_M(V(\Omega)) &= A(V(\Omega)). \end{aligned}$$

Then (2.5) implies

$$\left.\frac{d^2 A^2}{dv^2}\right|_{\epsilon=0} \le -2 \inf_M \mathfrak{Ric},$$

which implies

$$\left.\frac{d^2}{dv^2}\left\{A^2(v) + 2\left(\inf_M \mathfrak{Ric}\right) v^2\right\}\right|_{\epsilon=0} \le 0,$$

which implies the claim. □

NOTE. The theorem remains valid when M has a boundary, because we can use a variation of $\partial\Omega$ compactly supported on $\partial\Omega \cap M$ with infinitesimal vector pointing into Ω.

2.4. F. Almgren's uniqueness proof.

THEOREM 2.3. *Assume Ω is an isoperimetric region in $\mathbb{R}^n$. Then Ω is a disk.*

PROOF. In order to keep the differential-geometric ideas in the foreground, we will assume the boundary is smooth everywhere. Set $\Gamma = \partial\Omega$, and for convenience, assume $V(\Omega) = \boldsymbol{\omega}_{\mathbf{n}}$. Then, since Ω is a minimizer, we have $A(\Gamma) \le \mathbf{c}_{\mathbf{n-1}}$.

We first claim that the the constant mean curvature $\mathfrak{h}$ of $\partial\Omega$ satisfies

$$\mathfrak{h} \le n - 1. \tag{2.6}$$

To this end, consider the isoperimetric functional

$$J(D) = \frac{A(\partial D)}{V(D)^{1-1/n}},$$

where D varies over bounded domains in $\mathbb{R}^n$ possessing C^2 boundary. Let Φ_t denote a 1-parameter group of diffeomorphisms of $\mathbb{R}^n$, with associated time-dependent vector field $X = X(x,t)$, $\xi(x) = X(x,0)$, $\eta = \xi|\Gamma$. Then the first variation formulae in this more general situation are:

$$\frac{d}{dt}V(\Phi_t(\Omega))\bigg|_{t=0} = \int_\Gamma \langle \eta, \nu\rangle\, dA, \qquad \frac{d}{dt}A(\Phi_t(\Gamma))\bigg|_{t=0} = \int_\Gamma \mathfrak{h}\langle \eta, \nu\rangle\, dA.$$

where ν is the exterior unit normal vector field along Γ. Direct calculation implies, because Ω is an extremal of the isoperimetric functional,

$$\begin{aligned} 0 &= \frac{d}{dt}J(\Phi_t(\Omega))\bigg|_{t=0} \\ &= \frac{1}{V(\Omega)^{1-1/n}}\int_\Gamma \mathfrak{h}\langle \eta, \nu\rangle\, dA + \left(\frac{1}{n} - 1\right)\frac{A(\Gamma)}{V(\Omega)^{2-1/n}}\int_\Gamma \langle \eta, \nu\rangle\, dA, \end{aligned}$$

which implies

$$\text{(2.7)} \qquad \frac{\int \mathfrak{h}\langle \eta, \nu\rangle\, dA}{\int \langle \eta, \nu\rangle\, dA} = \frac{n-1}{n}\frac{A(\Gamma)}{V(\Omega)} \le n-1$$

(both integrals are over Γ). For any $w_o \in \Gamma$, let ϕ be a nonnegative C^∞ function compactly supported on a neighborhood of w_o in $\mathbb{R}^n$, and choose $X(x,t) = \phi(x)\nu_{w_o}$(so this particular vector field is time-independent). Then by picking ϕ with sufficiently small support about w_o one obtains (2.6). (The argument of this paragraph, applied to (2.7), implies that $\mathfrak{h} = \text{const.}$, namely $\mathfrak{h} = (n-1)A(\Gamma)/nV(\Omega)$.)

Next, we have $\overline{\Omega} = \Omega \cup \Gamma$, and let $\mathcal{C}$ denote the convex hull of $\overline{\Omega}$, $\Sigma = \partial\mathcal{C}$ the boundary of $\mathcal{C}$. One can easily show that every point of Σ has a unique supporting hyperplane, that is, Σ is everywhere differentiable.

Since Σ is convex, everywhere differentiable, then every point of Σ has a unique exterior unit normal vector, and the Gauss map

$$\mathfrak{G} : \Sigma \to \mathbb{S}^{n-1}$$

which takes every point of $w \in \Sigma$ to its exterior unit normal vector $\mathsf{n}(w)$ is well defined. The Gauss map is onto; furthermore, the Gauss map restricted to $\Sigma\backslash(\Sigma\cap\Gamma)$ adds no new points to the image in $\mathbb{S}^{n-1}$ that are not already in the image of $\Sigma\cap\Gamma$, namely,

$$\mathfrak{G}(\Sigma \setminus (\Sigma\cap\Gamma)) \subseteq \mathfrak{G}(\Sigma\cap\Gamma).$$

Therefore, to calculate the image of the area of the Gauss map, that is, the area in $\mathbb{S}^{n-1}$, we restrict the integral to $\Sigma\cap\Gamma$.

In a coordinate system on Γ, as described in the calculations above, the Jacobian transformation of $\mathfrak{G}$, $\mathfrak{G}_*$, is given in local coordinates by the linear transformation described by the Weingarten map $-\mathcal{G}^{-1}\mathcal{B}$, where $\mathcal{G}$ denotes the first fundamental form of Γ and $\mathcal{B}$ the second fundamental form. For points of $\Sigma\cap\Gamma$ we have $-\mathcal{B}$ positive semi-definite, which implies by the arithmetic-geometric mean inequality:

$$\begin{aligned} \mathbf{c_{n-1}} &= \int_{\Sigma\cap\Gamma} \det(-\mathcal{G}^{-1}\mathcal{B})\, dA \le \int_{\Sigma\cap\Gamma}\left\{\frac{\operatorname{tr}(-\mathcal{G}^{-1}\mathcal{B})}{n-1}\right\}^{n-1} dA = \cdots \\ \cdots &= \int_{\Sigma\cap\Gamma}\left(\frac{\mathfrak{h}}{n-1}\right)^{n-1} dA \le A(\Sigma\cap\Gamma) \le A(\Gamma), \end{aligned}$$

which implies $\mathbf{c_{n-1}} \le A(\Gamma)$. Therefore the two are equal.

But equality then implies $\Sigma = \Gamma$. Also, we have equality in the arithmetic-geometric mean inequality, which implies all points are umbilics (that is, $\mathcal{G}^{-1}\mathcal{B}$ is a scalar matrix at every point of Γ), which implies $-\mathcal{G}^{-1}\mathcal{B}$ is the identity. Therefore by a standard argument, there exists a constant vector $\mathbf{b}$ such that $\nu_x = x - \mathbf{b}$ for all x in Γ. Thus $|x - \mathbf{b}| = |\nu_x| = 1$, which implies Γ is a unit sphere. $\square$

2.5. M. Gromov's uniqueness proof. Here we assume that Ω is an isoperimetric domain (possibly with singularities as described above) of volume $\boldsymbol{\omega_n}$. Then (2.6) and (2.7) imply that the constant mean curvature $\mathfrak{h}$ of the smooth part of the boundary has curvature less than or equal to $n-1$. Also, we are assuming that $A(\partial\Omega) \leq \mathbf{c_{n-1}}$.

Because the codimension of the singular points of $\partial\Omega$ is at least 7, we restrict our attention to the regular points only. We construct Fermi coordinates in Ω based on $\partial\Omega$. For every $q \in \partial\Omega$ let $\mathbf{n}_q$ denote the *interior* unit normal vector field along $\partial\Omega$, and

$$v(t;q) = \exp_q t\mathbf{n}_q.$$

Then the geodesic (in this case: a line) $\gamma_q(t) = v(t;q)$ minimizes distance from $\partial\Omega$ to $\gamma_q(t)$ up to some positive value $\rho(q)$. The point $\gamma_q(\rho(q))$ is known as the *focal cut point* of q along γ_q. If $\kappa_1(g), \ldots, \kappa_{n-1}(q)$ denote the principal curvatures of the second fundamental form of $\partial\Omega$ at q, with respect to $\mathbf{n}_q$ (recall that the mean curvature is the sum of the principal curvatures), then it is standard that

$$\rho(q) \leq 1/\max_{j=1,\ldots,n-1} \kappa_j(q).$$

It is also standard that the volume of Ω is given by

$$V(\Omega) = \int_{\partial\Omega} dA_q \int_0^{\rho(q)} \prod_{j=1}^{n-1} \{1 - \kappa_j(q)t\}\, dt.$$

See Chavel (1994). Then for $V(\Omega) = \boldsymbol{\omega_n}$ we have

$$\begin{aligned}
\boldsymbol{\omega_n} &\leq \int_{\partial\Omega} dA_q \int_0^{1/\max \kappa_j(q)} \prod_{j=1}^{n-1} \{1 - \kappa_j(q)t\}\, dt \\
&\leq \int_{\partial\Omega} dA_q \int_0^{1/\max \kappa_j(q)} \left[1 - \frac{\mathfrak{h}t}{n-1}\right]^{n-1} dt \\
&\leq \int_{\partial\Omega} dA_q \int_0^{(n-1)/\mathfrak{h}} \left[1 - \frac{\mathfrak{h}t}{n-1}\right]^{n-1} dt \\
&= \frac{n-1}{n\mathfrak{h}} A(\partial\Omega) \leq \frac{A(\partial\Omega)}{n} \leq \frac{\mathbf{c_{n-1}}}{n} = \boldsymbol{\omega_n}.
\end{aligned}$$

Therefore we have equality in the arithmetic-geometric mean inequality, which implies every point of $\partial\Omega$ is umbilic, so $\partial\Omega$ is a sphere and Ω a disk. $\square$

3. Manifolds with curvature bounded from above

Surfaces. Here we assume that M is 2-dimensional and orientable. We use A and L for V and A, respectively, and write the isoperimetric function as $I(a)$. Note that according to our conventions, on the positively oriented boundary of a domain, the geodesic curvature κ_g and the scalar mean curvature $\mathfrak{h}$ are related by $\mathfrak{h} = \kappa_g$.

T. Carleman (1921) first showed that a disk imbedded in $\mathbb{R}^3$ as a minimal surface (that is, its mean curvature identically equal to 0) satisfies the 2-dimensional isoperimetric inequality

$$L^2 - 4\pi A \geq 0 \tag{3.1}$$

— his methods were those of complex analysis. The first to free the argument from complex analysis was W.T. Reid (1959), who based his argument on a formula of J.H. Jellet (1853). Definitive results for minimal varieties of arbitrary co-dimension in Euclidean spaces were obtained, via geometric measure theory, by F. Almgren (1986). We discussed his heuristic, above, in §2.4.

A. Weil (1926) generalized Carleman's result (minimal surfaces in Euclidean space have nonpositive Gauss curvature) to prove that for a simply connected surface with complete Riemannian metric of nonpositive Gauss curvature one also has (3.1). And E.F. Beckenbach–T. Rado (1933) noted that $L^2 \geq 4\pi A$ for all simply connected domains on a surface if and only if the Gauss curvature was nonpositive (for the necessary condition, just use Puiseux's and Diquet's formulae for the length and area of geodesic circles bounding disks of small radii; see Berger (1988, p. 382)).

I. Chavel (1978) combined Jellet's formula with A. Hurwitz' (1901), and using an argument of H.F. Weinberger (1956), gave a unified proof of the Carleman and Weil result. This approach was extended further in P. Li–R. Schoen–S. T. Yau (1984) and J. Choe (1990).

The G. Bol (1941) and F. Fiala (1940) inequalities. We now are given an oriented complete 2-dimensional *real analytic* Riemannian manifold M, and a relatively compact domain $\mathcal{D}$ bounded by a finite number of simple closed real analytic Jordan curves $\mathcal{C}$. Each component curve of $\mathcal{C}$ is given by a map

$$\omega : \mathbb{R} \to M, \quad |\omega'| = 1, \quad \omega \text{ is } L_o\text{-periodic}.$$

Let q denote a point in $\mathcal{C}$, as well as an arc length parameter along each component of $\mathcal{C}$. Also, let $\mathcal{C}'$ denote the unit velocity vector field along $\mathcal{C}$.

Let $\mathbf{n}(q)$ denote the interior unit normal vector field along $\mathcal{C}$ at $q \in \mathcal{C}$, and ι denote the rotation of tangent vectors by $\pi/2$-radians. The orientation of $\mathcal{C}$ is chosen as usual, that is, so that $\mathbf{n} = \iota\mathcal{C}'$. In particular,

$$\nabla_q \mathcal{C}' = \kappa_g \mathbf{n},$$

where κ_g denotes the geodesic curvature of ω.

Consider Fermi coordinates on M, given by

$$v(t;q) = \exp t\mathbf{n}(q), \qquad t \in \mathbb{R}, \quad q \in \mathcal{C}.$$

Then standard calculation yields

$$|\partial_t v| = 1, \qquad \nabla_t \partial_t v = 0, \qquad \langle \partial_t v, \partial_q v \rangle = 0.$$

Also, one has

$$\partial_q v(0,q) = \mathcal{C}'(q), \quad \nabla_t \partial_q v(0,q) = \nabla_q \mathbf{n}(q) = -\kappa_g \mathcal{C}'(q),$$

and the vector Jacobi's equation,

$$\nabla_t{}^2 \partial_q v + K \partial_q v = 0,$$

where K denotes the Gauss curvature of M. One can write

$$\partial_q v := \eta\,(-\iota(\partial_t v));$$

therefore

$$ds^2 = dt^2 + \eta^2\, dq^2$$

for the Riemannian metric, and η satisfies the *scalar* Jacobi's equation:

$$\partial_t{}^2\eta + K\eta = 0, \qquad \eta(0,q) = 1, \quad (\partial_t\eta)(0,q) = -\kappa_g(q).$$

For every $q \in \mathcal{C}$, we let $P(q)$ denote the first zero of $t \mapsto \partial_q v(t;q)$, that is, $v(P(q);q)$ *is the first focal point of* $\mathcal{C}$ *along the geodesic*

$$\gamma_q(t) = \exp t\mathbf{n}(q).$$

If no such zero exists then we set $P(q) = +\infty$. Since η and $\partial_t\eta$ cannot vanish simultaneously, then $P(q)$, when finite, is a real analytic function (by the implicit function theorem), with $P'(q)$ bounded on any strip $0 \le t \le r$.

Immediate results are

PROPOSITION 3.1. *If, for any given $r_o > 0$, the equation $P(q) = r_o$ has infinitely many solutions in $\mathcal{C}$, then $\mathcal{D}$ is a disk (possibly less its center).*

Similarly, if $P'(q) = 0$ has infinitely many solutions, then either (i) $\mathcal{D}$ is a disk (possibly less its center), or (ii) for every accumulation point q_o of solutions of $P'(q) = 0$ we have $P(q_o) = +\infty$.

Let Z_o denote the set of r for which $P(q) = r$ has at least one solution q_o satisfying $P'(q_o) = 0$ at $P(q_o) = r$; then Z_o is discrete.

PROPOSITION 3.2. (HÜLLCURVENSATZ) *For each component of $\mathcal{C}$, along the path $q \mapsto v(P(q);q)$ where $P(q)$ is finite, the Riemannian metric reads as*

$$ds = |P'(q)|\, dq;$$

so when $P' \neq 0$ on an interval $[q_1, q_2]$ we have that the length of the path along the first focal locus, from q_1 to q_2, is given by $|P(q_2) - P(q_1)|$.

DEFINITION. Given the geodesic $\gamma_q(t) := v(t;q)$ we let

$$\rho(q) = \sup\{\tau :\ d(\mathcal{C}, \gamma_q(t)) = t\ \forall\ t \in (0,\tau]\}.$$

So $\rho(q) < +\infty$ is the distance along γ_q after which γ_q ceases to minimize distance to $\mathcal{C}$. Then $\gamma_q(\rho(q))$ is referred to as *the focal cut point of q along γ_q*.

PROPOSITION 3.3. *A focal cut point $p = \exp \rho(q)\mathbf{n}(q)$ is either the first focal point of $\mathcal{C}$ along γ_q or is the intersection of at least two geodesic arcs minimizing distance from $\mathcal{C}$ to p. Note that these two arcs need not emanate from the same component of the boundary.*

So $\rho(q) \le P(q)$. Furthermore, if $\rho(q_o) = P(q_o) < \infty$ then $P'(q_o) = 0$.

PROOF. The proof of the first claim is standard, so we only consider the last claim.

If $P'(q_o) \neq 0$ then there exists q close to q_o with $P(q) < P(q_o)$. One can then travel from q to $v(P(q_o);q_o)$ by going along the geodesic $\gamma_q(t)$ from q to $\gamma_q(P(q))$ and then travelling along the focal locus to $v(P(q_o);q_o)$. By Proposition 3.2, the full path has length $P(q_o)$, which implies $\gamma_{q_o}(t)$ does not minimize distance from $\mathcal{C}$ to $\gamma_{q_o}(P(q_o))$, which is a contradiction. □

DEFINITION. A focal cut point p along the geodesic γ_q is *normal* if it is not a focal point of $\mathcal{C}$ along γ_q, and it is the endpoint of precisely two distance minimizing arcs (one of them γ_q) from $\mathcal{C}$ to p.

A cut point which is not normal will be referred to as *anormal.*

PROPOSITION 3.4. *Let p_o be a normal cut point along the two geodesics γ_{q_1} and γ_{q_2} at distance ρ_o. Then the function $\rho = \rho(q)$ is real analytic on neighborhoods of q_1 and q_2.*

Also, there exists a local real analytic function $\mu(q)$, from a neighborhood of q_1 to a neighborhood of q_2, satisfying

$$\mu(q_1) = q_2, \qquad v(\rho(q); q) = v(\rho(q)); \mu(q)).$$

Furthermore, the locus $v(\rho(q); q)$ near q_1 (or q_2) bisects the angle between the minimizing geodesics γ_q and $\gamma_{\mu(q)}$ meeting at $v(\rho(q); q)$.

Assume q_1 and q_2 are in the same component of $\mathcal{C}$, and $q_1 < q_2 < q_1 + L$, where L is the length of their common component. Then the map $q \mapsto \mu(q)$ is strictly decreasing.

PROOF. Let U_1 be a neighborhood of (ρ_o, q_1) such that $v_1 := v|U_1$ is a diffeomorphism, U_2 be a neighborhood of (ρ_o, q_2) such that $v_2 := v|U_2$ is a diffeomorphism, and satisfying $v_1(U_1) = v_2(U_2)$. Set

$$(\tau, \sigma) = ({v_2}^{-1} \circ v_1)(t, s)$$

(so we let s replace q in a neighborhood of q_1, and σ replace q in a neighborhood of q_2). Then write

$$\tau := \phi_1(t, s) \qquad \sigma := \phi_2(t, s).$$

If $\gamma_s(t) = \gamma_\sigma(\tau)$ denote two distinct geodesics producing the normal cut point at $\gamma_s(\rho) = \gamma_\sigma(\rho)$ then

$$\rho = \phi_1(\rho, s) \qquad \sigma = \phi_2(\rho, s).$$

Therefore, solve $t = \phi_1(t, s)$ for t as a function of s on some neighborhood of $s = q_1$. To use the implicit function theorem it suffices to show that $\partial\phi_1/\partial t < 1$ at $(t, s) = (\rho_o, q_1)$. But

$$(\partial\phi_1/\partial t)(\rho_o, q_1) = \langle \partial_t v(\rho_o, q_1), \partial_\tau v(\rho_o, q_2)\rangle < 1,$$

which is the first claim. The function $\mu(q)$ is given by $\mu(q) = \phi_2(\rho(q), q)$.

Thus, the cut locus is given, locally by $q \mapsto v(\rho(q); q)$. Its velocity vector is given by

$$\rho'(q)\partial_t v_{|v(\rho(q);q)} + \partial_q v_{|v(\rho(q);q)} = \rho'(q){\gamma_q}'(\rho(q)) + \partial_q v_{|\gamma_q(\rho(q))};$$

so the cosine of its angle with ${\gamma_q}'(\rho(q))$ is $\rho'(q)$. And the same is true for the geodesic $\gamma_{\mu(q)}$, since $\rho(\mu(q)) = \rho(q)$. This implies the cut locus is the angle bisector of the intersecting geodesics.

For the last claim it suffices to note that the arcs $\omega|[q_1, q_2]$, $\gamma_{q_1}|[0, \rho_o]$, and $\gamma_{q_2}|[0, \rho_o]$ are the boundary of a bounded domain in $\mathcal{D}$. □

COROLLARY 3.1. *In the above, we have $\rho'(q) = 0$ if and only if ${\gamma_q}'(\rho(q)) = -{\gamma_{\mu(q)}}'(\rho(q))$, that is, the two geodesics meet smoothly at the cut point.*

If $\rho'(q) \equiv 0$ on one component of $\mathcal{C}$ then either $\mathcal{D}$ is a geodesic disk, or a geodesic annulus (possibly degenerate).

PROPOSITION 3.5. *It is known that the function $q \mapsto \rho(q)$ is continuous.*

If $v(\rho(q_o); q_o)$ is a normal cut point, then q_o has a neighborhood for which $v(\rho(q); q)$ is a normal cut point.

PROOF. If there is a sequence of $q_j \to q_o$ such that all $v(\rho(q_j); q_j)$ are all focal cut points, then so is $v(\rho(q_o); q_o)$.

If one has a sequence $q_j \to q_o$ so that all $v(\rho(q_j); q_j)$ are all nonfocal, anormal cut points, then the "normality" of $v(\rho(q_o); q_o)$ comes from the strict decrease of the number of distinct geodesics associated with each q_j, which implies that the exponential map — more precisely, v — is not locally 1-1 on the preimage of $v(\rho(q_o); q_o)$, which implies $v(\rho(q_o); q_o)$ is a focal cut point along some geodesic, and is therefore anormal. □

The arguments of Propositions 3.4 and 3.5 also imply

PROPOSITION 3.6. *Let v_o be a nonfocal, anormal point. Then there exist a finite number of q-values* (mod L_o), $q_1, q_2, \ldots q_n$, $n \geq 3$, *satisfying*

$$v_o = v(\rho(q_j); q_j), \qquad j = 1, \ldots n.$$

Furthermore, for each j, $\rho(q)$ is C^1 on a small closed interval ending at q_j and real analytic on the interior of that interval; and the set of (normal!) cut points in a neighborhood of v_o consist of n arcs of class C^1 ending at v_o. At each point of these arcs, the arc bisects the angle of the geodesics meeting there.

COROLLARY 3.2. *The anormal cut points are isolated. In particular, there are at most a finite number of them.*

DEFINITION. A number $t > 0$ is called *exceptional* if there exists $q \in \mathcal{C}$ such that $v(t; q)$ is an anormal cut point, or is a normal cut point with $\rho'(q) = 0$.

COROLLARY 3.3. *The set of exceptional values is finite.*

For any given r_o, if $\rho(q)$ is not identically equal to r_o on a component of $\mathcal{C}$, then the set of solutions to $\rho(q) = r_o$ is finite.

PROOF. The first claim is obvious.

For the second, first recall that if $\rho(q) \equiv r_o$ on a component, then $\mathcal{D}$ must be a geodesic disk or annulus, and $\rho(q) \equiv r_o$ everywhere on $\mathcal{C}$.

So assume $\rho(q)$ is not identically equal to r_o, but is equal to r_o at infinitely many points of $\mathcal{C}$. Then $\rho(q) = r_o$ at infinitely many normal cut points of $\mathcal{C}$. But since $\rho(q)$ is real analytic away from afocal, anormal cut points, this implies $\rho(q) \equiv r_o$ on a component, which implies $\rho(q) \equiv r_o$ everywhere, which we assumed is not the case. □

NOTATION. Let t be an nonexceptional value, and define

$$\begin{aligned} \mathcal{C}_t &= \{p \in \mathcal{D} : \ d(p, \mathcal{C}) = t\}, \\ \Omega_t &= \{p \in \mathcal{D} : \ d(p, \mathcal{C}) < t\}, \\ \mathcal{D}_t &= \{p \in \mathcal{D} : \ d(p, \mathcal{C}) > t\}. \end{aligned}$$

Then $\mathcal{C}_t$ consists of a finite number of simply closed curves, piecewise real analytic, written as

$$p_t(q) = v(t; q), \qquad q \in \bigcup_{j=1}^{m(t)} [\alpha_j(t), \beta_j(t)],$$

where the $m(t)$ denotes the number of intervals required for $\mathcal{C}_t$. The value of $\alpha_j(t)$ *increases* with respect to t, and the value of $\beta_j(t)$ *decreases* with respect to t. The points $v(t, \alpha_j(t))$ and $v(t, \beta_j(t))$ are normal cut points, and both functions $\alpha_j(t)$ and $\beta_j(t)$ are local analytic inverse functions of $t = \rho(q)$.

If t_o is an exceptional value, then one has the same result with some of the simply closed curves possibly being degenerate (that is, a point), or two of them touching.

To consider what happens when $t \uparrow t_o$, first note that $\lim \alpha_j(t) := \alpha_o$ and $\lim \beta_j(t) := \beta_o$ exist as $t \uparrow t_o$. What might happen is that $\alpha_o = \beta_o$, in which case the "interval" disappears as t goes above t_o. Or, the interval $[\alpha_o, \beta_o]$ contains finitely many new points of the cut locus, in which case the number of "intervals" increases as t goes above t_o.

Assume that at t_o the limit of the set

$$\text{“}\lim_{t \uparrow t_o}\text{”}\, [\alpha_j(t), \beta_j(t)] = [\alpha_o, \delta_o] \cup [\delta_o, \beta_o].$$

Even if α_o denotes an anormal cut point, $\alpha_j(t)$ has an increasing continuous extension beyond t_o; and a corresponding remark applies for β_o. For δ_o, there exist continuous functions $\mathfrak{D}(t), \mathfrak{D}^*(t)$ defined for $t > t_o$ so that

$$\lim_{t \downarrow t_o} \mathfrak{D}(t) \uparrow \delta_o \qquad \text{and} \qquad \lim_{t \downarrow t_o} \mathfrak{D}^*(t) \downarrow \delta_o,$$

where $\mathcal{C}_t \cap [\alpha_o, \beta_o] = [\alpha_j(t), \mathfrak{D}(t)] \cup [\mathfrak{D}^*(t), \beta_j(t)]$.

NOTATION. For a point $q \in M$ and positive number $r > 0$, we let $B(q; R)$ denote the metric disk centered at q with radius r.

Also, we let τ denote the furthest distance from $\mathcal{C}$ into $\mathcal{D}$, that is,

$$\tau = \sup \{d(q, \mathcal{C}) : \; q \in \mathcal{D}\}.$$

Note that τ is the *inradius of* $\mathcal{D}$. That is, there exists $q_o \in \mathcal{D}$ such that $d(q_o, \mathcal{D}) = \tau$; $B(q_o; \tau) \subseteq \mathcal{D}$; and for any $t > \tau$, $q \in \mathcal{D}$, the disk $B(q; \tau)$ is *not* contained in $\mathcal{D}$.

THEOREM 3.1. *For every value t, let $L(t)$ and $A(t)$ denote*

$$L(t) = L(\mathcal{C}_t), \qquad A(t) = A(\mathcal{D}_t).$$

Then $L(t)$ and $A(t)$ are continuous with respect to t. Moreover, at all nonexceptional values of t we have

$$A'(t) = -L(t). \tag{3.2}$$

If the Gauss curvature K satisfies $K \leq \kappa$, then for all nonexceptional values of t we have

$$L'(t) \leq -2\pi\chi(\mathcal{D}) + \kappa A(\mathcal{D}_t). \tag{3.3}$$

Equations (3.2) *and* (3.3) *are valid even if $\mathcal{D}$ has a finite number of components.*

PROOF. We work only with $L(t)$. The argument for $A(t)$ is easier.

For any $t \in [0, \tau]$, $\mathcal{C}_t$ is a finite union of circles (some of them may be degenerate, that is, points), given by

$$\mathcal{C}_t = \exp t\mathfrak{C}_t, \qquad \mathfrak{C}_t = \bigcup_{j=1}^{m(t)} [\alpha_j(t), \beta_j(t)].$$

Then

$$L(t) = \int_{\mathcal{C}_t} ds = \int_{\mathfrak{C}_t} \eta(t; q)\, dq,$$

which implies

$$\begin{aligned}|L(t)-L(t_o)| &= \left|\int_{\mathfrak{C}_t}\eta(t;q)\,dq-\int_{\mathfrak{C}_{t_o}}\eta(t_o;q)\,dq\right| \\ &\le \int_{\mathfrak{C}_t\cap\mathfrak{C}_{t_o}}|\eta(t;q)-\eta(t_o;q)|\,dq+\sup_{[0,\tau]\times C}|\eta(t;q)|\int_{\mathfrak{C}_t\triangle\mathfrak{C}_{t_o}}dq,\end{aligned}$$

which implies that $L(t)$ is continuous.

As there are at most only a finite number of exceptional values of t, a similar argument shows that $L(t)$ is real analytic everywhere except possibly at the exceptional values.

For the non-exceptional value t, the loci

$$a_j(t) := v(t;\alpha_j(t)), \qquad b_j(t) := v(t;\beta_j(t)), \qquad j=1,\dots m(t),$$

consist of normal cut points along C_t, and are locally real analytic paths with respect to t. Let

$$\vartheta_{a_j(t)} = \angle(\partial_t v \to a_j{}'(t)), \qquad \vartheta_{b_j(t)} = \angle(\partial_t v \to b_j{}'(t)).$$

Then

$$\begin{aligned} a_j{}' &= \partial_t v + \alpha_j{}'\partial_q v = \partial_t v + (-\alpha_j{}'\eta)\iota\partial_t v, \\ b_j{}' &= \partial_t v + \beta_j{}'\partial_q v = \partial_t v + (-\beta_j{}'\eta)\iota\partial_t v, \end{aligned}$$

which implies

$$\alpha_j{}'(t)\eta(t;\alpha_j(t)) = -\tan\vartheta_{a_j(t)} \qquad \beta_j{}'(t)\eta(t;\beta_j(t)) = -\tan\vartheta_{b_j(t)}.$$

Therefore, using the Gauss–Bonnet theorem, we have

$$\begin{aligned} L(t) &= \sum_{j=1}^{m(t)}\int_{\alpha_j(t)}^{\beta_j(t)}\eta(t;q)\,dq, \\ L'(t) &= \sum_{j=1}^{m(t)}\left\{\beta_j{}'(t)\eta(t;\beta_j(t)) - \alpha_j{}'(t)\eta(t;\alpha_j(t)) + \int_{\alpha_j(t)}^{\beta_j(t)}(\partial_t\eta)(t;q)\,dq,\right\} \\ &= \sum_{j=1}^{m(t)}\left\{-\tan\vartheta_{b_j(t)} + \tan\vartheta_{a_j(t)} + \int_{\alpha_j(t)}^{\beta_j(t)} -\kappa_g(t;q)\eta(t;q)\,dq,\right\} \\ &= \sum_{j=1}^{m(t)}\{\vartheta_{b_j(t)} - \tan\vartheta_{b_j(t)} - (\vartheta_{a_j(t)} - \tan\vartheta_{a_j(t)})\} + 2\pi\chi(\Omega_t) \\ &\quad -\int_{\partial\mathcal{D}}\kappa_g\,ds - \iint_{\Omega_t}K\,dA \\ &\le -\int_{\partial\mathcal{D}}\kappa_g\,ds - \iint_{\Omega_t}K\,dA \\ &= -2\pi\chi(\mathcal{D}) + \iint_{\mathcal{D}_t}K\,dA, \end{aligned}$$

— to pass from the third equality to the inequality one uses the facts that $\vartheta_{b_j(t)}$ is positive, $\vartheta_{a_j(t)}$ is negative, and $\chi(\Omega_t)$ is nonpositive. That is,

$$L'(t) \le -2\pi\chi(\mathcal{D}) + \iint_{\mathcal{D}_t}K\,dA.$$

□

THEOREM 3.2. *If $\mathcal{D}$ is a relatively compact region in the complete real analytic surface M, with Gauss curvature K less or equal to the constant κ, then*

$$L^2(\partial\mathcal{D}) \geq 4\pi\chi(\mathcal{D})A(\mathcal{D}) - \kappa A^2(\mathcal{D}). \tag{3.4}$$

If $\mathcal{D}$ has n components all of which are, topologically, disks, then

$$L^2(\partial\mathcal{D}) \geq 4\pi n A(\mathcal{D}) - \kappa A^2(\mathcal{D}). \tag{3.5}$$

PROOF. We have

$$L(t)L'(t) \leq -2\pi\chi(\mathcal{D})L(t) + \kappa L(t)A(\mathcal{D}_t).$$

Integrate from 0 to τ; then

$$\frac{1}{2}\{L^2(\tau) - L^2(0)\} \leq 2\pi\chi(\mathcal{D})\{A(\tau) - A(0)\} - \kappa\{A^2(\tau) - A^2(0)\},$$

that is,

$$-\frac{1}{2}L^2(\partial\mathcal{D}) \leq -2\pi\chi(\mathcal{D})A(\mathcal{D}) + \kappa A^2(\mathcal{D}),$$

which implies (3.4). □

NOTE. If M is a Cartan–Hadamard surface, that is, it is complete, simply connected, with $K \leq \kappa \leq 0$ — and is therefore diffeomorphic to the plane — then one has

$$L^2(\partial\mathcal{D}) \geq 4\pi A(\mathcal{D}) - \kappa A^2(\mathcal{D})$$

for all regions in M. Indeed, it is certainly true for every region Ω diffeomorphic to a disjoint union of disks. If the region $\mathcal{D}$ is obtained from Ω by the removal of a finite number of diffeomorphic disks, then $A(\mathcal{D}) < A(\Omega)$, which implies (since $\kappa \leq 0$)

$$4\pi A(\mathcal{D}) - \kappa A^2(\mathcal{D}) < 4\pi A(\Omega) - \kappa A^2(\Omega).$$

But $L(\partial\mathcal{D}) > L(\partial\Omega)$. Since the inequality is valid for Ω, it is automatically valid for $\mathcal{D}$. The case of equality is easily handled.

P. Hartman (1964) extended the Bol–Fiala inequalities to the case of minimal differentiability assumptions by direct analysis of the continuity and differentiability of the function $L(t)$. I. Chavel & E.A. Feldman (1980) extended the inequalities to the differentiable case by approximating the differentiable case by the real analytic one. R. Osserman (1979) refined the original result in the real analytic case, and derived a Bonnesen type Bol–Fiala inequality, as follows:

NOTATION. Let inj denote the injectivity radius (see Chavel (1994, Chapter III)). Thus, for a point it denotes the injectivity radius of a point, and for M it denotes the infimum of inj taken over all points of M.

PROPOSITION 3.7. (Yu. D. Burago (1978)) *Assume M is complete, and homeomorphic to the plane, with $K \leq \kappa$. Then for $p_o \in \mathcal{D}$ satisfying $d(p_o, \partial\mathcal{D}) = \tau$, we have*

$$\operatorname{inj} p_o \geq \pi/\sqrt{\kappa}.$$

(If $\kappa \leq 0$ then $\pi/\sqrt{\kappa}$ denotes $+\infty$.)

PROOF. (C. Bavard (1984)) If $\kappa \leq 0$ then $\operatorname{inj} M = +\infty$ by the Cartan–Hadamard theorem. So assume $\kappa > 0$. First note that $K \leq \kappa$ implies that the first conjugate point of the initial point of a geodesic cannot occur prior to distance $\pi/\sqrt{\kappa}$ along the geodesic. This is a standard Sturmian argument.

Let q_o denote the closest cut point of p_o in M; then one has a geodesic loop starting and ending at p_o of length $2(\operatorname{inj} p_o)$, and q_o bisects the curve. Let $\mathfrak{D}$ denote the disk bounded by the loop. Then one can give an argument based on those of Propositions 3.4 and 3.6 that there is a geodesic emanating from q_o into $\mathfrak{D}$ with a conjugate point q in $\mathfrak{D}$. In particular, $d(q, q_o) \geq \pi/\sqrt{\kappa}$.

For every $w \in \partial\mathcal{D}$ consider a minimizing geodesic connecting w to q, and intersecting the geodesic loop at x. Then

$$d(w,q) = d(w,x) + d(x,q), \qquad d(p_o, q_o) = d(p_o, x) + d(x, q_o),$$

which implies

$$d(w,q) + d(p_o, q_o) \geq d(w, p_o) + d(q, q_o),$$

which implies

$$d(w,q) + \operatorname{inj} p_o \geq \tau + \pi/\sqrt{\kappa}.$$

Therefore, by minimizing $d(w,q)$ over $\partial\mathcal{D}$, we obtain

$$\tau + \operatorname{inj} p_o \geq d(\partial\mathcal{D}, q) + \operatorname{inj} p_o \geq \tau + \pi/\sqrt{\kappa},$$

which implies the claim. □

LEMMA 3.1. *Consider the real numbers A, a, L, ℓ, and κ, satisfying $A, a > 0$, $L, \ell \geq 0$, and*

$$0 = \ell^2 - 4\pi a + \kappa a^2, \tag{3.6}$$

$$\ell L \geq 2\pi(a + A) - \kappa a A, \tag{3.7}$$

If $\kappa A < 4\pi$ then $\ell > 0$, and

$$L^2 - 4\pi A + \kappa A^2 \geq \left\{\frac{2\pi}{\ell}(A - a)\right\}^2. \tag{3.8}$$

PROOF. From (3.6) we have $4\pi - \kappa a \geq 0$. Therefore $4\pi - \kappa A > 0$ implies

$$\frac{1}{a} + \frac{1}{A} > \frac{\kappa}{2\pi},$$

which implies from (3.7) that $\ell > 0$. Now square both sides of (3.7) and use (3.6). □

NOTATION. For any $q \in M$ and $r > 0$ we write $A(q;r) = A(B(q;r))$. In the complete simply connected 2-dimensional space form of constant curvature κ we let $\mathbb{B}_\kappa(r)$ denote a metric disk of radius r, $A_\kappa(r)$ its area, and $L_\kappa(r)$ the length of its boundary circle.

THEOREM 3.3. *Assume $\mathcal{D}$ is, topologically a disk, with boundary $\mathcal{C}$. Then*

$$L^2 - 4\pi A + \kappa A^2 \geq \left\{\frac{2\pi}{L_k(\tau)}(A - A_\kappa(\tau))\right\}^2 \geq 0. \tag{3.9}$$

We have $L^2 - 4\pi A + \kappa A^2 = 0$ if and only if $\mathcal{D} = \mathbb{B}_\kappa(\tau)$.

PROOF. Set

$$\mathcal{A}(t) = A(\Omega_t) = A - A(\mathcal{D}_t).$$

Then (3.3) reads as:

$$\mathcal{A}'' + \kappa\mathcal{A} \leq -2\pi + \kappa A, \qquad \mathcal{A}(0) = 0, \quad \mathcal{A}'(0) = L.$$

We want to compare $\mathcal{A}(t)$ to the solution $\mathcal{B}(t)$ of:

$$\mathcal{B}'' + \kappa\mathcal{B} = -2\pi + \kappa A, \qquad \mathcal{B}(0) = 0, \quad \mathcal{B}'(0) = L.$$

To this end consider b_ϵ given by:

$$b_\epsilon'' + \kappa b_\epsilon = -2\pi + \kappa A, \qquad b_\epsilon(0) = 0, \quad b_\epsilon'(0) = L + \epsilon,$$

and $a_\epsilon := b_\epsilon - \mathcal{A}$; thus,

$$a_\epsilon'' + \kappa a_\epsilon \geq 0, \qquad a_\epsilon(0) = 0, \quad a_\epsilon'(0) = \epsilon.$$

A standard Sturmian argument implies $a_\epsilon(t) \geq \epsilon \mathbf{S}_\kappa(t)$, which implies

$$b_\epsilon(t) \geq \mathcal{A}(t) + \epsilon \mathbf{S}_\kappa(t) \qquad \forall \epsilon > 0,$$

which implies $\mathcal{B}(t) \geq \mathcal{A}(t) \;\; \forall t \in [0, \pi/\sqrt{\kappa}]$. One checks explicitly that

$$\mathcal{B}(t) = \frac{(-2\pi + \kappa A)}{2\pi} A_\kappa(t) + \frac{L}{2\pi} L_\kappa(t);$$

therefore,

$$L L_\kappa(t) \geq 2\pi(\mathcal{A}(t) + A_\kappa(t)) - \kappa A A_\kappa(t).$$

Let $\tau_\kappa = \min\{\tau, \pi/\sqrt{\kappa}\}$. Then Burago's theorem implies

$$A = \mathcal{A}(\tau) \geq A(q_o; \tau) \geq A_\kappa(\tau_\kappa),$$

with equality in the first inequality if and only if $\mathcal{D} = B(q_o; \tau)$, and in the second inequality if and only if $\tau \leq \pi/\sqrt{\kappa}$ and $K|B(q_o;\tau) = \kappa$.

On the other hand, our interest in $\mathcal{D}$ is only for $A < 4\pi/\kappa$ — otherwise the inequality yields no information. For this case we have

$$L L_\kappa(\tau) \geq 2\pi(A + A_\kappa(\tau)) - \kappa A A_\kappa(\tau),$$

which implies (3.9).

The case of equality follows easily. $\square$

The benefit of the Bonnesen type estimate is that it preserves the characterization of the case of equality even when approximating the differentiable case by the analytic one.

Subsequent generalizations of the Bol–Fiala inequalities were given by A.D. Alexandrov (1948) and A. Huber (1954), and summarized in a unified presentation in Yu.D. Burago & V.A. Zalgaller (1988). Further generalization was given in P. Topping (1998, 1999).

Higher dimensions. C. Croke (1984) proved the following: Assume M is a 4-dimensional Riemannian manifold with smooth boundary, so that $\overline{M}$ is compact. Assume that M has nonpositive sectional curvature, and every geodesic in M minimizes distance from its initial point until it hits the boundary. Then

$$\frac{A(\partial M)}{V(M)^{3/4}} \geq \frac{A(\mathbb{S}^3)}{V(\mathbb{B}^4)^{3/4}},$$

with equality if and only if M is isometric to a disk in $\mathbb{R}^n$.

B. Kleiner (1992) then considered the 3-dimensional case: Let M be a 3-dimensional complete simply connected Riemannian manifold, with sectional curvature less than or equal to a constant $\kappa \leq 0$, and let $\Omega \subset\subset M$ with smooth boundary, and $\mathbf{B}_\kappa(\Omega)$ the disk in the 3-dimensional model space of constant curvature κ, having the same volume as Ω. Then

$$A(\partial\Omega) \geq A(\partial(\mathbf{B}_\kappa(\Omega))), \tag{3.10}$$

with equality if and only if Ω is isometric to $\mathbf{B}_\kappa(\Omega)$. His argument is an adaptation of Almgren's argument above. The reason it does not work in arbitrary dimensions is that in variable curvature one does not have a Gauss map. Therefore the equality

$$\mathbf{c_{n-1}} = \int_{\Sigma\cap\Gamma} \det\left(-\mathcal{G}^{-1}\mathcal{B}\right) dA$$

requires replacement. In the 3-dimensional manifold, one can replace this by

$$4\pi = \int_{\Sigma} K\, dA = \int_{\Sigma\cap\Gamma} K\, dA$$

by the Gauss–Bonnet theorem, unavailable in higher dimensions for the Gauss–Killing curvature $\det\left(-\mathcal{G}^{-1}\mathcal{B}\right)$.

References

L. V. Ahlfors, *Conformal Invariants*, McGraw-Hill, New York, 1973.

A. D. Alexandrov, *The Intrinsic Geometry of Convex Surfaces*, Gosudarstv. Izdat Tehn–Teor. Lit., Moscow–Leningrad, 1948 (Russian); German transl., Die Innere Geometrie der konvexen Flächen, Akademie Verlag, Berlin, 1955.

______, *A characteristic property of spheres*, Ann. di Math. Pura Appl. **58** (1962), 303–315.

F. Almgren, *Existence and Regularity Almost Everywhere of Solutions to Elliptic Variational Problems with Constraints*, Vol. 4, Memoir Amer. Math. Soc., Amer. Math. Soc., Providence, RI, 1976.

______, *Optimal isoperimetric inequalities*, Ind. U. Math. J. **35** (1986), 451–547.

T. Aubin, *Problèmes isopérimétriques et espaces de Sobolev*, J. Diff. Geom. **11** (1976), 573–598.

A. Baernstein, *Symmetrization in Analysis*, 2004 (to appear).

A. Baernstein and B. A. Taylor, *Spherical arrangements, subharmonic functions, and *-functions in n-space*, Duke. Math. J. **43** (1976), 245–268.

C. Bavard, *Le rayon d'injectivité des surfaces à courbure majoree*, J. Diff. Geom. **20** (1984), 137–142.

C. Bavard and P. Pansu, *Sur le volume minimal de R^2*, Ann. École Norm. Sup. Paris **19** (1986), 479–490.

E. F. Beckenbach and T. Radó, *Subharmonic functions and surfaces of negative curvature*, Trans. Amer. Math. Soc. **35** (1933), 662–674.

M. Berger, *A Panoramic View of Riemannian Geometry*, Springer Verlag, Berlin, 2003.

M. Berger and B. Gostiaux, *Differential Geometry: Manifolds, Curves, and Surfaces*, Springer Verlag, New York, 1988.

F. Bernstein, *Über die isoperimetrische Eigenschaft des Kreises auf der Kugeloberfläche und in der Ebene*, Math. Ann. **60** (1905), 117–136.

W. Blaschke, *Kreis und Kugel*, (2nd ed.), de Gruyter, Berlin, 1956.

G. Bol, *Isoperimetrische Ungleichung für Bereiche auf Flächen*, Jber. Deutsch. Math.-Verein. **51** 1941, 219–257.

Y. Burago, *The radius of injectivity on the surfaces whose curvature is bounded from above*, Ukrain. Geom. Sb. **21** (1978), 10–14. (Russian).

Y. Burago and V. Zalgaller, *Geometric Inequalities*, Springer Verlag, Berlin, 1988; Original Russian edition, *Geometricheskie neravenstva*, Leningrad, 1980.

X. Cabré, The isoperimetric inequality and the principal eigenvalue via the ABP method, preprint, 2003.

C. Carathéodory and E. Study, *Zwei Beweise des Satzes dass der Kreis unter alle Figuren gleichen Umgangs den grössten Inhalt hat*, **68** (1909), 133–144.

T. Carleman, *Zur theorie der Minimalflächen*, Math. Zeit. **9** (1921), 154–160.

I. Chavel, *On A. Hurwitz' method in isoperimetric inequalities*, Proc. Amer. Math. Soc. **71** (1978), 275–279.

______, *Riemannian Geometry: A Modern Introduction*, Cambridge Univ. Press, Cambridge, 1994.

______, *Isoperimetric Inequalities: Differential Geometric and Analytic Perspectives*, Cambridge Univ. Press, New York, 2001.

I. Chavel and E. A. Feldman, *Isoperimetric inequalities in curved surfaces*, Adv. in Math. **37** (1980), 83–98.

J. Choe, *The isoperimetric inequality for a minimal surface with radially connected boundary*, Ann. Scuola Norm. Sup. Pisa **17** (1990), 583–593.

C. B. Croke, *A sharp 4-dimensional isoperimetric inequality*, Comment. Math. Helv. **59** (1984), 187–192.

F. Fiala, *Le problème des isopérimètres sur les surfaces ouvertes à courbure positive*, Comment. Math. Helv. **13** (1940), 293–346.

R. J. Gardner, The Brunn–Minkowski inequality. *Bull. Amer. Math. Soc.* **39** (2002), 355–405.

M. Gromov, *Paul Levy's isoperimetric inequality*, preprint, 1980.

______, *Structures métriques pour les Variétés Riemannienes*, Cedic/Ferdnand Nathan, Paris, 1981.

______, *Isoperimetric inequalities in Riemannian manifolds*, Asymptotic Theory of Finite Dimensional Normed Spaces, Lecture Notes Math., vol. 1200, Springer Verlag, Berlin, 1986, pp. Appendix I, 114–129.

______, *Metric Structure for Riemannian and Non-Riemannian Spaces*, Appendices by M. Katz, P. Pansu, S.Semmes. (J. La Fontaine & P. Pansu, eds.), Engl. trans., S.M. Bates, Birkhauser, Boston, 1999.

H. Hadwiger, *Vorlesungungen über Inhalt, Oberfläche und Isoperimetrie*, Springer Verlag, Berlin, 1957.

P. Hartman, *Geodesic parallel coordinates in the large*, Amer. J. Math. **86** (1964), 705–727.

H. Howards, M. Hutchings and F. Morgan, *The isoperimetric problem on surfaces*, Amer. Math. Monthly **106** (1999), 430–439.

W. Y. Hsiang, *do Carmo Symposium*, Pitman Monographs Surveys Pure & Appl. Math., vol. 52, Longman Scientific & Technical, Essex, 1991, pp.˜229–240.

A. Huber, *On the isoperimetric inequality on surfaces of variable curvature*, Ann. Math. **60** (1954), 237–247.

A. Hurwitz, *Sur le problème des isopérimétres*, C. R. Acad. Sc. Paris **132** (1901), 401–403.

J. H. Jellet, *Sur la surface dont la courbure moyenne est constante*, J. Math. Pure Appl. **18** (1853), 163–167.

B. Kleiner, *An isoperimetric comparison theorem*, Invent. Math. **108** (1992), 37–47.

P. Li, R. Schoen and S. T. Yau, *On the isoperimetric inequality for minimal surfaces*, Ann. Sc. Norm. Sup. Pisa **11** (1984), 237–244.

E. H. Lieb and M. Loss, *Analysis*, Amer. Math. Soc., Providence, R.I., 1996.

D. Ohman, *Über den Brunn–Minkowskischen Satz*, Comment. Math. Helvitici **29** (1955), 215–222.

R. Osserman, *The isoperimetric inequality*, Bull. Amer. Math. Soc. **84** (1978), 1182–1238.

———, *Bonnesen-style inequalities*, Amer. Math. Monthly **86** (1979), 1–29.

———, *A Survey of Minimal Surfaces*, (2nd ed.), Dover, New York, 1986.

W. T. Reid, *The isoperimetric inequality and associated boundary problems*, J. Math. Mech. **8** (1959), 571–581.

A. Ros, *The isoperimetric problem*, Proceedings of the Clay Mathematical Institute–MSRI Summer School on Minimal Surfaces, Berkeley, 2001 (to appear). `http://www.ugr.es/~aros/isoper.htm`

E. Schmidt, *Der Brunn–Minkowskische Satz und sein Spiegel-theorem sowie die isoperimetrische Eigenschaft der Kugel in der euklidischen und nichteuklidischen Geometrie I, II*, Math. Nach. **1** (1948), 81–157; **2** (1949), 171–244.

H. A. Schwarz, *Beweis des Satzes, dass die Kugel kleinere Oberfläche besizt als jeder andere Körper gleichen Volumens*, Nach. Ges. Wiss. Göttingen (1884), 1–13; reprinted: *Gesammelte Werke*, Chelsea Publ. Co., Bronx, NY, 1971 (reprint of 1881–1882 ed.), pp. 327–340.

J. Steiner, *Einfache Beweise der isoperimetrische Hauptsätze*, J. Reine Angew. Math. **18** (1838), 281–296; reprinted: *Gesammelte Werke*, Chelsea Publ. Co., Bronx, NY, 1971 (reprint of 1881–1882 ed.).

———, *Über Maximum und Minimum*, reprinted: *Gesammelte Werke*, Chelsea Publ. Co., Bronx, NY, 1971 (reprint of 1881–1882 ed.) Vol. II, (1842), pp. 245–308.

E. Stredulinsky and W. P. Ziemer, *Area minimizing sets subject to a volume constraint in a convex set*, J. Geom. Anal. **7** (1997), 653–677.

P. Topping, *Mean curvature flow and geometric inequalities*, J. Reine Angew. Math. **503** (1998), 47–61.

———, *The isoperimetric inequality on a surface*, Manuscr. Math. **100** (1999), 23–33.

A. Weil, *Sur les surfaces a courbure negative*, C. R. Acad. Sc. Paris **182** (1926), 1069–1071.

H. F. Weinberger, *An isoperimetric inequality for the n–dimensional free membrane*, J. Rat. Mech. Anal. **5** (1956), 633–636.

B. White, *Existence of smooth embedded surfaces of prescribed genus that minimize parametric even elliptic functionals on 3–manifolds*, J. Diff. Geom. **33** (1991), 413–433.

Department of Mathematics, The City College of CUNY, New York, NY 10031, U.S.A.

E-mail address: `ichavel@gc.cuny.edu`

Contemporary Mathematics
Volume **387**, 2005

Hidden Symmetries and Arithmetic Manifolds

Benson Farb and Shmuel Weinberger

Dedicated to the memory of Robert Brooks.

ABSTRACT. A "hidden symmetry" of a Riemannian manifold M is an isometry of a finite-sheeted cover of M which is not a nontrivial lift. In this paper we characterize the locally symmetric metric(s) on an arithmetic manifold as the unique (up to homothety) metric which has an infinite number of hidden symmetries. This paper also serves to expose some of the ideas in a sequel paper, where we classify the isometry groups of the universal covers of all closed, aspherical Riemannian manifolds.

1. Introduction

Let M be a closed, locally symmetric Riemannian manifold of nonpositive curvature with no local torus factors; for example take M to be a hyperbolic manifold. Equivalently, $M = K\backslash G/\Gamma$ where G is a semisimple Lie group, K is a maximal compact subgroup, and Γ is a torsion-free, cocompact lattice in G. For simplicity, we will always assume that Γ is *irreducible*, or equivalently that M is not finitely covered by a smooth product; we will also assume for simplicity that $\dim(M) > 2$.

Let g_{loc} denote a locally symmetric metric on M. The Mostow Rigidity Theorem states that g_{loc} is unique up to homothety of the direct factors of $\widetilde{M}$ [1]. As the following result (essentially due to Borel) shows, the metric g_{loc} is also special because it reflects every symmetry of every other Riemannian metric on M.

PROPOSITION 1.1 (Most symmetry). *Let M be a closed, irreducible, locally symmetric Riemannian n-manifold, $n > 2$, of nonpositive curvature with no local torus factors. Then for every Riemannian metric h on M, the group* $\mathrm{Isom}(M, h)$ *is isomorphic to a subgroup of* $\mathrm{Isom}(M, g_{loc})$.

REMARKS.

(1) By an old result of Bochner, the group $\mathrm{Isom}(M, g_{loc})$ is finite.
(2) It is *not* true in general that $\mathrm{Isom}(M, h)$ is topologically conjugate to a subgroup of $\mathrm{Isom}(M, g_{loc})$: there are examples of hyperbolic n-manifolds, $n > 4$, with periodic diffeomorphisms not topologically conjugate into

2000 *Mathematics Subject Classification.* Primary 53C35, 53C23.
Both authors are supported in part by the NSF.

[1] We will henceforth abuse notation and call g_{loc} the "unique locally symmetric metric on M".

Isom(M, g_{loc}); we give examples in §4.1. By averaging, one sees that these diffeomorphisms preserve some Riemannian metric.

(3) We conjecture that Proposition 1.1 remains true when M has finite volume, and the metric h is any finite volume Riemannian metric. However, we only know how to prove this when M admits a locally symmetric metric of negative curvature.

In §2 below we give three proofs of Proposition 1.1: one using coarse geometry and Smith theory, one using harmonic maps, and one (essentially due to Borel) using algebraic topology. These proofs are meant as a warm-up for the proof of our main theorem (Theorem 1.4 below); they are also meant as an introduction to some of the ideas in [**3**], where we will make use of the interplay among all three proofs.

We remark that both the harmonic maps proof and the algebraic-topological proof apply *mutatis mutandis* to any aspherical manifold with a degree one map to M that is an isomorphism on fundamental group. We also remark that each of the proofs applies to give a slightly modified version of Proposition 1.1 in dimension 2.

Hidden symmetries. Does the property of g_{loc} given in Proposition 1.1 characterize the locally symmetric metric up to homothety? The answer is obviously no: since Isom(M, g_{loc}) is finite, it is easy to equivariantly perturb the metric (say in the neighborhood of a single free orbit, for simplicity) to give a non-isometric manifold with the same isometry group. Indeed, there is an infinite dimensions worth of such metrics.

The main discovery of this paper is that we can actually detect g_{loc} exactly by looking at the isometry groups of finite-sheeted covers of M. To make this more precise, we will need the following.

DEFINITION 1.2 (Hidden symmetry). Let M be a finite volume Riemannian manifold. A *hidden symmetry* of M is an isometry ϕ of a d-sheeted, $1 < d < \infty$, Riemannian cover of M with the property that ϕ is not the lift of an isometry.

So, for example, the deck transformations on any finite cover of M, which are just lifts of the identity, are not hidden symmetries of M.

To find examples of locally symmetric manifolds with a lot of hidden symmetry, we must consider *arithmetic manifolds.* These are manifolds $M = \Gamma\backslash G/K$ with G semisimple, $K \subset G$ a maximal compact subgroup, and Γ an irreducible, arithmetic lattice. Borel showed that these exist in abundance in every semisimple Lie group G, and Margulis's Arithmeticity Theorem (see, e.g., [**6**]) states that if $\text{Rank}_{\mathbf{R}}(G) > 1$ and if Γ is irreducible, then every such M is arithmetic.

Recall that the *commensurator* of a lattice Γ in a Lie group G is defined to be

$$\text{Comm}_G(\Gamma) := \{g \in G : g\Gamma g^{-1} \cap \Gamma \text{ has finite index in both } g\Gamma g^{-1} \text{ and } \Gamma\}$$

Borel and Harish-Chandra proved that arithmetic groups have infinite index in their commensurators. Using this we will prove the following.

PROPOSITION 1.3 (Arithmetic metrics have hidden symmetry). *Let M be a closed, arithmetic manifold. Then (M, g_{loc}) has infinitely many hidden symmetries; these can be taken to occur on irregular covers.*

We remark that, while every lift to $\widetilde{M} = G/K$ of a hidden symmetry lies in the commensurator $\text{Comm}_G(\Gamma)$, the set of hidden symmetries is a rather small subset of $\mathrm{Comm}_G(\Gamma)$; see §3.1 below. Margulis proved that irreducible, arithmetic lattices in semisimple Lie groups are precisely those irreducible lattices that have infinite index in their commensurator (see [**6**]).

The following theorem gives our promised characterization of the locally symmetric metric on an arithmetic manifold.

THEOREM 1.4 (Characterizing the arithmetic metric). *Let M be a closed arithmetic manifold with $\dim(M) > 2$. Then the locally symmetric metric g_{loc} is the unique metric (up to homothety) on M having infinitely many hidden symmetries.*

Conversely, if $M = \Gamma\backslash G/K$ with G semisimple and Γ an irreducible, torsion-free, cocompact non-arithmetic *lattice, then there exists $d = d(M)$ so that any Riemannian metric on M has at most d hidden symmetries, and this maximal bound is achieved by uncountably many non-homothetic Riemannian metrics.*

Theorem 1.4 characterizes the locally symmetric metric via a countable amount of data, namely the (finite) isometry groups of the finite sheeted covers of M. In the special case when (M, h) is negatively curved, Theorem 1.4 can be deduced from results of Furman in [**5**]. In the case that (M, h) is nonpositively curved, then Theorem 1.4 is a consequence of theorems of Eberlein [Eb] and Margulis (see below). A different approach from ours to proving Theorem 1.4 might be to use Furman's main theorem in [**4**] classifying which locally-compact topological groups contain a lattice in a semisimple Lie group as a lattice.

In [**3**] we will prove the following much stronger theorem:

THEOREM 1.5 (Characterization of arithmetic manifolds). *Let (M, h) be any closed, aspherical Riemannian manifold with* $\dim(M) > 2$. *Suppose that $\pi_1(M)$ contains no normal infinite abelian subgroup, and that M is "irreducible" in the sense that no finite cover of M is a smooth product. If M has an infinitely many hidden symmetries, then (M, h) is isometric to an arithmetic manifold.*

REMARKS.

(1) Each of the hypotheses in Theorem 1.5 is necessary: every irreducible lattice in a semisimple Lie group is not virtually a product, and does not contain any normal infinite abelian subgroup.
(2) Note that Theorem 1.4 above requires the assumption that the fundamental group of M is arithmetic; we make no such a priori assumption in Theorem 1.5.
(3) In [**3**] we give a similar characterization of (not necessarily arithmetic) locally symmetric manifolds in terms of symmetries that potentially remain hidden until one gets to the universal cover.
(4) The special role of normal abelian subgroups can be seen by considering, for example, any 3-dimensional solvmanifold M; this is a manifold whose universal cover is the unique 1-connected non-nilpotent, solvable Lie group Sol. Every such M has an infinite number of hidden symmetries, but M admits many non-locally homogeneous metrics that have all of these symmetries; indeed any $\mathbf{R}^2$-invariant perturbation of the Sol-metric has all of these symmetries. Similar examples can be constructed on torus bundles that are very far from locally homogeneous metrics.
(5) Theorem 1.5 has an extension to all closed, aspherical Riemannian manifolds; in particular, normal abelian subgroups are allowed. See [**3**].

Theorem 1.4 can be extended to non-aspherical manifolds; in the end it is really a theorem about closed manifolds whose fundamental group is arithmetic. For a lattice Δ, we say that a manifold M has *Δ-fundamental hidden symmetries* if there is a homomorphism $\pi_1(M) \to \Delta$ and hidden symmetries of M that are intertwined with hidden symmetries of Δ. When $\Delta = \pi_1(M)$ then we call these *fundamental* hidden symmetries.

THEOREM 1.6 (Hidden symmetry theorem, nonaspherical case). *Let Δ be an irreducible, cocompact lattice in a semisimple Lie group $G \not\approx \mathrm{SL}(2, \mathbf{R})$ with maximal compact subgroup K. If a closed Riemannian manifold M has infinitely many Δ-fundamental hidden symmetries, then M is a Riemannian fibration over the arithmetic manifold $\Delta\backslash G/K$.*

In §4.2 we discuss a "near counterexample" to Theorem 1.5 and Theorem 1.6, namely a closed, aspherical manifold with infinitely many "topological hidden symmetries". We also state in §4.2 a conjecture on how metrics with greater and greater hidden symmetry must degenerate.

It is a pleasure to thank David Fisher, Karin Melnick, Kevin Whyte, and the referee for their useful comments and suggestions.

2. Three proofs of Proposition 1.1

Each element $\phi \in \mathrm{Isom}(M, h)$ induces an automorphism of $\pi_1(M)$, which (since ϕ may move a given basepoint) is only defined up to conjugacy, thus giving a map $\psi : \mathrm{Isom}(M, h) \to \mathrm{Out}(\pi_1(M))$. Mostow Rigidity states that, under the hypothesis of Proposition 1.1, any isomorphism between fundamental groups of two locally symmetric manifolds is induced by a unique isometry. We then have

$$\mathrm{Isom}(M, h) \xrightarrow{\psi} \mathrm{Out}(\pi_1 M) \to \mathrm{Isom}(M, g_{loc})$$

with the last map an isomorphism by Mostow Rigidity. Hence to prove Proposition 1.1, it is enough to prove that ψ is injective.

Suppose $\ker(\psi) \neq 0$. We claim there exists $\phi \in \ker(\psi)$ with order some prime $p > 1$. To prove this, first note that $\mathrm{Isom}(M, h)$ is a compact (since M is compact) Lie group, by the classical Steenrod-Myers Theorem. Any element of the connected component $\mathrm{Isom}^0(M, h)$ of the identity of $\mathrm{Isom}(M, h)$ is homotopically trivial, and so lies in $\ker(\psi)$. Now every nontrivial, connected, compact Lie group contains a circle, and so contains elements of any prime order. If $\mathrm{Isom}^0(M, h)$ is trivial, then $\mathrm{Isom}(M, h) \supseteq \ker(\psi)$ is a finite group, and so contains an element g of order some prime $p > 1$, and the claim is proved.

So, we have an element $g \in \mathrm{Isom}(M, h)$ of order $p > 1$, with p prime, so that g is homotopically trivial. We now give three different proofs that $g = 1$, proving that ψ must indeed be injective.

2.1. Smith theory proof. Let $F =< g >$, and consider the group Λ of lifts of every element of F to $\widetilde{M}$. There is an exact sequence

$$1 \to \pi_1(M) \to \Lambda \to F \to 1 \tag{2.1}$$

with the kernel acting on $\widetilde{M}$ by deck transformations. The fact that g is homotopically trivial gives that the action $F \to \mathrm{Out}(\pi_1(M))$ is trivial. Further, since M is locally symmetric, nonpositively curved, and has no local torus factors, it follows that the center $Z(\pi_1(M)) = 0$, so in particular $H^2(G, Z(\pi_1(M))) = 0$. These two facts imply that the exact sequence (2.1) splits, so that the F-action on M lifts to an action by isometries on $(\widetilde{M}, h)$, where by abuse of notation h denotes the lift to $\widetilde{M}$ of the metric h on M.

Now, the lift of any two metrics on a compact manifold are quasi-isometric (Effremovich-Milnor-Svarc), so the lift of each element of F induces a quasi-isometry of $\widetilde{M}$ endowed with the metric g_{loc}. But each element of F is homotopically trivial on M, and since homotopies are compact, there exists $D > 0$ so that

$$d(x, f(x)) \leq D \text{ for all } f \in F, x \in \widetilde{M}$$

where d is the distance on $X := (\widetilde{M}, g_{loc})$.

Now X has a compactification $\overline{X} = X \cup \partial X$ by Hausdorff equivalence classes of geodesic rays. In this compactification, $\overline{X}$ is homeomorphic to a closed n-ball, $n = \dim(M)$, and each isometry of X extends to a homeomorphism of this ball. Since each $f \in F$ moves points by at most D, and since ∂X is defined by Hausdorff equivalence classes of rays, it follows that the homeomorphic action of any $f \in F$ on $\overline{X}$ restricts to the identity on ∂X.

Now, since $f = g$ has prime order p, we may apply Smith Theory to the action of f on the homology n-ball $(\overline{X}, \partial X)$, where $n = \dim(M)$. This gives that the pair $(\text{Fix}(f), \text{Fix}(f|_{\partial X}))$ is a mod-p homology r-ball for some $0 \leq r \leq n$. Since f is the identity on ∂X, in other words since $\text{Fix}(f|_{\partial X}) = \partial X$, it must be that $r = n$. Further, if $\text{Fix}(f) \cap X \neq X$, then radial projection away from a point $z \notin \text{Fix}(f) \cap X$ would give a homotopy equivalence of $(\text{Fix}(f), \text{Fix}(f|_{\partial X}))$ with $(\partial X, \partial X)$, contradicting that the former is a mod-p homology n-disk with $n > 1$. Hence $\text{Fix}(f) \cap X = X$, so that f is trivial on X, and hence on M.

2.2. Harmonic maps proof. We recall that a map $f : N \to M$ between Riemannian manifolds is *harmonic* if it minimizes the energy functional

$$E(f) = \int_N ||Df_x||^2 d\text{vol}_N$$

The key properties of harmonic maps between closed Riemannian manifolds which we will need are the following (see, e.g., [**7**], which is indeed the inspiration for this proof):

- (Eells-Sampson) When the target manifold has nonpositive sectional curvatures, a harmonic map exists in each homotopy class.
- (Hartman) The image of any two harmonic maps in a given homotopy class are the boundary of an isometric product with an interval; in particular the harmonic map in a given homotopy class is unique when the domain and target have the same dimension.
- (easy) The precomposition and postcomposition of a harmonic map with an isometry gives a harmonic map.

Now let $F =< g >$. Since (M, g_{loc}) is nonpositively curved, there is a unique harmonic map ϕ in the homotopy class of the identity map $(M, h) \to (M, g_{loc})$. Precomposing ϕ with any element $f \in F$ then gives, since f is homotopically trivial, that

$$\phi \circ f = \phi$$

so that ϕ is F-invariant. This principal of "uniqueness implies invariance" will be quite useful for us later.

As ϕ is F-invariant and F is finite, the map ϕ factors through a map $\phi_2 : M \to M/F$. Now, restricting to orientation-preserving isometries if necessary, the fixed set of F is a submanifold of M of codimension at least two. Hence the quotient orbifold M/F is a pseudomanifold, and so one can apply degree theory to M/F. Since ϕ factors through ϕ_2, and since ϕ_2 has degree $|F|$, it follows that ϕ has degree at least $|F|$. But ϕ is homotopic to the identity map, and so has degree one. Hence $|F| = 1$, so that g is the identity.

2.3. Algebraic-topological proof. We consider the action of the orbifold fundamental group of $M/\operatorname{Isom}(M)$ on the universal cover $\widetilde{M}$. Note that this group, by (the algebraic form of) Mostow rigidity, has a representation into the algebraic group G. Note that the symmetric space $X = G/K$ is an example of the universal space $\underline{EG}$ for proper G-actions (see, e.g., [BCH]), and thus there is a well-defined equivariant homotopy class intertwining these actions. As in the proof in Subsection 2.2, considering the group generated by $\pi_1(M)$ and by g gives a contradiction.

REMARK. The difference between this proof and the harmonic maps proof is that in the latter one can pick out a canonical unique map, while in the algebraic proof there is merely a contractible space of choices. The proof of our main result (Theorem 1.4) will use the extra rigidity of the analytic proof.

3. Proof of the main results

In this section we prove the main results of this paper.

3.1. Hidden symmetries in arithmetic manifolds. We begin this subsection with the following.

PROOF OF PROPOSITION 1.3.
Let $M = \Gamma\backslash G/K$, endowed with the locally symmetric metric. The source of the hidden symmetries in the arithmetic case is intersections of conjugates of Γ. As M is arithmetic, we have that $[\operatorname{Comm}_G(\Gamma) : \Gamma] = \infty$. Let $g \in \operatorname{Comm}_G(\Gamma) \setminus N_G(\Gamma)$, where $N_G(\Gamma)$ denotes the normalizer of Γ in G (which is finite modulo Γ, by Bochner). The subgroup $\Delta_g := g\Gamma g^{-1} \cap \Gamma$ then contains a subgroup of finite index H_g which is normal in $g\Gamma g^{-1}$ but is not normal in Γ. The cover of M corresponding to H_g then has a hidden symmetry. □

It would be interesting to explore manifolds with hidden symmetries occurring in *regular* covers.

Hidden symmetries are thin in $\operatorname{Comm}_G(\Gamma)$. We now give[2] substance to the claim that the set of hidden symmetries of $M = \Gamma\backslash G/K$ is a rather thin subset of $\operatorname{Comm}_G(\Gamma)$.

First, note that $g \in G$ is (the lift to G/K of) a hidden symmetry of M if and only if g normalizes some finite index subgroup of Γ. Given $g \in \operatorname{Comm}_G(\Gamma)$, if g normalizes Γ', then some power g^k of g is an inner automorphism of Γ' since the outer automorphism group of any cocompact lattice in a simple $G \not\approx \mathrm{SL}(2, \mathbf{R})$ is finite by a result of Bochner (this also follows from Mostow Rigidity). Thus multiplying g^k by an element of Γ' we obtain an element which commutes with Γ', and is therefore the identity. Thus $g^k \in \Gamma$ for some k, which is certainly not all of $\operatorname{Comm}_G(\Gamma)$. For an easy example (in the noncocompact case), most elements of

[2]We would like to thank Kevin Whyte for this clean description.

$\mathrm{SL}_n(\mathbf{Q})$ do not have any power in $\mathrm{SL}_n(\mathbf{Z})$, for example the diagonal matrix with entries $(1/2, 2, 1, 1, 1, \ldots)$.

3.2. Proof of Theorem 1.4.

Non-arithmetic case. First suppose that $M = \Gamma\backslash G/K$ is not arithmetic. By a theorem of Margulis, Γ has finite index in $\Delta := \mathrm{Comm}_G(\Gamma)$. Since every lift to $\widetilde{M}$ of every hidden symmetry of M lies in Δ, it follows that M can only have finitely many hidden symmetries. The universal bound d follows from Proposition 1.1. Now any perturbation of the metric on M which is invariant with respect to the finite group Δ/Γ gives a metric on M with the same number of hidden symmetries as (M, g_{loc}).

Arithmetic case. Let X denote the universal cover of (M, h). For any lattice Δ which intersects $\Gamma = \pi_1(M)$ in a subgroup of finite index, we may consider the harmonic map f_Δ in the homotopy class of the identity map from the cover of M corresponding to Δ with the metric h to the corresponding cover with the metric g_{loc}. Uniqueness of harmonic maps gives us, as in the harmonic maps proof of Proposition 1.1 in §2.2, that $f_\Delta = f_\Gamma$.

Noting this for all such Δ, we see that the lift $\widetilde{f}$ of the harmonic representative f of the identity map $(M, h) \to (M, g_{loc})$ will be g-equivariant with respect any g lying in the union of such Δ in G. Further, again by uniqueness of harmonic maps, we see that $\widetilde{f}$ will be equivariant with respect to the group U generated by all such Δ. The given implies that U contains Γ as a subgroup of infinite index. The (topological) closure of U is a closed, hence Lie, subgroup of G. By the Borel Density theorem (see, e.g., [Ma]), G is the smallest algebraic subgroup containing Γ; from this it follows that U is dense in G.

Using the Arzela-Ascoli theorem, for any $g \in G$, we can find an isometry ϕ of X which f intertwines g: one simply takes any limit of elements of U converging to g. We claim that such a ϕ is unique. The only possible non-uniqueness comes from the stabilizer of a point, a compact group. Hence if ϕ were not unique one could factor the map $\widetilde{f} : (X, h) \to (X, g_{loc})$ through some isometry. If this isometry were of finite order, the order would divide $\deg(f) = 1$, a contradiction; if it were of infinite order, then f would factor through a lower-dimensional quotient, also contradicting $\deg(f) = 1$. This proves the claim.

Hence we have shown that X admits an isometric G-action, and that $\widetilde{f}$ intertwines this action with the standard isometric G-action on the symmetric space (X, g_{loc}). Thus the harmonic map f is actually a homothety, and the theorem is proven.

As the proof of Theorem 1.6 is similar, we leave it as an exercise to the reader.

4. Examples

4.1. An exotic periodic diffeomorphism. According to an old example, for any $n > 4$ the group $\mathbf{Z}/p\mathbf{Z}$ has a smooth action on the n-sphere S^n with fixed-set a knotted mod-p homology sphere K, with (S^n, K) not homeomorphic to (S^n, S^m) for any $0 \leq m \leq n$. To obtain such an action, take a $(k+1)$-dimensional contractible manifold X whose product with an interval is a ball, for example Mazur manifolds for $k = 3$. Then for $n - k$ even, the manifold $X \times D^{n-k}$ has a $\mathbf{Z}/p\mathbf{Z}$

action with fixed-point set X. Restricting to the boundary gives an action an a sphere with fixed-point set ∂X, which is definitely non-simply connected.

Now if one takes a $\mathbf{Z}/p\mathbf{Z}$ action on an n-manifold M and if the fixed-set of this action is nonempty, then one can equivariantly connect-sum with one of the knotted (S^n, K) as above to get a new action on M whose fixed set is definitely different, for example the fundamental group of the fixed set won't inject into $\pi_1(M)$.

4.2. A near counterexample. We now give what seems like a "near counterexample" to Theorem 1.5 and Theorem 1.6, namely a closed, aspherical manifold with infinitely many "topological hidden symmetries".

Let M be a noncocompact, arithmetic hyperbolic manifold, say of dimension 3, and say with one end, or "cusp". This end is homeomorphic to the product of a 2-torus T^2 with $[0, \infty)$. Let V be the closed manifold obtained from M by first deleting $T^2 \times (N, \infty)$ for any large N, and then doubling the resulting manifold. We will not need this, but Leeb has shown that such V can be given metrics of nonpositive curvature.

Since M is arithmetic, it has infinitely many hidden symmetries. It is easy to see that any hidden symmetry of M induces a finite order diffeomorphism of some finite cover of V; we call this a "hidden topological symmetry". Hence V has infinitely many "topological hidden symmetries", but it is easy to see that M has no locally symmetric metric, and indeed is not a fiber bundle over any manifold. Note that any topological hidden symmetry preserves some Riemannian metric on the corresponding finite cover (by averaging); of course this metric depends (as it must, by Theorem 1.5) on the cover of M.

The reason this example ends up not being a counterexample to the theorems is that the core of a cover is not the cover of the core. Equivalently, think about the cover of a cusp. This will be disconnected and different components are likely to cover it different numbers of times. These different covering cusps will be interchanged in a hidden symmetry; this necessitates isogenies of the cusp with shifted versions of itself that go further down the tube.

Given this situation, we make the following conjecture.

CONJECTURE 4.1 (Stretching cusps). *Let M be the double of a cusped arithmetic hyperbolic n-manifold N, or more generally the double of a negatively curved, arithmetic locally symmetric manifold N which is noncompact but has finite volume. Suppose g_i is a sequence of Riemannian metrics on M with the property that (M, g_i) has at least i hidden symmetries. Then the sequence $\{(M, g_i)\}$ Gromov-Hausdorff converges (in the appropriate sense) to a disjoint union of two isometric copies of N.*

Here we are using a notion of convergence where one chooses two (sequences of) basepoints. It would be interesting to consider the special case of N hyperbolic and each of the metrics g_i being nonpositively curved.

References

1. P. Baum, A. Connes, N. Higson, *Classifying space for proper actions and K-theory of group C^*-algebras*, C^*-algebras: 1943–1993 (San Antonio, TX, 1993), Contemp. Math., vol. 167, 1994, pp. 240–291.
2. P. Eberlein, *Lattices in spaces of nonpositive curvature*, Annals of Math. (2) **111** (1980), no. 3, 435–476.

3. B. Farb and S. Weinberger, The isometry groups of covers of closed aspherical manifolds, in preparation.
4. A. Furman, *Mostow-Margulis rigidity with locally compact targets*, Geom. Funct. Anal. **11** (2001), no. 1, 30–59.
5. ———, *On minimal, strongly proximal actions of locally compact group*, Israel J. Math., to appear.
6. G.A. Margulis, *Discrete subgroups of semisimple Lie groups*, Ergebnisse der Mathematik und ihrer Grenzgebiete (3), 17, Springer-Verlag, Berlin, 1991.
7. R. Schoen and S.T. Yau, *Compact group actions and the topology of manifolds with nonpositive curvature*, Topology **18** (1979), no. 4, 361–380.

Dept. of Mathematics, University of Chicago, 5734 University Ave, Chicago, Il 60637, U.S.A.
E-mail address: farb@math.uchicago.edu

Dept. of Mathematics, University of Chicago, 5734 University Ave, Chicago, Il 60637, U.S.A.
E-mail address: shmuel@math.uchicago.edu

Contemporary Mathematics
Volume **387**, 2005

Variants of the $3N + 1$ Conjecture and Multiplicative Semigroups

Hershel M. Farkas

Dedicated to the memory of Robert Brooks

ABSTRACT. In this note we show that a variant of the $3N + 1$ conjecture has some interesting combinatorial consequences of the form every odd positive integer is expressible as a product of elements in the set of rational numbers $\left\{\frac{4k+1}{2k+1},\ k = 0, 1, \ldots\right\}$ These results can be rephrased in the language of multiplicative semigroups generated by sets of rational numbers.

1. Introduction

The famous $3N+1$ conjecture or problem is the following. Let N be a positive odd integer. Then $3N+1$ is necessarily an even integer; and we can divide out the highest power of 2 in it, obtaining $\frac{3N+1}{2^{k(N)}}$. This number is now once again an odd number and we can proceed as before. The $3N+1$ conjecture says that you always end up with the periodic sequence $1, 1, 1, \ldots$

Despite a great deal of work by many people (see [**3**], [**4**]), we do not know whether the conjecture is true or false. We do not settle this conjecture in this paper but rather indicate possible connections with some other problems. We do, however, show that if we change the hypotheses a bit, we can in fact solve the problem. In fact, it is known that there are variants of the problem which are undecidable [**1**].

Let us begin with an example which shows some of the subtleties of the problem. If N is congruent to $1 \mod 4$ say $N = 2^m j + 1$ where $m \geq 2$, and $(j, 2) = 1$, then $3N + 1$ is divisible by 4; and the result is an odd number congruent to $1 \mod 4$ provided $m \geq 4$. Hence if $m = 2k$ is even, then repeating this process k times brings us to the number $3^k j + 1$, which is clearly even but about which there is no information concerning the highest power of 2 which divides it. We can, however, say that the number we end up with is surely less than the number we started with. Moreover, when we come to this point and divide out the highest power of 2, whatever it may be, the new odd number may be congruent to $3 \mod 4$. If m were odd in this process, say $m = 2k + 1$, then the number after k steps would

2000 *Mathematics Subject Classification.* Primary 20M25; Secondary 11K31.

be $3^k \cdot 2 \cdot j + 1$, which is of course an odd number which is always congruent to $3 \mod 4$.

If, on the other hand, N is congruent to $3 \mod 4$, say $N = 2^m j + 3$ where $m \geq 2$, $(j, 2) = 1$, then $3N + 1$ is at most divisible by 2 and $\frac{3N+1}{2} = 3 \cdot 2^{m-1} j + 5$, which is congruent to $3 \mod 4$ only when $m = 2$ and otherwise is congruent to $1 \mod 4$. The point is that when the number is congruent to $1 \mod 4$ the process lowers the number and when the number is congruent to $3 \mod 4$ the process raises the number. The $3N + 1$ conjecture says that the reducers win in the end since the number ultimately goes down to 1.

In order to prove the result, it is sufficient to show that beginning with the odd number N the process always leads to a number less than N. If one could show this then an induction would finish the proof. This remark already shows that the problem is with the numbers congruent to $3 \mod 4$, since we have already observed that starting with a number congruent to $1 \mod 4$ always leads to a smaller number.

To give the reader a feeling for the problem we indicate the orbit of the number 27 under this process:

$$(27, 41, 31, 47, 71, 107, 161, 121, 91, 137, 103, 155, 233, 175, 263, 345, 593, 445, 167, 251, 377, 282, 425, 319, 479, 719, 1079, 1619, 2429, 911, 1367, 2051, 3077, 577, 433, 325, 61, 23, 35, 53, 5, 1, 1, 1, \ldots).$$

2. A perturbation of the problem

In order to formulate a problem that we can solve, we consider the following variant. Recognizing that the real problem seems to be the numbers congruent to $3 \mod 4$, we change the rules as follows. If the odd number N is congruent to $3 \mod 4$, we do as before. If the odd number N is congruent to $1 \mod 4$ we replace it with $\frac{N+1}{2}$. The virtue of our new game over the preceding one is that we always stay in the realm of odd numbers. We do not have to divide out the highest powers of 2. If we look at the orbit of 27 now under the new rules, we get

$$(27, 41, 21, 11, 17, 9, 5, 3, 5, 3, \ldots)$$

once again an orbit which ends in a periodic sequence $5, 3, 5, 3, \ldots$. There are also quite long orbits here as the reader can see if one begins with the number 55. While we believe that with these rules you always will end up with the periodic sequence given, we have not been able to prove it.

What can we do? We continue with the process just described but allow ourselves now to divide out powers of 3. In this case, the orbit of 27 is now trivial

$$(27, 1, 1, 1, \ldots)$$

while the orbit of 55 is

$$(55, 83, 125, 63, 7, 11, 17, 9, 1, 1, 1, \ldots)$$

We can prove the following

THEOREM 1. *Let $F(n)$ satisfy*

$$F(n) = \begin{cases} \frac{n}{3}, & n \equiv 0 \mod 3, \\ \frac{3n+1}{2}, & n \not\equiv 0 \mod 3 \text{ and } n \equiv 3 \mod 4, \\ \frac{n+1}{2}, & n \not\equiv 0 \mod 3 \text{ and } n \equiv 1 \mod 4. \end{cases}$$

Then for any odd number N, the sequence $F(N), F(F(N)), F(F(F(N))), \ldots$ terminates with $1, 1, 1, \ldots$. In other words a modified $3N+1$ conjecture is true.

PROOF. The result is proven by induction. The result is clearly true for $N = 1$. Assume we have proven it for all odd j which are less than or equal to the odd number N. We show it is also true for $N+2$. If $N+2$ is congruent to $1 \mod 4$ then $\frac{N+3}{2}$ is an odd number less than $N+2$ and its orbit terminates with a sequence of 1's by the induction hypothesis. There remains the case of $N+2$ congruent to $3 \mod 4$. Denote the odd number $N+2$ by N_1 and consider the sequence

$$N_1, N_2, N_3, \ldots, N_k, N_{k+1}$$

where

$$N_j = \frac{3N_{j-1}+1}{2}, \qquad j = 2, \ldots, k+1$$

and N_j is congruent to $3 \mod 4$ for $j = 1, \ldots, k$ while N_{k+1} is congruent to $1 \mod 4$. Notice immediately that

$$N_j + 1 = \frac{3}{2}(N_{j-1}+1) = \left(\frac{3}{2}\right)^{j-1}(N_1+1)$$

so that in particular choosing $j = k+1$ yields

$$N_{k+1} + 1 = \left(\frac{3}{2}\right)^{k}(N_1+1).$$

As a consequence of this we see that choosing k as the largest positive integer such that (N_1+1) is divisible by 2^{k+1}, forces N_{k+1} to be congruent to $1 \mod 4$ since

$$N_{k+1} = \left(\frac{3}{2}\right)^{k}(N_1+1) - 1 = 3^k\frac{N_1+1}{2^k} - 1.$$

The last expression is clearly congruent to $1 \mod 4$ since it is of the form $(3^k \cdot 2 \cdot l) - 1$ with l odd.

It thus follows that the next step in the process leads to $\frac{N_{k+1}+1}{2} = \frac{3^k(N_1+1)}{2^{k+1}}$. We now divide out as is permissible by 3^k and have arrived at a number of the form $\frac{N_1+1}{2^{k+1}}$ which is less than N_1 so the proof is complete by induction. □

REMARK. We have thus succeeded in proving this variant of the $3N+1$ conjecture where we have changed the rules slightly by reducing in the case of odd integers congruent to 1 mod 4 differently. In this way we always obtain odd numbers without having to divide out powers of 2. The price, however, was dividing out powers of 3. We believe that the result is true even without dividing out these powers of 3 but then you get a periodic sequence $(5, 3, 5, 3, \ldots)$ rather than the trivial sequence of $(1, 1, \ldots)$. The interesting thing though may be that we have in fact also solved another problem.

3. A combinatorial problem

Let S denote the set of rational numbers of the form $\left\{\frac{4n+1}{2n+1}\right\}$ with $0 \leq n \leq \infty$. What are the positive integers which are expressible as products of rational numbers taken from the set S?

THEOREM 2. *Every positive odd integer is expressible as a product of elements in S.*

PROOF. The proof is by induction. 1 is obviously expressible as a product of elements in S since it is in S. Assume now that the result has been proven for all odd integers $j \leq N$. We prove it now for $j = N+2$.

Note if $j = N+2$ is composite then we factor it and we are done since each factor is an odd number less than N and is therefore by induction hypothesis expressible as a product of elements in S. If $N+2$ is a prime then we as before need consider the two cases of the prime congruent to 1 or 3 mod 4. If the prime is congruent to 1 mod 4 then we can easily write $\frac{N+2}{\frac{N+3}{2}}$ which is our set S and then

$$N+2 = \frac{N+2}{\frac{N+3}{2}} \cdot \frac{N+3}{2}.$$

$\frac{N+3}{2}$ is an odd integer less than N and is thus representable as a product of elements in S hence we have represented $N+2$ as a product as well.

If the prime $N+2$ is congruent to 3 mod 4 one has to work a bit harder. We consider once again the sequence

$$N_1, N_2, .., N_k, N_{k+1}$$

as in the proof of the previous theorem but now form the product

$$\frac{3N_1}{N_2} \cdot \frac{3N_2}{N_3} \cdot \ldots \cdot \frac{3N_k}{N_{k+1}} \cdot \frac{N_{k+1}}{\frac{N_{k+1}+1}{2}}.$$

It is clear that each factor in the product is an element of the set S. However, the product is clearly equal to

$$\frac{3^k N_1}{\frac{N_{k+1}+1}{2}} = \frac{N_1}{\frac{N_1+1}{2^{k+1}}}.$$

Once again $\frac{N_1+1}{2^{k+1}} < N_1$, so we are finished by induction. □

4. Elementary number theory

The above problem, namely, determining which positive integers appear as products of elements in the set S can be phrased differently in terms of elementary ideas in number theory. Recall that the arithmetical function $d(n)$ counts the number of divisors of n including the trivial divisors 1 and n. Thus $d(p) = 2$ whenever p is a prime. Consider therefore the sequence $\left\{\frac{d(n^2)}{d(n)}\right\}$. This is a sequence of rational numbers. The question is what are the positive integers which appear in this sequence. It is easy to see that the function $d(n)$ is always even unless n is a square. Hence $d(n^2)$ will always be an odd integer while if n is not a square $d(n)$ will be even. Hence if n is not a square the quotient will not be an integer. The only way the quotient can possible be an integer therefore is when n is actually a square and thus n^2 is a 4th power. Namely it is sufficient to look at the subsequence $\left\{\frac{d(n^4)}{d(n^2)}\right\}$. If the primary decomposition of n is given by $n = p_1^{\alpha_1} \cdots p_k^{\alpha_k}$, then

$$\frac{d(n^4)}{d(n^2)} = \prod_{l=1}^{k} \frac{(4\alpha_l + 1)}{(2\alpha_l + 1)}$$

namely a product of elements in the set S. Hence as a consequence of the proof of the previous theorem, we have also proven

THEOREM 3. *The set of integers contained in the multiplicative semigroup generated by the rationals* $\left\{\frac{d(n^2)}{d(n)}\right\}$ *contains the positive odd integers.*

REMARK. We have now seen that the variant of the $3N+1$ problem has led us to some simple number theoretic problems which can of course be generalized. As an example of such a generalization consider the question of which positive integers appear in the sequence $\left\{\frac{d(n^3)}{d(n)}\right\}$.

By analogy with the above, we prove

THEOREM 4. *The set of integers contained in the multiplicative semigroup generated by* $\left\{\frac{d(n^3)}{d(n)},\ n \geq 0\right\}$, *consists of all positive integers not congruent to* $0 \mod 3$.

PROOF. We immediately observe that since $d(n^3) = \prod_{l=1}^{k}(3\alpha_l+1)$, no integers congruent to $0 \mod 3$ can possibly occur. In fact what we are asking is what positive integers can be expressed as a product of elements of the set T where T is the set of rational numbers of the form $\frac{3n+1}{n+1}$. We shall show that every positive integer not congruent to $0 \mod 3$ is so expressible and thus prove the theorem.

Once again the tool is induction. We shall show that every prime congruent to 1 or $2 \mod 3$ is so expressible. We have to consider the cases of primes congruent to 1 or $2 \mod 3$ separately. Assume therefore we have proven the result for all positive integers j less than or equal to N. We of course mean for all integers not congruent to $0 \mod 3$. Let $3l+1$ be the first prime greater than N. We show that it too is expressible as such a product.

Consider the rational number $\frac{3l+1}{l+1}$. If $l+1$ is not congruent to $0 \mod 3$, it has a decomposition into primes congruent to 1 or $2 \mod 3$; and since each of these primes is less than N, the induction hypothesis finishes the proof. If $l+1$ is congruent to $0 \mod 3$, so that l is in fact congruent to $2 \mod 3$, we proceed as follows:

Denote $3l+1$ by N_1 and consider the sequence

$$N_1, N_2, \ldots, N_k$$

where

$$N_j = \frac{4N_{j-1}+2}{3}, \qquad j = 2, \ldots, k$$

and k is the largest integer for which N_1+2 is divisible by 3^k. The product

$$\frac{4N_1}{N_2} \cdot \frac{4N_2}{N_3} \cdot \ldots \cdot \frac{4N_{k-1}}{N_k} \cdot \frac{N_k}{\frac{N_k+2}{3}}$$

has the property that each term in the product is an element of our set T. Hence this product is representable as a product of elements of T.

It is easy to see that

$$N_k + 2 = \left(\frac{4}{3}\right)^{k-1}(N_1+2)$$

so that the product in fact equals

$$\frac{4^{k-1}N_1}{\frac{N_k+2}{3}} = \frac{N_1}{\frac{N_1+2}{3^k}}.$$

Since $\frac{N_1+2}{3^k}$ is an integer less than N and not congruent to $0 \mod 3$ it is expressible as a product of elements of T by the induction hypothesis and we are done.

It remains to treat the case of the first prime larger than N being of the form $3l+2$. Here the result is a little bit trickier than before.

We begin by making the observation that

$$\frac{2(3l+2)}{2l+2} = \frac{3(2l+1)+1}{2l+2}$$

is in fact in our set T so that the term $\frac{3l+2}{l+1}$ is in fact also in our set T. Since $l+1 < 3l+2$, it is representable as a product of elements of T provided l is not congruent to $2 \mod 3$. Hence once again the obstruction is when l is congruent to $2 \mod 3$.

We observe that

$$\frac{5(3l+2)}{5l+4} = \frac{3(5l+3)+1}{5l+4}$$

so that $\frac{5(3l+2)}{5l+4}$ is in our set T and furthermore if l was congruent to $2 \mod 3$, the same is true for $5l+4$. We define the sequence

$$(N_1, N_2, \ldots, N_k)$$

where $N_1 = 3l+2, N_j = \frac{5N_{j-1}+2}{3} \quad j = 2, .., k$ where k is the largest integer so that 3^k divides N_1+1. It is clear that

$$N_j + 1 = \left(\frac{5}{3}\right)^{j-1} (N_1+1)$$

so that in particular we have

$$N_k + 1 = \left(\frac{5}{3}\right)^{k-1} (N_1+1).$$

Consider now the product

$$\frac{5N_1}{N_2} \cdot \frac{5N_2}{N_3} \cdot \ldots \cdot \frac{5N_{k-1}}{N_k} \cdot \frac{2N_k}{\frac{2N_k+2}{3}}.$$

It clearly equals $\frac{N_1}{\frac{N_1+1}{3^k}}$. The term $\frac{N_1+1}{3^k}$ is not divisible by 3 and is less than N_1 so is representable as a product and this concludes the proof. □

5. Concluding remarks

Our purpose in this paper has been to try and tie the $3N+1$ conjecture or problem to some other problems in combinatorics thus perhaps giving some new idea how to attack it. We observe in this final section how to convert the $3N+1$ problem into one where we are asking what is the set of integers represented by taking products from a given set.

Let $Q = \left\{\frac{n}{\frac{3n+1}{2}}\right\}$ for all odd positive integers n. Adjoin to this set all positive powers of 2. Does Q contain all positive integers? Alternatively we can rephrase this as: let Q denote the multiplicative semigroup generated by the rationals $\left\{\frac{2n+1}{3n+2}, \ n \geq 0\right\}$ and 2. Does Q contain all positive integers?

If the $3N+1$ conjecture is true, then the answer to the above question is yes. However, it may be that the answer is yes but that the $3N+1$ conjecture is false, since the proposed question is weaker than the original conjecture.

References

1. J. H. Conway, *Unpredictable Iiterations*, Proc. 1972 Number Theory Conference, University of Colorado, pp. 49-52.
2. G. H. Hardy and E. M. Wright, *The Theory of Numbers*, Clarendon Press, 1979.
3. J. C. Lagarias, *The $3x+1$ problem and its generalizations*, American Math. Monthly **92** (1985), 3-23.
4. G. J. Wirsching, *The Dynamical System Generated by the 3N+1 Problem*, Lecture Notes in Mathematics, no. 1681, Springer Verlag, 1998.

INSTITUTE OF MATHEMATICS, THE HEBREW UNIVERSITY OF JERUSALEM, JERUSALEM 91904, ISRAEL

E-mail address: farkas@math.huji.ac.il

Contemporary Mathematics
Volume **387**, 2005

Energy Capacity Inequalities via an Action Selector

Urs Frauenfelder, Viktor Ginzburg, and Felix Schlenk

ABSTRACT. An action selector for a symplectic manifold (M, ω) associates with each compactly supported Hamiltonian function H on M an action value of H in a suitable way. Action selectors are known to exist for a broad class of symplectic manifolds. We show how the existence of an action selector leads to sharp energy capacity inequalities between the Gromov width, the Hofer–Zehnder capacity, and the displacement energy. We also obtain sharp lower bounds for the smallest action of a closed characteristic on contact type hypersurfaces.

1. Introduction and main results

Consider an arbitrary symplectic manifold (M, ω). We set $I = [0, 1]$ and denote, for each subset A of M, by $\mathcal{H}(I \times A)$ the set of smooth functions $H \colon I \times M \to \mathbb{R}$ whose support is compact and contained in $I \times \operatorname{Int} A$. We abbreviate $\mathcal{H} = \mathcal{H}(I \times M)$, and denote by $\mathcal{H}(A)$ the set of those functions in $\mathcal{H}(I \times A)$ which do not depend on $t \in I$. The Hamiltonian vector field of $H \in \mathcal{H}$ defined by

$$\omega\left(X_{H_t}, \cdot\right) = -dH_t\left(\cdot\right) \tag{1.1}$$

generates a flow φ_H^t with time-1-map φ_H. For $H \in \mathcal{H}$, the set of contractible 1-periodic orbits of φ_H^t is denoted $\mathcal{P}^\circ(H)$. Given $x \in \mathcal{P}^\circ(H)$, let $\mathcal{D}(x)$ be the set of smooth discs $\bar{x} \colon D^2 = \{z \in \mathbb{C} \mid |z| \leq 1\} \to M$ satisfying $\bar{x}(e^{it}) = x(t)$. We shall identify the map $\bar{x}$ with its oriented image and write $\int_{\bar{x}} \omega = \int_{D^2} (\bar{x})^* \omega$. For $H \in \mathcal{H}$, the action functional $\mathcal{A}_H$ on $\bar{\mathcal{P}}^\circ(H) = \{(x, \bar{x}) \mid x \in \mathcal{P}^\circ(H), \bar{x} \in \mathcal{D}(x)\}$ is defined as

$$\mathcal{A}_H\left(\bar{x}\right) = -\int_{\bar{x}} \omega + \int_0^1 H\left(t, x(t)\right) dt,$$

and its *action spectrum* is

$$\Sigma^\circ(H) = \left\{\mathcal{A}_H\left(\bar{x}\right) \mid (x, \bar{x}) \in \bar{\mathcal{P}}^\circ(H)\right\}.$$

Note that for $a \in \Sigma^\circ(H)$ the set $a + \omega(\pi_2(M))$ also belongs to $\Sigma^\circ(H)$. As a consequence, $\Sigma^\circ(H)$ need not be closed and is dense in $\mathbb{R}$ if $\omega(\pi_2(M))$ is dense. For

---

2000 *Mathematics Subject Classification.* Primary 53D40; Secondary 37J10.

This work is supported in part by the Swiss National Science Foundation (UF), the US National Science Foundation and the UCSC faculty research funds (VG), and the Deutsche Forschungsgemeinschaft (FS).

$H \in \mathcal{H}$ we abbreviate

$$E^+(H) = \int_0^1 \max_{x \in M} H(t,x)\, dt,$$

and we recall that for $H, K \in \mathcal{H}$ the composition $\varphi_H \circ \varphi_K$ is generated by

$$(H\#K)(t,x) = H(t,x) + K\left(t, \left(\varphi_H^t\right)^{-1}(x)\right).$$

We say that a Hamiltonian $H \in \mathcal{H}(M)$ is *simple* and write $H \in \mathcal{S}(M)$ if

(P1) $H \geq 0$,

(P2) $H|_U = \max H$ for some open non-empty set $U \subset M$,

(P3) the only critical values of H are 0 and $\max H$.

We emphasize that simple Hamiltonians are, as is clear from (P3), "normalized" to have minimum equal to 0. Furthermore, we say that $H \in \mathcal{S}(M)$ is *Hofer–Zehnder admissible* and write $H \in \mathcal{S}^\circ_{\mathrm{HZ}}(M)$ if the flow φ_H^t has no non-constant, contractible in M, T-periodic orbit with period $T \leq 1$.

1.1. Axioms for an action selector. A *weak action selector* σ for (M, ω) is a map $\sigma \colon \mathcal{H} \to \mathbb{R}$ satisfying the following axioms.

(AS1) $\sigma(H) \in \Sigma^\circ(H)$ for all $H \in \mathcal{H}$;

(AS2) $\sigma(H) > 0$ for all $H \in \mathcal{S}(M)$ with $H \not\equiv 0$;

(AS3) $\sigma(H) \leq E^+(H)$ for all $H \in \mathcal{H}$;

(AS4) σ is continuous with respect to the C^0-topology on $\mathcal{H}$;

(AS5) $\sigma(H\#K) \leq \sigma(H) + E^+(K)$ for all $H, K \in \mathcal{H}$.

An *action selector* σ for (M, ω) is a weak action selector which *in addition* to (AS2) satisfies

(AS2$^+$) $\sigma(H) = \max H$ for all $H \in \mathcal{S}^\circ_{\mathrm{HZ}}(M)$.

A symplectic manifold (M, ω) is called *weakly exact* if $[\omega]$ vanishes on $\pi_2(M)$.

Remark 1.1. If (M, ω) is weakly exact, a weak action selector for (M, ω) is an action selector for (M, ω).

Proof. Given $H \in \mathcal{S}^\circ_{\mathrm{HZ}}(M)$ on a weakly exact symplectic manifold (M, ω), we have $\Sigma^\circ(H) = \{0, \max H\}$, and hence axioms (AS1) and (AS2) imply $\sigma(H) = \max H$. □

Remark 1.2. If σ is an action selector for (M, ω), then (AS3) is a consequence of the other axioms.

Proof. Axioms (AS5) and (AS2$^+$) applied to $H = 0$ yield

$$\sigma(K) = \sigma(0\#K) \leq \sigma(0) + E^+(K) = E^+(K)$$

for all $K \in \mathcal{H}$. □

An *exhaustion* of a symplectic manifold (M, ω) is an increasing sequence of submanifolds $M_i \subset M$ exhausting M, that is,

$$M_1 \subset M_2 \subset \cdots \subset M_i \subset \cdots \subset M \qquad \text{and} \qquad \bigcup_i M_i = M.$$

By a (weak) action selector for an exhaustion (M_i) of (M, ω), we mean a collection σ_i of (weak) action selectors for (M_i, ω). While a (weak) action selector for (M, ω) obviously restricts to a (weak) action selector for any of its exhaustions, the restriction of σ_{i+1} to M_i is not assumed to agree with σ_i, whence it is unclear whether every (weak) action selector for an exhaustion of (M, ω) fits together to a (weak) action selector for (M, ω); see 3. (ii) and (iii) in Appendix A. The existence of a (weak) action selector for exhaustions of (M, ω) as defined here is sufficient for our purposes in this paper and can sometimes be established even when it is unknown whether a global (weak) action selector for (M, ω) exists; see 3. (iii) in Appendix A.

EXAMPLES OF (WEAK) ACTION SELECTORS.

1. $(\mathbb{R}^{2n}, \omega_0)$. Let $\omega_0 = \sum_i dq_i \wedge dp_i$ be the standard symplectic form on $\mathbb{R}^{2n}$. Action selectors σ_{V} and σ_{HZ} for $(\mathbb{R}^{2n}, \omega_0)$ have been constructed by Viterbo in [**43**] and by Hofer and Zehnder in [**20, 24**]. The constructions of σ_{V} and σ_{HZ} are outlined in Appendix A.

2. Weakly exact closed symplectic manifolds. After work on selector-like invariants for standard cotangent bundles over a closed base by Oh [**31, 32**], an action selector σ_{PSS} was constructed for weakly exact closed symplectic manifolds by Schwarz in [**39**] by making use of the Piunikhin–Salamon–Schwarz isomorphism. Examples of such manifolds are (products of) closed surfaces of positive genus. The construction of σ_{PSS} is outlined in Appendix A.

3. Weakly exact convex symplectic manifolds. A compact symplectic manifold (M, ω) with boundary ∂M is said to be *convex* if there exists a Liouville vector field X (i.e., $\mathcal{L}_X\omega = d\iota_X\omega \overset{!}{=} \omega$) which is defined near ∂M and is everywhere transverse to ∂M, pointing outward. A non-compact symplectic manifold (M, ω) is *convex* if it admits an exhaustion by compact convex submanifolds. Examples of weakly exact convex symplectic manifolds are cotangent bundles (T^*B, ω_0) over a closed base B endowed with the standard symplectic form $\omega_0 = \sum_i dp_i \wedge dq_i$, and unit-ball bundles therein, and, more generally, Stein manifolds and Stein domains, see 3. (i) in Appendix A. Other examples are twisted cotangent bundles over closed orientable surfaces of genus at least 2, see [**9, 11**]. Building on [**39**] and [**45**], an action selector σ_{PSS} for (exhaustions of) weakly exact convex symplectic manifolds was constructed in [**9**].

4. Rational strongly semi-positive closed symplectic manifolds. A $2n$-dimensional symplectic manifold (M, ω) is *strongly semi-positive* if it satisfies one of the following conditions.

(SP1) $\omega(A) = \lambda\, c_1(A)$ for every $A \in \pi_2(M)$ where $\lambda \geq 0$;

(SP2) $c_1(A) = 0$ for every $A \in \pi_2(M)$;

(SP3) the minimal Chern number $N \geq 0$ defined by $c_1(\pi_2(M)) = N\mathbb{Z}$ is at least $n - 1$.

Here, $c_1 = c_1(\omega)$ is the first Chern class of any almost complex structure on TM compatible with ω, see [**28**, Section 4.1]. Examples are symplectic manifolds of dimension ≤ 4, complex Grassmannians, and Calabi–Yau manifolds. A symplectic manifold (M, ω) is *rational* if $\omega(\pi_2(M))$ is a discrete subset of $\mathbb{R}$. The action selector σ_{PSS} from [**39**] for weakly exact closed symplectic manifolds has been extended to a *weak* action selector σ_{PSS} for rational strongly semi-positive closed symplectic manifolds in [**29**].

5. Rational closed symplectic manifolds. A construction of an action selector σ_{Oh} for all rational closed symplectic manifolds was proposed by Oh in [**33, 34, 35**].

REMARKS 1.3.

(1) In view of the action selector σ_{Oh}, our concept of a weak action selector appears to be superfluous. The verification of (AS2$^+$) for σ_{Oh} is, however, difficult already for rational strongly semi-positive closed symplectic manifolds. We thus find it interesting to see which results can be formally derived from the existence of a weak action selector.

(2) We have chosen a minimal set of axioms for a (weak) action selector required for its applications that we have in mind: The existence of a (weak) action selector for a given symplectic manifold will imply Theorems 1 and 2 below. The selector-like invariants constructed in [**31, 32**] and the (weak) action selectors σ_V, σ_{HZ}, σ_{PSS} and σ_{Oh} have many further properties, and they lead to many other results in Hamiltonian dynamics [**9, 10, 12, 24, 31, 32, 34, 35, 37, 39, 43**] as well as to some insight into the algebraic structure of the groups of Hamiltonian and symplectic diffeomorphisms of certain closed symplectic manifolds [**7, 8, 37**]. An example of an additional property is sub-additivity:

$$\sigma(H\#K) \leq \sigma(H) + \sigma(K) \quad \text{for all } H, K \in \mathcal{H}.$$

This together with (AS3) is stronger than the axioms (AS3) and (AS5), which are sufficient for our purposes. Sub-additivity holds for the selectors σ_V, σ_{PSS} and σ_{Oh}, but has not been established for σ_{HZ}. An additional property shared by all known (weak) action selectors and useful for intuition is the monotonicity property

$$\sigma(H) \leq \sigma(K) \quad \text{for all } H, K \in \mathcal{H} \text{ with } H \leq K.$$

OPEN PROBLEMS 1.4.

(1) It is not known whether every (exhaustion of a non-compact) symplectic manifold admits a (weak) action selector. The simplest symplectic manifold for which no weak action selector has yet been constructed is the convex strongly semi-positive symplectic manifold $\left(T^*S^2, \omega_0 + \pi^*\sigma\right)$, where σ is an area form on S^2.

(2) Is every weak action selector an action selector?

(3) Do axioms (AS1)–(AS5) uniquely determine (weak) action selectors? This would imply that every (weak) action selector for an exhaustion of (M, ω) fits together to a (weak) action selector for (M, ω). More specifically, the action selectors σ_V, σ_{HZ}, σ_{PSS} and σ_{Oh} are all defined by a variational procedure (see Appendix A), and so it should be possible to compare them: Do σ_V, σ_{HZ} and σ_{PSS} agree on $\left(\mathbb{R}^{2n}, \omega_0\right)$? Does σ_{PSS} agree with σ_{Oh} on weakly exact or strongly semi-positive closed symplectic manifolds?

1.2. A sharp energy capacity inequality. Given a symplectic manifold (M, ω) one can associate to each subset A of M various symplectic invariants.

1. The Gromov width. The Gromov width of A is defined as

$$c_G(A) = \sup\left\{\pi r^2 \mid B^{2n}(r) \text{ symplectically embeds into } (M, \omega)\right\}.$$

Here, $B^{2n}(r)$ denotes the open ball in $\left(\mathbb{R}^{2n}, \omega_0\right)$ of radius r. The Gromov width, which was introduced by Gromov in [**15**], measures the symplectic size of (M, ω) in a geometric way; it corresponds to the injectivity radius of a Riemannian manifold. Darboux's theorem states that every point of a symplectic manifold has an open neighborhood which is symplectomorphic to some ball $B^{2n}(r)$, and hence $c_G(A)$ vanishes only if A has empty interior.

2. Hofer–Zehnder capacities. Hofer–Zehnder capacities, in contrast with the Gromov width, measure the symplectic size of a set from the perspective of Hamiltonian dynamics on this set. We consider two variants. For each subset $A \subset M$ let $\mathcal{S}(A)$ be the set of simple functions in $\mathcal{H}(A)$. We say that a function $H \in \mathcal{S}(A)$ is HZ-*admissible* if the flow φ_H^t has no non-constant T-periodic orbit with period $T \leq 1$, and as before $H \in \mathcal{S}(A)$ is HZ$^\circ$-*admissible* if the flow φ_H^t has no non-constant T-periodic orbit with period $T \leq 1$ which is contractible in M. Set

$$\begin{aligned} \mathcal{S}_{\mathrm{HZ}}(A) &= \{H \in \mathcal{S}(A) \mid H \text{ is HZ-admissible}\}, \\ \mathcal{S}_{\mathrm{HZ}}^\circ(A, M) &= \{H \in \mathcal{S}(A) \mid H \text{ is HZ}^\circ\text{-admissible}\}. \end{aligned}$$

Following [**23, 24**] and [**26, 39**], we define the Hofer–Zehnder capacity and the π_1-sensitive Hofer–Zehnder capacity of $A \subset (M, \omega)$ as

$$\begin{aligned} c_{\mathrm{HZ}}(A) &= \sup\{\max H \mid H \in \mathcal{S}_{\mathrm{HZ}}(A)\}, \\ c_{\mathrm{HZ}}^\circ(A, M) &= \sup\{\max H \mid H \in \mathcal{S}_{\mathrm{HZ}}^\circ(A, M)\}. \end{aligned}$$

Both $c_{\mathrm{HZ}}(A)$ and $c_{\mathrm{HZ}}^\circ(A, M)$ vanish if and only if A has empty interior.

REMARKS 1.5.

(1) The definition of the Hofer–Zehnder capacities $C_{\mathrm{HZ}}(A)$ and $C_{\mathrm{HZ}}^\circ(A, M)$ given in [**23, 24**] and [**26, 39**] uses the larger class

$$\mathcal{F}(A) = \{H \in \mathcal{H}(A) \mid H \text{ satisfies (P1) and (P2)}\}.$$

Of course, $c_{\mathrm{HZ}}(A) \leq C_{\mathrm{HZ}}(A)$ and $c_{\mathrm{HZ}}^\circ(A, M) \leq C_{\mathrm{HZ}}^\circ(A, M)$. It follows from [**23**] that equalities hold for convex subsets of $(\mathbb{R}^{2n}, \omega_0)$, and we do not know examples with $c_{\mathrm{HZ}}(A) < C_{\mathrm{HZ}}(A)$ or $c_{\mathrm{HZ}}^\circ(A, M) < C_{\mathrm{HZ}}^\circ(A, M)$.

(2) The main virtue of the Hofer–Zehnder capacity $C_{\mathrm{HZ}}(A)$ is that $C_{\mathrm{HZ}}(A) < \infty$ implies almost existence of periodic orbits near any compact regular energy level of an autonomous Hamiltonian system on A, and similarly for $C_{\mathrm{HZ}}^\circ(A, M)$. As we shall show in Appendix B, this continues to hold for $c_{\mathrm{HZ}}(A)$ and $c_{\mathrm{HZ}}^\circ(A, M)$.

(3) Corollary 1 below remains valid for C_{HZ} and C_{HZ}° when (AS2$^+$) is required to hold for the larger classes $\mathcal{F}_{\mathrm{HZ}}(A)$ and $\mathcal{F}_{\mathrm{HZ}}^\circ(A, M)$ of Hofer–Zehnder admissible functions from $\mathcal{F}(A)$. The action selectors σ_{V}, σ_{HZ}, σ_{PSS} and σ_{Oh} from Examples 1, 2, 3 and 5 do satisfy this stronger axiom.

3. Displacement energy. An invariant with both geometric and dynamical features is the *displacement energy* introduced in [**19, 25**]. The Hofer norm $\|H\|$ of $H \in \mathcal{H}$ is defined as

$$\|H\| = \int_0^1 \left(\sup_{x \in M} H(t, x) - \inf_{x \in M} H(t, x) \right) dt,$$

and the displacement energy $e(A, M) = e(A, M, \omega) \in [0, \infty]$ is defined as

$$e(A, M) = \inf\{\|H\| \mid H \in \mathcal{H}, \varphi_H(A) \cap A = \emptyset\}$$

if A is compact and as

$$e(A, M) = \sup\{e(K, M) \mid K \subset A \text{ is compact}\}$$

for a general subset A of M.

Since the invariants c_{G}, c_{HZ} and c°_{HZ}, and e are defined in different ways, relations between them lead to many applications. It is easy to see that $c_{\mathrm{G}}(A) \le c_{\mathrm{HZ}}(A)$, see e.g. [**24**], and it follows from definitions that $c_{\mathrm{HZ}}(A) \le c^{\circ}_{\mathrm{HZ}}(A,M)$. In order to compare $c^{\circ}_{\mathrm{HZ}}(A,M)$ with $e(A,M)$, we introduce further invariants. Following [**43**], we make the

DEFINITION 1.6. Assume that (M,ω) admits a weak action selector σ. For each subset A of M the *spectral capacities* $c_\sigma(A,M)$ and $c^\sigma(A,M)$ are defined as

$$\begin{aligned} c_\sigma(A,M) &= \sup\{\sigma(H) \mid H \in \mathcal{S}(A)\}, \\ c^\sigma(A,M) &= \sup\{\sigma(H) \mid H \in \mathcal{H}(I \times A)\}. \end{aligned}$$

Of course, $c_\sigma(A,M) \le c^\sigma(A,M)$, and $c_\sigma(A,M) = c^\sigma(A,M)$ provided that σ is monotone and that $A \subsetneq M$ if M is closed. If M is closed, $c^\sigma(M,M) = \infty$, because then $\mathcal{H}(I \times M)$ contains the constant functions and $\sigma(const) = const$. (Indeed, $\sigma(0) \le 0$ by (AS3) and $\sigma(0) \ge 0$ by (AS2) and (AS4), so that $\sigma(0) = 0$. This, together with (AS1), (AS4) and the fact that $\omega(\pi_2(M))$ is countable, implies that $\sigma(const) = const$.) Since simple Hamiltonians have minimum 0, this argument does not apply to c_σ. However, for all closed M for which we know $c_\sigma(M,M)$, even this capacity is infinite.

The following result, proved in Section 2, is an elaboration of an observation in [**13**].

THEOREM 1. *Consider a symplectic manifold (M,ω) and an arbitrary subset A of M.*

(i) *If σ is a weak action selector for (M,ω), then*

$$c_\sigma(A,M) \le c^\sigma(A,M) \le e(A,M).$$

(ii) *If σ is an action selector, then $c^{\circ}_{\mathrm{HZ}}(A,M) \le c_\sigma(A,M)$, so that*

$$c_{\mathrm{G}}(A) \le c_{\mathrm{HZ}}(A) \le c^{\circ}_{\mathrm{HZ}}(A,M) \le c_\sigma(A,M) \le c^\sigma(A,M) \le e(A,M).$$

COROLLARY 1. *Assume that an exhaustion of (M,ω) admits an action selector. Then*

$$c_{\mathrm{G}}(A) \le c_{\mathrm{HZ}}(A) \le c^{\circ}_{\mathrm{HZ}}(A,M) \le e(A,M)$$

for every subset A of M.

PROOF OF COROLLARY 1. The first two inequalities in the corollary are clear. Let (M_i) be an exhaustion of (M,ω) with action selectors σ_i. For every i with $A \subset M_i$, Theorem 1 (ii) yields

$$c^{\circ}_{\mathrm{HZ}}(A,M_i) \le c_\sigma(A,M_i) \le e(A,M_i),$$

and so we readily find $c^{\circ}_{\mathrm{HZ}}(A,M) \le e(A,M)$. □

REMARK 1.7.

(1) Theorem 1 and Corollary 1 are sharp. Indeed, let

$$\varphi \colon B^{2n}(3r) \hookrightarrow (M,\omega)$$

be a Darboux ball. For $A = \varphi\left(B^{2n}(r)\right)$ we have $c_G(A) = e(A,M) = \pi r^2$.

(2) It would be interesting to know whether in the situation of Theorem 1 (ii) there exists an example with $c_{\mathrm{HZ}}^{\circ}(A,M) < c_\sigma(A,M)$. A more specific version of this problem is posed in 1.12 below.

(3) (i) The assertion of Corollary 1 was first obtained by Hofer, [**20**], for $(\mathbb{R}^{2n},\omega_0)$, see also [**24**, Section 5.5]; in fact, our axioms for a (weak) action selector are extracted from [**24**], and our proof of Theorem 1 closely follows [**24**]. Later on, the energy-capacity inequality

$$c_{\mathrm{HZ}}^{\circ}(A,M) \le 2\,e\,(A,M)$$

was established for subsets of weakly exact symplectic manifolds which are closed [**39**] or convex [**9**], and it was pointed out in [**13**] how to remove the factor 2 in this inequality at least for open manifolds M.

(ii) The energy-capacity inequality $c_G(A) \le 2\,e(A,M)$ was proved in [**25**] for every subset A of *any* symplectic manifold (M,ω). This inequality implies that the Hofer norm on the group of compactly supported Hamiltonian diffeomorphisms is non-degenerate. In view of Corollary 1, it is conceivable that the factor 2 can always be omitted. ◊

Let $Z^{2n}(r)$ be the standard symplectic cylinder $B^2(r)\times\mathbb{R}^{2n-2}\subset(\mathbb{R}^{2n},\omega_0)$. Another consequence of Theorem 1 (ii) is

COROLLARY 1.8. *Assume that σ is an action selector for $(\mathbb{R}^{2n},\omega_0)$. Then for all $r>0$,*

$$c_\sigma\left(B^{2n}(r)\right)=c_\sigma\left(Z^{2n}(r)\right)=\pi r^2 \quad \textit{and} \quad c^\sigma\left(B^{2n}(r)\right)=c^\sigma\left(Z^{2n}(r)\right)=\pi r^2.$$

1.3. An estimate for the smallest spectral value. A *hypersurface* S in (M,ω) is a smooth compact connected orientable codimension 1 submanifold without boundary contained in $M\setminus\partial M$. A *closed characteristic* on S is an embedded circle in S all of whose tangent lines belong to the distinguished line bundle

$$\mathcal{L}_S = \{(x,\xi)\in TS \mid \omega(\xi,\eta)=0 \text{ for all } \eta\in T_xS\}.$$

Denote by $\mathcal{P}(S)$ the set of closed characteristics on S, and by $\mathcal{P}^\circ(S)$ the set of those closed characteristics on S which are contractible in M. For $x\in\mathcal{P}^\circ(S)$ let $\mathcal{D}(x)$ be the set of smooth discs $\bar{x}\colon D^2\to M$ with boundary x. The *contractible action spectrum* of S is the set

$$\Sigma^\circ(S) = \left\{ \int_{\bar{x}}\omega \;\middle|\; x\in\mathcal{P}^\circ(S),\, \bar{x}\in\mathcal{D}(x)\right\}.$$

If $\omega|_S = d\lambda$ for some 1-form λ on S, the *full action spectrum* is defined as

$$\Sigma(S,\lambda) = \left\{ \int_x \lambda \;\middle|\; x\in\mathcal{P}(S)\right\}.$$

Note that $\Sigma(S,\lambda)$ is independent of the choice of λ if $H^1(S;\mathbb{R})=0$. The sets $\Sigma^\circ(S)$ and $\Sigma(S,\lambda)$ are important collections of numerical invariants of S, see [**4, 5**]. Here, we are interested in the "smallest" spectral value. If $\mathcal{P}^\circ(S)$ or $\mathcal{P}(S)$ is non-empty, we define

$$\alpha_1^\circ(S)=\inf\{|\alpha| \mid \alpha\in\Sigma^\circ(S)\} \quad\text{and}\quad \alpha_1(S,\lambda)=\inf\{|\alpha|\mid\alpha\in\Sigma(S,\lambda)\}.$$

If (M,ω) is not rational, then $\alpha_1^\circ(S) = 0$ for any hypersurface $S \subset (M,\omega)$ with $\mathcal{P}^\circ(S) \neq \emptyset$, and our results for $\alpha_1(S,\lambda)$ will deal with hypersurfaces in weakly exact symplectic manifolds. We shall thus assume in this paragraph that (M,ω) is rational. We shall also assume that S is a hypersurface of *contact type*. This means that there exists a Liouville vector field X which is defined near S and is transverse to S. Equivalently, there exists a contact form λ on S (i.e., a 1-form λ such that $d\lambda = \omega|_S$ and $\lambda \wedge (d\lambda)^{n-1}$ is a volume form on S). The equivalence is given by $\iota_X\omega = \lambda$. A hypersurface S is said to be of *restricted contact type* if it is transverse to a Liouville vector field X defined *on all* of M. The contact form $\lambda = \iota_X\omega$ is then globally defined and $d\lambda = d\iota_X\omega = \omega$ so that (M,ω) is exact. If (M,ω) is exact, then every hypersurface S of contact type with $H^1(S;\mathbb{R}) = 0$ is of restricted contact type.

EXAMPLES 1.9.

(1) Consider a hypersurface S of $\left(\mathbb{R}^{2n},\omega_0\right)$. It bounds a bounded domain U. Then

$$\begin{aligned} & U \text{ is convex} \\ \Longrightarrow\ & U \text{ is starshaped} \\ \Longrightarrow\ & S \text{ is of restricted contact type} \\ \Longrightarrow\ & S \text{ is of contact type.} \end{aligned}$$

Examples show that none of these arrows can be inverted, see [**2, 18**].

(2) Let T^*B be the cotangent bundle over a closed base endowed with the symplectic form $\omega_0 = d\lambda_0$, where $\lambda_0 = \sum_i p_i dq_i$.

(i) Assume that B is endowed with a Riemannian metric. Any regular energy level $S_c = \{H = c\}$ of a classical Hamiltonian $H(q,p) = \frac{1}{2}|p|^2 + V(q)$ is of contact type, see [**1**, Theorem 1.2.2], and if $c > \max V$, then S_c is of restricted contact type, see (iii) below. If $c > \max V$, then $\Sigma(S_c) \neq \emptyset$, and if $c < \max V$, then $\Sigma^\circ(S_c) \neq \emptyset$; see [**24**, p. 131] for a brief history of this existence problem and for further references.

(ii) Assume that S is a hypersurface of contact type in (T^*B,ω_0) and that $\dim B \geq 2$. Then $\Sigma(S) \neq \emptyset$ if the bounded component of $T^*B \setminus S$ contains B, see [**21**], and $\Sigma^\circ(S) \neq \emptyset$ if B is simply connected, [**44**].

(iii) Assume that S is a hypersurface in T^*B such that for each $q \in B$ the intersection $S \cap T_q^*B$ bounds a strictly convex domain in T_q^*B containing 0. Then S is transverse to the Liouville vector field $X(q,p) = \sum_i p_i \frac{\partial}{\partial p_i}$, and hence S is of restricted contact type. Moreover, there exists a unique Finsler metric $F\colon TB \to \mathbb{R}$ on B such that S is the unit cosphere bundle $\{x \in T^*B \mid F^*(x) = 1\}$, where $F^*\colon T^*B \to \mathbb{R}$ is the "dual norm" in each fibre, defined by

$$F^*(q,p) = \sup_{0 \neq \dot{q} \in T_qB} \frac{p(\dot{q})}{F(q,\dot{q})},$$

see [**4**, Section 4.a)]. The projection $\pi\colon T^*B \to B$ induces a bijection between $\mathcal{P}(S)$ and the non-empty set of prime geodesics in the Finsler metric F, and the action $\left|\int_x \lambda_0\right|$ of $x \in \mathcal{P}(S)$ equals the F-length of $\pi(x)$. In particular, $\alpha_1(S,\lambda_0)$ is the length of the shortest closed F-geodesic on B.

(3) Let S be a smoothly embedded loop in the 2-sphere S^2 endowed with an area form. Then S is of contact type, but not of restricted contact type. It bounds two discs of areas a_1 and a_2, and $\alpha_1^\circ(S) = \min(a_1, a_2)$. ◊

The following two propositions seem to be well known; proofs are given for the reader's convenience in Appendix C.

PROPOSITION 1.10. *Assume that S is a hypersurface of contact type in a weakly exact symplectic manifold (M, ω).*

(i) *If S is simply connected or if S is of restricted contact type, then $0 \notin \Sigma^\circ(S)$ and $\Sigma^\circ(S)$ is closed. If $\mathcal{P}^\circ(S) \neq \emptyset$, we thus have $\alpha_1^\circ(S) > 0$.*

(ii) *If S is of contact type, then $0 \notin \Sigma(S, \lambda)$ and $\Sigma(S, \lambda)$ is closed for any contact form λ on S. Thus, $\alpha_1(S, \lambda) > 0$ if $\mathcal{P}(S) \neq \emptyset$.*

PROPOSITION 1.11. *Assume that (M, ω) is a symplectic manifold, and that $U \subset M$ is a relatively compact domain whose boundary S is a hypersurface of restricted contact type. If $\mathcal{P}^\circ(S) \neq \emptyset$, then*

$$0 < \alpha_1^\circ(S) \leq c_{\mathrm{HZ}}^\circ(U, M),$$

and if $\mathcal{P}(S) \neq \emptyset$, then

$$0 < \alpha_1(S, \lambda) \leq c_{\mathrm{HZ}}(U)$$

for any globally defined contact form λ.

For a convex bounded domain with smooth boundary in $(\mathbb{R}^{2n}, \omega_0)$, it holds that $\alpha_1(S) = c_{\mathrm{HZ}}(U)$, see [**23**], and for the full Bordeaux bottle (which is star–shaped) it holds that $\alpha_1(S) < c_{\mathrm{HZ}}(U)$, see [**24**, p. 99].

OPEN PROBLEM 1.12. Assume that a hypersurface $S \subset B^4(1) \subset (\mathbb{R}^{2n}, \omega_0)$ bounds a star-shaped domain U. Then $c_{\mathrm{HZ}}(U) \leq c_\sigma(U)$ for any action selector σ for $B^4(1)$ in view of Theorem 1 (ii). Does equality hold for σ_V, σ_{HZ} and σ_{PSS}? Since it is known that $c_\sigma(U)$ belongs to $\Sigma^\circ(S)$ for every monotone action selector σ, the first step toward a solution of this problem would be to see whether $c_{\mathrm{HZ}}(U) \in \Sigma^\circ(S)$. ◊

Our next goal is to find upper bounds of $\alpha_1^\circ(S)$ for contact type hypersurfaces S which might not be of restricted contact type and which might not bound a domain U. Since the interior of a hypersurface S is empty, $c_\sigma(S, M) = 0$. We thus consider the *outer spectral capacity*

$$\hat{c}_\sigma(S, M) = \inf\{c_\sigma(U, M) \mid S \subset U,\, U \text{ open in } M\}.$$

The *index of rationality* $\rho(M, \omega) \in (0, \infty]$ of a rational symplectic manifold (M, ω) is set ∞ if (M, ω) is weakly exact and is defined to be the positive generator of the cyclic subgroup $\omega(\pi_2(M))$ of $\mathbb{R}$ if (M, ω) is not weakly exact. The following result improves Corollary 11.2 in [**9**].

THEOREM 2. *Assume that (M, ω) is a rational symplectic manifold admitting a weak action selector σ, and consider a hypersurface $S \subset (M, \omega)$ of contact type. Then $\hat{c}_\sigma(S, M) \leq e(S, M)$, and if $\hat{c}_\sigma(S, M) < \rho(M, \omega)$, then $\mathcal{P}^\circ(S) \neq \emptyset$ and*

$$\alpha_1^\circ(S) \leq \hat{c}_\sigma(S, M) \leq e(S, M).$$

COROLLARY 2. *Assume that an exhaustion of the rational symplectic manifold (M, ω) admits a weak action selector, and consider a hypersurface S of contact type such that $e(S, M) < \rho(M, \omega)$. Then $\mathcal{P}^\circ(S) \neq \emptyset$ and $\alpha_1^\circ(S) \leq e(S, M)$.*

REMARKS 1.13.

(1) Again, Theorem 2 and Corollary 2 are sharp: For the circle $S \subset (\mathbb{R}^2, \omega_0)$ bounding $B^2(r)$ we have $\alpha_1^\circ(S) = e(S, \mathbb{R}^2) = \pi r^2$.
(2) Theorem 2 and Corollary 2 establish the Weinstein conjecture for a large class of hypersurfaces of contact type. We refer to [**13, 38**] for the state of the art of the Weinstein conjecture.
(3) Consider a contact hypersurface S in $(\mathbb{R}^{2n}, \omega_0)$. Then S is contained in a ball of radius $\operatorname{Diam}(S)$, where $\operatorname{Diam}(S)$ is the diameter of S. Since $e\left(B^{2n}(r), \mathbb{R}^{2n}\right) = \pi r^2$, we find $e(S, \mathbb{R}^{2n}) \leq \pi \operatorname{Diam}(S)^2$, and hence

$$\alpha_1(S) \leq \pi \operatorname{Diam}(S)^2,$$

improving the estimates in [**9, 22**]. ◊

We finally consider a billiard table $\overline{U} \subset \mathbb{R}^n$ with smooth boundary, i.e., U is a bounded domain in $\mathbb{R}^n(q)$ with smooth connected boundary. The length of a billiard trajectory on $\overline{U}$ is measured with respect to the Euclidean length, and the Euclidean volume is $\operatorname{vol}(U)$. Let D^n be the closed unit ball in $\mathbb{R}^n(p)$. Combining a generalization of Corollary 2 with the construction in [**46**], we shall obtain the following result of Viterbo, whose proof in [**46**] uses the fact that $\overline{U} \times D^n \subset (\mathbb{R}^{2n}, \omega_0)$ can be approximated by domains with boundary of restricted contact type.

PROPOSITION 1.14 (Viterbo). *Let U be a bounded domain in $\mathbb{R}^n$ with smooth boundary. Then there exists a periodic billiard trajectory on $\overline{U}$ of length*

$$l \leq e\left(\overline{U} \times D^n, \mathbb{R}^{2n}\right) \leq C_n \left(\operatorname{vol}(U)\right)^{1/n},$$

where C_n is a constant depending only on n.

Acknowledgements. We are grateful to Peter Albers, David Hermann and Matthias Schwarz for most helpful discussions. We thank Dietmar Salamon for sending us Chapter 12 of the forthcoming book [**29**]. Part of this note was written when the third author was visiting Hokkaido University; he wishes to thank Kaoru Ono for his warm hospitality. The first and third author thank Hokkaido University and Leipzig University for their support.

2. Proofs

2.1. Proof of Theorem 1. (i) Assume that (M, ω) is a symplectic manifold admitting a weak action selector σ. The support $\operatorname{supp} H$ of $H \in \mathcal{H}$ is defined as the closure of the set

$$\bigcup_{t \in [0,1]} \{x \in M \mid H_t(x) \neq 0\}.$$

The main ingredient in the proof of Theorem 1 is

PROPOSITION 2.1. *Assume that $H, K \in \mathcal{H}$ are such that φ_K displaces* $\operatorname{supp} H$. *Then $\sigma(H) \leq \|K\|$.*

PROOF. We follow the argument in [**9, 13, 24, 39**]. Given a smooth function $\lambda\colon [0,1] \to [0,1]$ with $\lambda(0) = 0$ and $\lambda(1) = 1$, we define $H^\lambda \in \mathcal{H}$ as

$$H^\lambda(t,x) = \lambda'(t)\, H(\lambda(t), x).$$

LEMMA 2.2. $\sigma\left(H^{\lambda}\right)=\sigma(H)$.

PROOF. The Hamiltonian flow of H^{λ} is a reparametrization of the flow of H:

$$\varphi_{H^{\lambda}}^{t}(x) = \varphi_{H}^{\lambda(t)}(x), \quad x \in M.$$

Since $\lambda(0) = 0$ and $\lambda(1) = 1$, this reparametrization gives rise to a one-to-one correspondence between one-periodic orbits of the flows: the orbit $x^{\lambda} \in \mathcal{P}^{\circ}\left(H^{\lambda}\right)$ corresponds to $x \in \mathcal{P}^{\circ}(H)$ when

$$x^{\lambda}(t) = x(\lambda(t)).$$

Moreover,

$$\int_0^1 H^{\lambda}\left(t, x^{\lambda}(t)\right) dt = \int_0^1 \lambda'(t)\, H(\lambda(t), x(\lambda(t)))\, dt = \int_0^1 H(t, x(t))\, dt,$$

and hence $\mathcal{A}_{H^{\lambda}}\left(\bar{x}^{\lambda}\right) = \mathcal{A}_H(\bar{x})$ for all $(x, \bar{x}) \in \bar{\mathcal{P}}^{\circ}(H) \cong \bar{\mathcal{P}}^{\circ}\left(H^{\lambda}\right)$. Consider now the smooth family of functions $\lambda_{\tau}\colon [0,1] \to [0,1]$ given by

$$\lambda_{\tau}(t) = (1-\tau)t + \tau\lambda(t), \quad \tau \in [0,1].$$

Then $H^{\lambda_0} = H$ and $H^{\lambda_1} = H^{\lambda}$. By the above, $\Sigma^{\circ}\left(H^{\lambda_{\tau}}\right) = \Sigma^{\circ}(H)$ for all $\tau \in [0,1]$, and according to [**24, 39**] and [**33**, Lemma 2.2], the Lebesgue measure of this set vanishes. In view of (AS4), the map

$$[0,1] \to \Sigma^{\circ}(H), \quad \tau \mapsto \sigma\left(H^{\lambda_{\tau}}\right)$$

is constant. In particular, $\sigma\left(H^{\lambda}\right) = \sigma(H)$. □

Utilizing Lemma 2.2, we can assume that $H_t = 0$ for $t \in [0, 1/2]$ and $K_t = 0$ for $t \in [1/2, 1]$. With this parametrization,

$$(\tau H \# K)(t,x) = \tau H(t,x) + K(t,x)$$

for all $\tau \in [0,1]$ and $(t,x) \in I \times M$, and the time-1-flow of $\tau H \# K$ is the time-1-flow of K followed by the time-1-flow of τH. Since φ_K displaces $\operatorname{supp} \tau H = \operatorname{supp} H$ for each $\tau \in (0,1]$, it follows that $x(0) \notin \operatorname{supp} \tau H$ for each $x \in \mathcal{P}^{\circ}(K)$ and each $x \in \mathcal{P}^{\circ}(\tau H \# K)$, and we conclude that $\mathcal{P}^{\circ}(\tau H \# K) = \mathcal{P}^{\circ}(K)$. Therefore, $\Sigma^{\circ}(\tau H \# K) = \Sigma^{\circ}(K)$ for each $\tau \in [0,1]$, and arguing as above we find $\sigma(H \# K) = \sigma(K)$. The inverse φ_K^{-1} of φ_K is generated by $K^{-}(t,x) = -K\left(t, \varphi_K^t(x)\right)$. Notice that $K \# K^{-} = 0$. Combining this with (AS5), we conclude

$$\sigma(H) = \sigma\left(H \# K \# K^{-}\right) \leq \sigma(H \# K) + E^{+}\left(K^{-}\right) = \sigma(K) + E^{+}\left(K^{-}\right).$$

Using

$$\max_{x \in M} K^{-}(t,x) = \max_{x \in M}\left(-K\left(t, \varphi_K^t(x)\right)\right) = -\min_{x \in M} K(t,x)$$

and (AS3), we finally obtain

$$\begin{aligned} \sigma(H) \leq \sigma(K) + E^{+}\left(K^{-}\right) &\leq \int_0^1 \left(\max_{x \in M} K(t,x) - \min_{x \in M} K(t,x)\right) dt \\ &= \|K\|, \end{aligned}$$

as desired. □

Consider again a symplectic manifold (M,ω) admitting a weak action selector σ, and let $A \subset M$. In order to show that $c^{\sigma}(A,M) \le e(A,M)$, we can assume that $e(A,M) < \infty$. Fix $\delta > 0$ and choose $K \in \mathcal{H}$ such that $\|K\| \le e(A,M) + \delta$ and φ_K displaces A. In particular, φ_K displaces $\operatorname{supp} H$ for any $H \in \mathcal{H}(I \times A)$. Proposition 2.1 thus yields $\sigma(H) \le \|K\| \le e(A,M) + \delta$. Taking the supremum over $H \in \mathcal{H}(I \times A)$, we find $c^{\sigma}(A,M) \le e(A,M) + \delta$, and since $\delta > 0$ is arbitrary, the claim follows.

(ii) Assume that σ is an action selector for (M,ω). In order to show that $c^{\circ}_{\mathrm{HZ}}(A,M) \le c_{\sigma}(A,M)$, we need to prove that $\max H \le \sigma(H)$ for every $H \in \mathcal{S}^{\circ}_{\mathrm{HZ}}(A,M)$, and this holds by (AS2$^+$).

2.2. Proof of Theorem 2. We consider an arbitrary hypersurface S of a symplectic manifold (M,ω). Examples show that $\Sigma^{\circ}(S)$ can be empty, see [**12, 14**]. We therefore follow [**22**] and consider parametrized neighborhoods of S. Since S is orientable and contained in $M \setminus \partial M$, there exists an open neighborhood I of 0 and a smooth diffeomorphism

$$\psi\colon S \times I \to U \subset M$$

such that $\psi(x,0) = x$ for $x \in S$. We call ψ a *thickening* of S and set $S_{\epsilon} = \psi(S \times \{\epsilon\})$.

THEOREM 2.3. *Assume that (M,ω) is a rational symplectic manifold admitting a weak action selector σ, and consider a hypersurface S of (M,ω). Then $\hat{c}_{\sigma}(S,M) \le e(S,M)$. If $\hat{c}_{\sigma}(S,M) < \rho(M,\omega)$, then for every thickening ψ of S and every $\delta > 0$ there exists $\epsilon \in [-\delta,\delta]$ such that*

$$\mathcal{P}^{\circ}(S_{\epsilon}) \neq \emptyset \quad \textit{and} \quad \alpha_1(S_{\epsilon}) \le \hat{c}_{\sigma}(S,M) + \delta \le e(S,M) + \delta.$$

PROOF. By Theorem 1 (i), $c_{\sigma}(U,M) \le e(U,M)$ for every open neighborhood U of S, and hence the first claim follows from taking the infimum. In order to prove the second claim, we can assume that $\delta > 0$ is so small that

$$C := \hat{c}_{\sigma}(S,M) + \delta < \rho := \rho(M,\omega).$$

Let $\tau \in (0,\delta)$ be so small that for the neighborhood $U_{\tau} = \psi(S \times (-\tau,\tau))$ of S we have $c_{\sigma}(U_{\tau},M) < C$. We choose a smooth function $f\colon \mathbb{R} \to [0,C]$ such that

$$\begin{aligned} f(t) &= 0 && \text{if } t \notin \left]-\tfrac{\tau}{2},\tfrac{\tau}{2}\right[, \\ f(t) &= C && \text{if } t \in \left[-\tfrac{\tau}{4},\tfrac{\tau}{4}\right], \\ |f'(t)| &\neq 0 && \text{if } t \in \left]-\tfrac{\tau}{2},-\tfrac{\tau}{4}\right[\cup \left]\tfrac{\tau}{4},\tfrac{\tau}{2}\right[. \end{aligned}$$

Define $H \in \mathcal{S}(M)$ by

$$H(x) = \begin{cases} f(t) & \text{if } x \in S_t, \\ 0 & \text{otherwise.} \end{cases}$$

By (AS2), we have $\sigma(H) > 0$, so that

$$0 < \sigma(H) \le c_{\sigma}(U_{\tau},M) < C < \rho. \tag{2.1}$$

By (AS1), there exists $(x,\bar{x}) \in \bar{\mathcal{P}}^{\circ}(H)$ such that $\sigma(H) = \mathcal{A}_H(\bar{x})$.

LEMMA 2.4. *The orbit x is not constant.*

PROOF. If x is constant, our choice of H yields $H(x) \in \{0, C\}$. If $H(x) = 0$, then

$$0 < \sigma(H) = \mathcal{A}_H(\bar{x}) = -\int_{\bar{x}} \omega < C, \tag{2.2}$$

and if $H(x) = C$, then

$$0 < \sigma(H) = \mathcal{A}_H(\bar{x}) = -\int_{\bar{x}} \omega + C < C. \tag{2.3}$$

Since $\bar{x}$ is a sphere, $\int_{\bar{x}} \omega \in \rho\mathbb{Z}$, and so both (2.2) and (2.3) contradict $C < \rho$. □

By construction of H and by Lemma 2.4, there exists $\epsilon \in (-\tau, \tau) \subset [-\delta, \delta]$ such that $(x, \bar{x}) \in \bar{\mathcal{P}}^\circ(S_\epsilon)$ and $0 < f(\epsilon) < C$ and

$$0 < \sigma(H) = -\int_{\bar{x}} \omega + f(\epsilon) < C.$$

We conclude that $\left|\int_{\bar{x}} \omega\right| \leq C$, so that $\alpha_1(S_\epsilon) < \hat{c}_\sigma(S, M) + \delta$, as claimed. □

Theorem 2 is an easy consequence of Theorem 2.3; we refer the reader to [**9**, Section 11] for a detailed argument.

REMARK 2.5. Assume that S is a hypersurface of contact type bounding a domain U. An obvious modification of the proof of Theorem 2.3 shows that Theorem 2 holds for $c_\sigma(U, M)$ as well. It is, however, easy to see that $\hat{c}_\sigma(S, M) \leq c_\sigma(U, M)$.

2.3. Proof of Proposition 2. Let $U \subset \mathbb{R}^n$ be a bounded domain with smooth boundary, and abbreviate $e = e(\overline{U} \times D^n, \mathbb{R}^{2n})$. We choose a smooth family W_t, $t \in (0, 1]$, of bounded domains in $\mathbb{R}^{2n}$ with smooth boundaries S_t such that

$$W_t \subset W_{t'} \text{ if } t \leq t' \quad \text{and} \quad \bigcap_{t \in (0,1]} W_t = \overline{U} \times D^n.$$

In view of Theorem 2.3, we can find a sequence $t_k \to 0$ such that S_{t_k} carries a closed characteristic γ_k with $\int_{\gamma_k} \lambda_0 \leq e + 1/k$. Choose a parametrization $\gamma_k(t) = (q_k(t), p_k(t)) : [0, 1] \to S_{t_k}$ of γ_k. Since $p_k(t) = f_k(t) \frac{\dot{q}_k(t)}{|\dot{q}_k(t)|}$ for a function $f_k \geq 1$,

$$\text{length}(q_k) = \int_0^1 |\dot{q}_k(t)|\, dt \leq \int_0^1 \langle p_k(t), \dot{q}_k(t) \rangle dt = \int_{\gamma_k} \lambda_0 \leq e + \tfrac{1}{k}.$$

We therefore find a subsequence q_{k_m} converging to a billiard trajectory q on U, see [**3**], and length$(q) \leq e$. The estimate $e \leq C_n \left(\text{vol}(U)\right)^{1/n}$ is proved in [**46**]. □

Appendix A. Constructions of action selectors

In this appendix we outline the constructions of the (weak) action selectors σ_{V}, σ_{HZ} and σ_{PSS}.

A.1. σ_{V} and σ_{HZ} for $(\mathbb{R}^{2n}, \omega_0)$.

The selector σ_{V}. The following description is taken from [**17, 43**]. The construction of the action selector σ_{V} makes use of generating functions. For $H \in \mathcal{H}(\mathbb{R}^{2n})$ the graph

$$\Gamma_H = \{(x, \varphi_H(x)) \mid x \in \mathbb{R}^{2n}\}$$

of φ_H is a Lagrangian submanifold of $(\mathbb{R}^{2n} \times \mathbb{R}^{2n}, -\omega_0 \oplus \omega_0)$. Under the symplectomorphism

$$(q, p, Q, P) \mapsto \left(\frac{q+Q}{2}, \frac{p+P}{2}, p-P, Q-q\right)$$

of $(\mathbb{R}^{2n} \times \mathbb{R}^{2n}, -\omega_0 \oplus \omega_0)$ with the cotangent bundle $T^*\Delta$ over the diagonal in $\mathbb{R}^{2n} \times \mathbb{R}^{2n}$, the graph Γ_H is mapped to a Lagrangian submanifold of $T^*\Delta$ which outside a compact set coincides with Δ. Adding a point ∞ to $\Delta \cong \mathbb{R}^{2n} \cong S^{2n} \backslash \{\infty\}$ one obtains a closed Lagrangian submanifold Γ of T^*S^{2n}, which is Hamiltonian isotopic to the zero section. According to a result of Sikorav, Γ admits a generating function quadratic at infinity. This means that for some $N \geq 0$ there exists a smooth function $S\colon S^{2n} \times \mathbb{R}^N \to \mathbb{R}$ such that

- 0 is a regular value of the fibre derivative $\partial_\xi S\colon S^{2n} \times \mathbb{R}^N \to \mathbb{R}^N$;
- $\Gamma = \{(q, \partial_q S(q,\xi)) \mid (q,\xi) \in S^{2n} \times \mathbb{R}^N \text{ satisfies } \partial_\xi S(q,\xi) = 0\}$;
- $S(q,\xi) = Q(\xi)$ away from a compact set for a non-degenerate quadratic form Q on $\mathbb{R}^N$.

Critical points (q,ξ) of S correspond to 1-periodic orbits $x \in \mathcal{P}^\circ(H)$, and if S is normalized such that $S(\infty, 0) = 0$, then $\mathcal{A}_H(x) = -S(q,\xi)$. For $a \in \mathbb{R}$ set

$$S^a = \{(q,\xi) \in S^{2n} \times \mathbb{R}^N \mid S(q,\xi) < a\}.$$

We denote by H_* homology with real coefficients. It is shown in [**40, 43**], see also [**42**], that the relative homology groups $\mathrm{H}_*(S^b, S^a)$ do not depend on the choice of S, and it is easily seen that $\mathrm{H}_*(S^a, S^{-a})$ does not depend on a for a large enough. Let i be the index of Q, and for a large enough let $1 \in \mathrm{H}_i(S^a, S^{-a})$ be the image of the positive generator of $\mathrm{H}_0(S^{2n})$ under the Thom isomorphism $\mathrm{H}_0(S^{2n}) = \mathrm{H}_i(S^a, S^{-a})$. Consider the map

$$j_S^\lambda\colon \mathrm{H}_i(S^a, S^{-a}) \to \mathrm{H}_i(S^a, S^\lambda)$$

induced by inclusion. The action selector σ_{V} is defined as

$$\sigma_{\mathrm{V}}(H) = -\inf\{\lambda \mid j_S^\lambda(1) = 0\}.$$

It follows from classical Lusternik–Schnirelmann theory that $\sigma_{\mathrm{V}}(H)$ is a critical value of $-S$ and hence belongs to $\Sigma^\circ(H)$. Similarly, the axioms (AS2) to (AS5) are translations of properties of special critical values of generating functions quadratic at infinity, which can be verified by elementary but by no means trivial topological arguments.

The selector σ_{HZ}. The following description is taken from Section 5.3 of [**24**]. The action selector σ_{HZ} is defined via a direct minimax on the space of loops

$$E = H^{1/2} = \left\{ x \in L^2\left(S^1; \mathbb{R}^{2n}\right) \,\middle|\, \sum_{k\in\mathbb{Z}} |k|\,|x_k|^2 < \infty \right\},$$

where

$$x = \sum_{k\in\mathbb{Z}} e^{k2\pi Jt} x_k, \qquad x_k \in \mathbb{R}^{2n},$$

is the Fourier series of x and $-J$ is the standard almost complex structure of $\mathbb{R}^{2n} \cong \mathbb{C}^n$. The space E is a Hilbert space with inner product

$$\langle x, y\rangle = \langle x_0, y_0\rangle + 2\pi \sum_{k\in\mathbb{Z}} |k|\,\langle x_k, y_k\rangle,$$

and there is an orthogonal splitting

$$E = E^- \oplus E^0 \oplus E^+, \qquad x = x^- + x^0 + x^+,$$

into the spaces of $x \in E$ having only Fourier coefficients for $k < 0$, $k = 0$, $k > 0$. The action functional $\mathcal{A}_H \colon C^\infty\left(S^1; \mathbb{R}^{2n}\right) \to \mathbb{R}$ extends to E as

$$\mathcal{A}_H(x) = a(x) + b(x), \qquad x \in E,$$

where $a(x) = -\int_{\bar{x}} \omega = \frac{1}{2}\left\|x^+\right\|^2 - \frac{1}{2}\left\|x^-\right\|^2$ and $b(x) = \int_0^1 H(t, x(t))\,dt$. The function $\mathcal{A}_H \colon E \to \mathbb{R}$ is differentiable, and its gradient is given by

$$\nabla \mathcal{A}_H(x) = x^+ - x^- + \nabla b(x).$$

Prompted by the structure of the gradient flows generated by such gradients, we consider the group G of homeomorphisms h of E satisfying

- h and h^{-1} map bounded sets to bounded sets;
- there exist continuous maps $\gamma^\pm \colon E \to \mathbb{R}$ and $k \colon E \to E$ mapping bounded sets to precompact sets and such that there exists $r = r(h)$ satisfying

  $$\gamma^\pm(x) = 0 \quad \text{and} \quad k(x) = 0 \quad \text{for all } x \in E \text{ with } \|x\| \geq r$$

  and

  $$h(x) = e^{\gamma^+(x)} x^+ + x^0 + e^{\gamma^-(x)} x^- + k(x) \quad \text{for all } x \in E.$$

The selector σ_{HZ} is defined as

$$\sigma_{\mathrm{HZ}}(H) = \sup_{F\in\mathcal{F}} \inf_{x\in F} \mathcal{A}_H(x)$$

where

$$\mathcal{F} = \left\{ h\left(E^+\right) \mid h \in G \right\}.$$

It follows from Schauder's fixed point theorem that

$$h\left(E^+\right) \cap \left(E^- \oplus E^0\right) \neq \emptyset \qquad \text{for all } h \in G.$$

This intersection result implies that $\sigma_{\mathrm{HZ}}(H)$ is finite. Using the fact that $\mathcal{A}_H$ satisfies the Palais–Smale condition one then shows that $\sigma_{\mathrm{HZ}}(H) \in \Sigma^\circ(H)$. The other axioms readily follow from definitions.

A.2. $\sigma_{\rm PSS}$ for weakly exact closed symplectic manifolds. The following description closely follows [**39**]. The construction of the action selector $\sigma_{\rm PSS}$ is along the lines of the construction of $\sigma_{\rm V}$, but instead of working with a finite dimensional reduction S, one works with the action functional $\mathcal{A}_H$ on the full space $\mathcal{L}^\circ$ of contractible loops. Notice that $\mathcal{P}^\circ(H)$ is the set of critical points of $\mathcal{A}_H$. Let $H \in \mathcal{H}$ be such that all $x \in \mathcal{P}^\circ(H)$ are non-degenerate in the sense that

$$\det\left(\mathrm{id} - d\varphi_H^1(x(0))\right) \neq 0. \tag{A.1}$$

Then $\mathcal{P}^\circ(H)$ is a finite set, and $\mathcal{A}_H$ is a Morse function on $\mathcal{L}^\circ$. The Morse indices are, however, all infinite and hence do not lead to a change of topology of the sublevel sets of $\mathcal{A}_H$. Floer overcame this problem by constructing a *relative* Morse theory for $\mathcal{A}_H$, from which he extracted a Morse-type homology isomorphic to the homology of M. It is called Floer homology and will serve as a substitute for the homology groups $\mathrm{H}_*\left(S^a, S^{-a}\right)$ with a large.

Consider the $\mathbb{Z}_2$-vector space $\mathrm{CF}(M,H)$ freely generated by the elements of $\mathcal{P}^\circ(H)$. In order to define a differential on it, fix $x, y \in \mathcal{P}^\circ(H)$. For a generic family J_t, $t \in S^1$, of almost complex structures on M for which $g_t = \omega \circ (\mathrm{id} \times J_t)$ is a Riemannian metric for each t, the set $\mathcal{M}(x,y)$ of solutions $u \in C^\infty\left(\mathbb{R} \times S^1, M\right)$ of the elliptic partial differential equation

$$\partial_s u + J_t(u)(\partial_t u - X_{H_t}(u)) = 0 \tag{A.2}$$

with asymptotic boundary conditions

$$\lim_{s \to -\infty} u(s,t) = x(t), \qquad \lim_{s \to \infty} u(s,t) = y(t), \tag{A.3}$$

is a smooth manifold. Notice that $u \in \mathcal{M}(x,y)$ is a downward gradient flow line of $\mathcal{A}_H$ with respect to the L^2-metric on $\mathcal{L}^\circ$ induced by the family g_t. Let $\mathcal{M}^1(x,y)$ be the union of the 1-dimensional components of $\mathcal{M}(x,y)$. The real numbers act on $\mathcal{M}^1(x,y)$ by shift in the s-variable. Since (M,ω) is weakly exact, $\mathcal{M}^1(x,y)/\mathbb{R}$ is compact, and so one can set

$$n(x,y) = \#\left\{\mathcal{M}^1(x,y)/\mathbb{R}\right\} \mod 2.$$

The linear map ∂ on $\mathrm{CF}(M,H)$ defined as the linear extension of

$$\partial x = \sum_{y \in \mathcal{P}^\circ(H)} n(x,y)\, y, \qquad x \in \mathcal{P}^\circ(H), \tag{A.4}$$

indeed satisfies $\partial \circ \partial = 0$, and the homology $\mathrm{HF}(M,H)$ of the resulting complex is independent of the choice of the family J_t. It is in fact independent of H as well, and agrees with the homology $\mathrm{H}(M;\mathbb{Z}_2)$ of M.

One way of constructing an isomorphism from the homology $\mathrm{H}(M;\mathbb{Z}_2)$ to the Floer homology $\mathrm{HF}(M,H)$ is as follows. The homology $\mathrm{H}(M;\mathbb{Z}_2)$ is canonically isomorphic to the Morse homology $\mathrm{HM}(M;\mathbb{Z}_2)$ of M. A chain complex $\mathrm{CM}(M,F)$ for this homology is generated by the critical points of a Morse function F on M, and the differential is as in (A.4), where now $n(c,c')$ is the number (mod 2) of negative gradient flow lines of F (with respect to a generic Riemannian metric) connecting critical points $c, c' \in \mathrm{Crit}\, F$. In order to relate $\mathrm{HM}(M;\mathbb{Z}_2)$ with $\mathrm{HF}(M,H)$, one defines a chain map $\phi\colon \mathrm{CM}(M,F) \to \mathrm{CF}(M,H)$ by

$$\phi(c) = \sum_{x \in \mathcal{P}^\circ(H)} n(c,x)\, x, \quad c \in \mathrm{Crit}(F),$$

where now $n(c,x)$ is the number (mod 2) of mixed trajectories (γ, u) from $c \in \operatorname{Crit} F$ to $x \in \mathcal{P}^\circ(H)$ as in Figure 1.

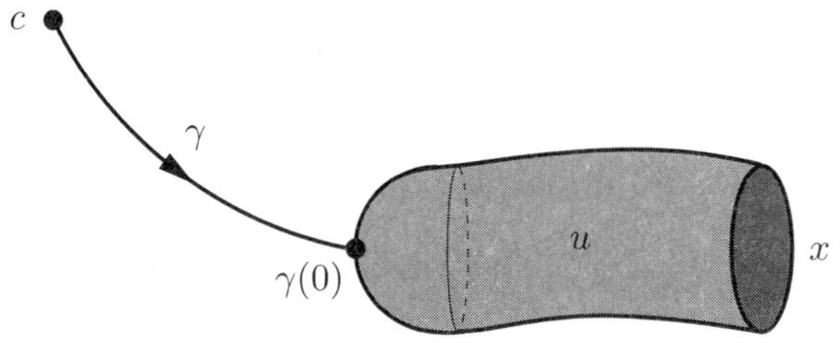

FIGURE 1. An element of $\mathcal{M}(c,x)$.

Here, $\gamma\colon (-\infty, 0] \to M$ is an integral curve of $-\nabla F$, i.e.,

$$\dot\gamma(s) = -\nabla F(\gamma(s)),$$

and $u\colon \mathbb{R} \times S^1 \to M$ is a solution of the equation

$$\partial_s u + J_t(u)(\partial_t u - X_{\beta(s)H_t}(u)) = 0,$$

where $\beta\colon \mathbb{R} \to [0,1]$ is a smooth cut off function such that

$$\beta(s) = 0,\ s \le 0; \quad \beta'(s) \ge 0,\ s \in \mathbb{R}; \quad \beta(s) = 1,\ s \ge 1,$$

and the boundary conditions are

$$\lim_{s\to-\infty} \gamma(s) = c, \quad \gamma(0) = \lim_{s\to-\infty} u(s,t), \quad \lim_{s\to\infty} u(s,t) = x(t).$$

The map ϕ indeed commutes with the differentials and thus induces a map

$$\Phi^H\colon \operatorname{HM}(M;\mathbb{Z}_2) \to \operatorname{HF}(M,H),$$

which turns out to be an isomorphism. The composition

$$\Phi^H_{\mathrm{PSS}}\colon \operatorname{H}(M;\mathbb{Z}_2) \cong \operatorname{HM}(M;\mathbb{Z}_2) \xrightarrow{\Phi^H} \operatorname{HF}(M,H)$$

is called the Piunikhin–Salamon–Schwarz (PSS) isomorphism.

The Floer chain complex $\operatorname{CF}(M,H)$ comes with a natural real filtration given by the action functional: For $\lambda \in \mathbb{R}$ consider the linear subspace

$$\operatorname{CF}^\lambda(M,H) = \left\{ \sum_{x\in\mathcal{P}^\circ(H)} \xi_x\, x \;\middle|\; \xi_x = 0 \text{ if } \mathcal{A}_H(x) > \lambda \right\}$$

of $\operatorname{CF}(M,H)$. Since $\mathcal{A}_H$ decreases along solutions of (A.2), the boundary operator ∂ preserves $\operatorname{CF}^\lambda(M,H)$ and hence descends to the quotient complex

$$\operatorname{CF}(M,H)/\operatorname{CF}^\lambda(M,H).$$

Its homology is denoted by $\operatorname{HF}^\lambda(M,H)$. Let $j^\lambda\colon \operatorname{HF}(M,H) \to \operatorname{HF}^\lambda(M,H)$ be the map induced by the natural projection $\operatorname{CF}(M,H) \to \operatorname{CF}(M,H)/\operatorname{CF}^\lambda(M,H)$, and let 1 be the generator of $\operatorname{H}_{2n}(M;\mathbb{Z}_2) \cong \operatorname{H}^0(M;\mathbb{Z}_2)$. Since Φ^H_{PSS} is an isomorphism,

$$\sigma_{\mathrm{PSS}}(H) = \inf\left\{\lambda \in \mathbb{R} \mid j^\lambda\left(\Phi^H_{\mathrm{PSS}}(1)\right) = 0\right\} \tag{A.5}$$

is finite. Furthermore, it is clear that $\sigma_{\mathrm{PSS}}(H)$ lies in $\Sigma^\circ(H)$. So far we have defined $\sigma_{\mathrm{PSS}}(H)$ for those $H \in \mathcal{H}$ that satisfy (A.1). The function σ_{PSS} is C^0-continuous on this set; this is proved by using the compatibility $\Phi_{\mathrm{PSS}}^K = \Phi^{KH} \circ \Phi_{\mathrm{PSS}}^H$ of the PSS-isomorphisms with the canonical isomorphism

$$\Phi^{KH} \colon \mathrm{HF}(M,H) \to \mathrm{HF}(M,K)$$

obtained by counting solutions of (A.2) with H replaced by $(1-\beta)H + \beta K$ and with asymptotic boundary conditions $x \in \mathcal{P}^\circ(H)$, $y \in \mathcal{P}^\circ(K)$. Since the set of H satisfying (A.1) is C^∞-dense in $\mathcal{H}$, the function σ_{PSS} C^0-continuously extends to all of $\mathcal{H}$, and one readily verifies that $\sigma_{\mathrm{PSS}}(H) \in \Sigma^\circ(H)$ for all $H \in \mathcal{H}$. The explicit nature of the isomorphism Φ_{PSS} is also important for the verification of the remaining axioms. The proof of (AS2) is difficult. One studies equation (A.2) with H replaced by λH for each $\lambda \in [0,1]$ and uses a cobordism argument similar to the one used in proving $\partial \circ \partial = 0$. Finally, the proof of (AS5) uses the product structure on Floer homology given by the pair of pants product and a sharp energy estimate for the pair of pants.

A.3. σ_{PSS} for weakly exact convex symplectic manifolds. The construction of σ_{PSS} for weakly exact convex symplectic manifolds in [**9**] follows the construction outlined above. We can assume that M is compact. Since M has boundary ∂M, the set $\mathcal{P}^\circ(H)$ contains infinitely many critical points for every $H \in \mathcal{H}$. This problem is overcome by using the geometry of M near ∂M. On a neighborhood $\partial M \times (-\epsilon, 0]$ with coordinates (x,r) the symplectic form ω is $d\,(e^r\alpha)$, where $\alpha = (\iota_X\omega)|_{\partial M}$. Let $\mathcal{H}_\partial$ be the set of Hamiltonians on M which satisfy (A.1) and which near ∂M are of the form $H(x,r) = f(r)$ with $f'(r)$ positive and so small that the flow φ_H^t has no 1-periodic orbits near ∂M. For $H \in \mathcal{H}_\partial$ the set $\mathcal{P}^\circ(H)$ is finite, and the strong maximum principle for uniformly elliptic operators implies that solutions of (A.2) and (A.3) stay away from ∂M. The Floer homology $\mathrm{HF}(M,H)$ can thus be defined and, by using a yet stronger version of the maximum principle, the isomorphism Φ^H from the Morse homology $\mathrm{HM}(M;\mathbb{Z}_2)$ to $\mathrm{HF}(M,H)$ can also be constructed. Here, the Morse homology is defined via Morse functions which are of the form $F(x,r) = e^{-r}$ near ∂M. After identifying $\mathrm{HM}_{2n}(M;\mathbb{Z}_2)$ with $\mathrm{H}_{2n}(M,\partial M;\mathbb{Z}_2) \cong \mathrm{H}^0(M;\mathbb{Z}_2)$, one defines $\sigma_{\mathrm{PSS}}(H) \in \Sigma^\circ(H)$ as in (A.5). Since $\mathcal{H}$ lies in the C^∞-closure of $\mathcal{H}_\partial$ and since σ_{PSS} is C^0-continuous on $\mathcal{H}_\partial$, we obtain a C^0-continuous function σ_{PSS} on $\mathcal{H}$ verifying axioms (AS1) and (AS2) for an action selector. The remaining axioms are also verified as in the closed case.

Consider now an exhaustion $M_1 \subset M_2 \subset \cdots$ of a non-compact convex weakly exact symplectic manifold (M,ω), and denote the action selector σ_{PSS} for M_i by σ_i. The following examples shed some light on the problem whether an action selector for an exhaustion of (M,ω) fits together to an action selector of (M,ω).

(i) Assume that $(M \setminus M_1, \omega)$ is symplectomorphic to $(\partial M_1 \times (0, r_\infty), d\,(e^r\alpha))$ for some $r_\infty \in (0,\infty]$ and that $M_i = M_1 \cup \partial M_1 \times [0, r_i]$, $0 = r_1 < r_2 < \cdots < r_\infty$. Then every $H_i \in \mathcal{H}_\partial(M_i)$ extends to $H_{i+1} \in \mathcal{H}_\partial(M_{i+1})$ with $H_{i+1}(x,r) = f(r)$ on $M_{i+1} \setminus M_i$. The chain complexes $\mathrm{CF}(M_i, H_i)$ and $\mathrm{CF}(M_{i+1}, H_{i+1})$ then agree, and so σ_{i+1} restricts to σ_i on M_i. Examples are $\left(\mathbb{R}^{2n}, \omega_0\right)$ exhausted by balls, cotangent bundles (T^*B, ω_0) over a closed base exhausted by ball bundles, and, more generally, Stein manifolds (M,J,f). In the latter example, (M,J) is an open complex manifold and $f \colon M \to \mathbb{R}$ is a smooth exhausting plurisubharmonic function without critical points off a compact subset of M. Here, "exhausting" means that f is proper and bounded from below, and "plurisubharmonic" means that $\omega = -d\,(df \circ J)$ is a symplectic form with $\omega\,(v, Jv) > 0$ for all $0 \neq v \in TM$.

Then the gradient $X = \nabla f$ with respect to the Kähler metric $\omega \circ (\mathrm{id} \times J)$ satisfies $\mathcal{L}_X \omega = \omega$. Furthermore, if the critical points of f are contained in, say, $\{f < 1\}$, the manifold M is exhausted by the exact convex symplectic manifolds $M_i = \{f \leq i\}$.

(ii) Assume that $M_{i+1} \setminus M_i$ is diffeomorphic but not symplectomorphic to $\partial M_i \times (r_i, r_{i+1}]$. Then every $H_i \in \mathcal{H}_\partial(M_i)$ extends to $H_{i+1} \in \mathcal{H}_\partial(M_{i+1})$ such that the vector spaces $\mathrm{CF}(M_i, H_i)$ and $\mathrm{CF}(M_{i+1}, H_{i+1})$ agree. Moreover, then the homology groups of M_i and M_{i+1} agree, so that the Floer homology groups $\mathrm{HF}\,(M_i, H_i)$ and $\mathrm{HF}\,(M_{i+1}, H_{i+1})$ also agree. However, solutions of (A.2) and (A.3) for H_{i+1} used to define the differential ∂_{i+1} of $\mathrm{CF}(M_{i+1}, H_{i+1})$ might enter $M_{i+1} \setminus M_i$, and so the differentials ∂_i and ∂_{i+1} might not agree. It is thus unclear whether $\sigma_i\,(H_i) = \sigma_{i+1}\,(H_{i+1})$, and so σ_{i+1} might not restrict to σ_i on M_i. An example of a symplectic manifold admitting an exhaustion of type (ii) but no exhaustion of type (i) is the camel space $M \subset (\mathbb{R}^4, \omega_0)$ defined as

$$M = \{y_1 < 0\} \cup \{y_1 > 0\} \cup \left\{x_1^2 + x_2^2 + y_2^2 < 1,\, y_1 = 0\right\},$$

cf. [**6**, Proposition 3.4.A]. It is not hard to find an exhaustion of type (ii): For $i \geq 1$ consider the union of the two closed 4-balls of radius i centered at $y_1 = \pm i$. Smoothing this set appropriately near $y_1 = 0$ for each i one obtains starshaped dumbbells M_i with smooth boundary forming an exhaustion of M of type (ii), see [**30**, Lemma 5.1] for details. Assume now that M admits an exhaustion (M_i) such that $(M \setminus M_1, \omega_0)$ is symplectomorphic to $(\partial M_1 \times (0, r_\infty), d(e^r \alpha))$. Since M has infinite volume, $r_\infty = \infty$. Since $M = M_1 \cup \partial M_1 \times (0, \infty)$ is diffeomorphic to the standard $\mathbb{R}^4$, the interior of M_1 is also diffeomorphic to the standard $\mathbb{R}^4$, and so $M \setminus M_1$ is diffeomorphic to $S^3 \times \mathbb{R}$. In particular, $H^1\,(\partial M_1; \mathbb{R}) = 0$. The primitive $\lambda = \iota_{\frac{\partial}{\partial r}} \omega_0$ of ω_0 on $M \setminus M_1$ therefore smoothly extends to all of M. The Liouville vector field X defined by $\iota_X \omega_0 = \lambda$ then agrees with $\frac{\partial}{\partial r}$ on $M \setminus M_1$ and hence integrates to a flow on M. A standard construction now shows that the identical embedding of a dumbbell as above extends to a symplectic embedding of $(\mathbb{R}^4, \omega_0)$ into M, but this contradicts the Symplectic Camel Theorem, see Lemma 4.1.7 and Proposition 4.1.8 in [**41**]. The same arguments apply to camel spaces in $(\mathbb{R}^{2n}, \omega_0)$ for $n \geq 3$, where one then uses the Symplectic Camel Theorem proved in [**43**].

(iii) If $M_{i+1} \setminus M_i$ is not diffeomorphic to $\partial M_i \times (0, 1]$, then any extension $H_{i+1} \in \mathcal{H}_\partial(M_{i+1})$ of $H_i \in \mathcal{H}_\partial(M_i)$ must have critical points on $M_{i+1} \setminus M_i$. Already the vector spaces $\mathrm{CF}(M_i, H_i)$ and $\mathrm{CF}(M_{i+1}, H_{i+1})$ are thus different, and so it is unclear whether $\sigma_i\,(H_i) = \sigma_{i+1}\,(H_{i+1})$. An example is the orientable surface M of infinite genus and with one end. More explicitly, M is the surface with exhaustion $M_1 \subset M_2 \subset \cdots$, where M_1 is a torus with an open disc removed and $M_{i+1} \setminus \mathrm{Int}\, M_i$ is a torus with two open discs removed, $i = 1, 2, \ldots$. For every area form on M we obtain an exact convex symplectic manifold, and it is unclear whether σ_{i+1} restricts to σ_i on M_i, $i = 1, 2, \ldots$. Higher-dimensional examples are products of such surfaces and also Stein manifolds of infinite topology.

A.4. σ_{PSS} for rational strongly semi-positive closed symplectic manifolds. The construction of σ_{PSS} for rational strongly semi-positive closed symplectic manifolds is similar to the weakly exact case. If (M, ω) is not weakly exact, the action functional $\mathcal{A}_H$ is well-defined only on a suitable infinite cover of $\mathcal{L}^\circ$, however. As a consequence, the set of critical points of $\mathcal{A}_H$ is infinite for every $H \in \mathcal{H}$. One therefore needs to define the Floer chain complex over a certain Novikov ring, and the Floer homology will thus be a module over this ring. The strong semi-positivity condition excludes bubbling off of pseudo-holomorphic spheres in the compactifications of the moduli spaces relevant to the definition of Floer homology, and it is also used in the construction of the PSS-isomorphism. The selector σ_{PSS} can then be defined as before. Since (M,ω) is rational, the spectrum $\Sigma^\circ(H)$ is a closed and nowhere dense subset of $\mathbb{R}$ for all $H \in \mathcal{H}$, and so the axioms (AS1), (AS3), (AS4) and (AS5) for a weak action selector can be verified as in the weakly exact case. The verification of (AS2), however, makes use of the Poincaré duality for Floer homology. It is available only if M is closed, and thus no weak action selector has yet been constructed for convex strongly semi-positive symplectic manifolds.

Appendix B. Almost existence via c_{HZ} and c_{HZ}°

In this appendix we explain why the finiteness of the symplectic capacities c_{HZ} and c_{HZ}° leads to almost existence results for periodic orbits. We shall only look at c_{HZ}°, the argument for c_{HZ} being the same. Let (M,ω) be a symplectic manifold. For each $\epsilon \in [0,1]$ and each $A \subset M$ we consider the set $\mathcal{F}^\epsilon(A)$ of functions in $\mathcal{H}(A)$ satisfying

(P1) $H \geq 0$,

(P2) $H|_U = \max H$ for some open non-empty set $U \subset A$,

(P3$^\epsilon$) the critical values of H lie in $[0, \epsilon \max H] \cup \max H$.

We denote by $\mathcal{F}_{\mathrm{HZ}}^{\circ,\epsilon}(A,M)$ the set of those $H \in \mathcal{F}^\epsilon(A)$ for which the flow φ_H^t has no non-constant, contractible in M, T-periodic orbits with period $T \leq 1$, and set

$$C_{\mathrm{HZ}}^{\circ,\epsilon}(A,M) = \sup\left\{\max H \mid H \in \mathcal{F}_{\mathrm{HZ}}^{\circ,\epsilon}(A,M)\right\}.$$

Notice that $C_{\mathrm{HZ}}^{\circ,1}(A,M) = C_{\mathrm{HZ}}^\circ(A,M)$ and $C_{\mathrm{HZ}}^{\circ,0}(A,M) = c_{\mathrm{HZ}}^\circ(A,M)$. Moreover, we claim that for every $\epsilon \in (0,1)$,

$$(1-\epsilon)C_{\mathrm{HZ}}^{\circ,\epsilon}(A,M) \leq c_{\mathrm{HZ}}^\circ(A,M) \leq C_{\mathrm{HZ}}^{\circ,\epsilon}(A,M). \tag{B.1}$$

The second inequality follows from definitions. To prove the first inequality, we need to show that for each $H \in \mathcal{F}_{\mathrm{HZ}}^{\circ,\epsilon}(A,M)$ and each $\delta > 0$ there exists $K \in \mathcal{S}_{\mathrm{HZ}}^\circ(A,M)$ with $\max K = (1-\epsilon)\max H - \delta$. The idea of the argument is to take as K only "the upper part of H". To be more precise, fix $H \in \mathcal{F}_{\mathrm{HZ}}^{\circ,\epsilon}(A,M)$ and $\delta \in \,]0,(1-\epsilon)\max H[$. Following [**14**, Section 6] and [**38**], we choose a surjective smooth map $f\colon [0,\max H] \to [0,(1-\epsilon)\max H - \delta]$ such that

$$f(t) = 0 \text{ if } t \in [0,\epsilon \max H] \quad \text{and} \quad 0 \leq f'(t) \leq 1 \text{ for all } t \in [0,\max H].$$

Set $K = f \circ H$. Since $|f'| \leq 1$ and since the flow of H has no non-trivial contractible periodic orbits with period $T \leq 1$, the same holds for the flow of K. Furthermore, since $H \in \mathcal{F}_{\mathrm{HZ}}^{\circ,\epsilon}(A,M)$, we have $K \in \mathcal{S}_{\mathrm{HZ}}^\circ(A,M)$, and by construction $\max K = (1-\epsilon)\max H - \delta$, which completes the proof of (B.1).

COROLLARY B.1. *Assume that A is a subset of a symplectic manifold (M,ω) such that $c_{\mathrm{HZ}}^\circ(A,M) < \infty$. Then almost every regular energy level of a proper function on A carries a periodic orbit which is contractible in M.*

PROOF. The corollary is proved for C_{HZ}° in [**24, 27**], and the argument given there applies to $C_{\mathrm{HZ}}^{\circ,\epsilon}$ for every $\epsilon \in (0,1]$. The corollary thus follows from the first inequality in (B.1). □

Appendix C. Proofs of Propositions 1.10 and 1.11

C.1. Proof of Proposition 1.10. (i) If S is simply connected, we choose a contact form λ on S; for $x \in \mathcal{P}^\circ(S)$ we can choose $\bar{x} \in \mathcal{D}(x)$ contained in S, so that $\int_{\bar{x}} \omega = \int_x \lambda$. If S is of restricted contact type, we choose a globally defined contact form λ; for $x \in \mathcal{P}^\circ(S)$ we find again that $\int_{\bar{x}} \omega = \int_x \lambda$. The Reeb vector field R on S associated with λ is defined by

$$\iota_R \omega = 0 \quad \text{and} \quad \lambda(R) = 1.$$

Choose a Riemannian metric on S such that the length of R is 1. Parametrizing x such that $\dot{x}(t) = R$, we see that $\int_x \lambda$ is the length of x and is hence positive. If x_j is a sequence in $\mathcal{P}^\circ(S)$ with $\int_{x_j} \lambda \to a$, then the lengths $\int_{x_j} \lambda$ are bounded, and hence a subsequence of x_j converges to a closed characteristic $x \in \mathcal{P}^\circ(S)$ of length a, see [**24**, p. 109]. This proves that $\Sigma^\circ(S)$ is closed. The proof of (ii) is similar. □

C.2. Proof of Proposition 1.11. We prove the first statement in Proposition 1.11; the second statement can be proved in a similar way. The inequality $0 < \alpha_1^\circ(S)$ follows from Proposition 1.10 (i). We will show

$$\alpha_1^\circ(S) \le c_{\mathrm{HZ}}^\circ(U, M) \tag{C.1}$$

for the capacity $C_{\mathrm{HZ}}^\circ(U, M)$. It will then be clear from the proof that (C.1) holds for all capacities $C_{\mathrm{HZ}}^{\circ,\epsilon}(U, M)$, $\epsilon \in (0, 1]$, introduced in Appendix B, and hence also for $c_{\mathrm{HZ}}^\circ(U, M)$ in view of (B.1).

Step 1. *A convenient thickening of S.* Let X be a Liouville vector field on M transverse to $S = \partial U$. Then X points outward. Let $\lambda = \iota_X \omega|_S$ be the associated contact form on S and let R be the Reeb vector field of λ on S. Using the flow φ_X^t of X, which exists in a neighborhood of $\overline{U}$ for small t, we see that a neighborhood of S is symplectomorphic to the thickening $(-\epsilon, \epsilon) \times S$ with coordinates (t, x) and symplectic form

$$\omega(t, x) = d((1 + t)\lambda(x)). \tag{C.2}$$

Then the Liouville vector field is $X(t, x) = (1 + t)\frac{\partial}{\partial t}$. Set $S_t = \{t\} \times S$. Using (1.1) and (C.2), we see that the Hamiltonian vector field X_H of $H\colon (-\epsilon, \epsilon) \times S$, $H(t, x) = t$, points along S_t and equals the *translate* of R on S_t. Thus, the periodic orbits of the flow of H on S_t are exactly the periodic orbits of the Reeb flow on S. Set $U_t = U \cup (-t, t) \times S$. Then $U_t = \varphi_X^{\ln(1+t)}(U)$, and hence the conformality of the symplectic capacity C_{HZ}° implies that

$$C_{\mathrm{HZ}}^\circ(U_t, M) = (1 + t)\, C_{\mathrm{HZ}}^\circ(U, M). \tag{C.3}$$

Step 2. Abbreviate $C = C_{\mathrm{HZ}}^\circ(U, M)$. Arguing by contradiction, we assume that $\alpha_1^\circ(S) > C$. Let $\Sigma^\circ(R)$ be the set of periods of those periodic orbits of the Reeb field R on S which are contractible in M. Since $\Sigma^\circ(S) = \Sigma^\circ(R)$, the number $\alpha_1^\circ(S)$ is the infimum of $\Sigma^\circ(R)$, and so we find $L > 1$ such that

$$[0, LC] \cap \Sigma^\circ(R) = \emptyset. \tag{C.4}$$

Fix a small $\delta > 0$. Using (C.4), we find $K \in \mathcal{F}_{\mathrm{HZ}}^\circ(U, M)$ such that

$$\max K = (1 - \epsilon - \delta)\, C.$$

Let F be a smooth function on the shell $(-\epsilon, 0) \times S$ such that $F(x,t) = f(t)$, where $f\colon (-\epsilon, 0) \to \mathbb{R}$ is a monotone decreasing function such that

$$\begin{aligned} f(t) &= l\epsilon C && \text{for } t \text{ near } -\epsilon, \\ f(t) &= 0 && \text{for } t \text{ near } 0, \\ |f'(t)| &< LC && \text{for all } t, \end{aligned}$$

where $l > 1$. We extend F to the ambient manifold M in the obvious way: $F \equiv l\epsilon C$ inside the shell and $F \equiv 0$ outside the shell. By the construction of the thickening $(-\epsilon, \epsilon) \times S$, by (C.4) and by the choice of f, the function F belongs to $\mathcal{F}^{\circ}_{\mathrm{HZ}}(U, M)$. Since $F \equiv l\epsilon C$ on the support of K, the Hamiltonian $H = K + F$ also belongs to $\mathcal{F}^{\circ}_{\mathrm{HZ}}(U, M)$. However,

$$\max H = (1 - \epsilon - \delta + l\epsilon)\, C.$$

Since $l > 1$, we can choose $\delta > 0$ so small that $\max H > C$. This contradicts $H \in \mathcal{F}^{\circ}_{\mathrm{HZ}}(U, M)$. □

REMARK C.1. In the above proof, the assumption that S is of *restricted* contact type was used only to obtain identity (C.3). If S is a hypersurface of contact type bounding U such that the functions $t \mapsto c^{\circ}_{\mathrm{HZ}}(U_t, M)$ and $t \mapsto c_{\mathrm{HZ}}(U_t)$ are 1-Lipschitz at 0, then Proposition 1.11 still holds for S.

References

1. C. Abbas and H. Hofer, *Holomorphic Curves and Global Questions in Contact Geometry*, Book in preparation.
2. S. M. Bates, *A capacity representation theorem for some non-convex domains*, Math. Z. **227** (1998), 571–581.
3. V. Benci and F. Giannoni, *Periodic bounce trajectories with a low number of bounce points*, Ann. Inst. H. Poincaré Anal. Non Linéaire **6** (1989), 73–93.
4. K. Cieliebak, *Symplectic boundaries: Closed characteristics and action spectra*, Diss. ETH No. 11610, Zürich, 1996.
5. K. Cieliebak, A. Floer, H. Hofer and K. Wysocki, *Applications of symplectic homology. II. Stability of the action spectrum*, Math. Z. **223** (1996), 27–45.
6. Ya. Eliashberg and M. Gromov, *Convex symplectic manifolds*, Several Complex Variables and Complex Geometry, Proceedings, Summer Research Institute, Santa Cruz, 1989, Part 2 (E. Bedford et al., eds.), Proc. Sympos. Pure Math. **52**, Amer. Math. Soc., Providence, 1991, pp. 135–162.
7. M. Entov, *Commutator length of symplectomorphisms*, Comment. Math. Helv. **79** (2004), 58–104.
8. M. Entov and L. Polterovich, *Calabi quasimorphism and quantum homology*, Int. Math. Res. Not. **2003**, 1635–1676.
9. U. Frauenfelder and F. Schlenk, *Hamiltonian dynamics on convex symplectic manifolds*, math.SG/0303282.
10. ______, *Slow entropy and symplectomorphisms of cotangent bundles*, math.SG/0404017.
11. V. Ginzburg, *On closed trajectories of a charge in a magnetic field. An application of symplectic geometry*, Contact and Symplectic Geometry (Cambridge, 1994), Publ. Newton Inst. **8**, Cambridge Univ. Press, Cambridge, 1996, pp. 131–148.
12. ______, *Hamiltonian dynamical systems without periodic orbits*, Northern California Symplectic Geometry Seminar, 35–48, Amer. Math. Soc. Transl. Ser. 2, **196**, Amer. Math. Soc., Providence, RI, 1999.
13. ______, *The Weinstein conjecture and theorems of nearby and almost existence*, The Breadth of Symplectic and Poisson Geometry, Progr. Math. **232**, Birkhäuser Boston, Boston, MA, 2005, pp. 139–172.
14. V. Ginzburg and B. Gürel, *Relative Hofer–Zehnder capacity and periodic orbits in twisted cotangent bundles*, Duke Math. J. **123** (2004), 1–47.

15. M. Gromov, *Pseudo-holomorphic curves in symplectic manifolds*, Invent. Math. **82** (1985), 307–347.
16. D. Hermann, *Holomorphic curves and Hamiltonian systems in an open set with restricted contact-type boundary*, Duke Math. J. **103** (2000), 335–374.
17. ———, *Inner and outer Hamiltonian capacities*, Bull. Soc. Math. France **132** (2004), 509–541.
18. ———, *Non-equivalence of symplectic capacities for open sets with restricted contact type boundary*, `http://www.math.u-psud.fr/ biblio/pub/1998/abs/ppo1998_32.html`.
19. H. Hofer, *On the topological properties of symplectic maps*, Proc. Roy. Soc. Edinburgh Sect. A **115** (1990), 25–38.
20. ———, *Estimates for the energy of a symplectic map*, Comment. Math. Helv. **68** (1993), 48–72.
21. H. Hofer and C. Viterbo, *The Weinstein conjecture in cotangent bundles and related results*, Ann. Scuola Norm. Sup. Pisa Cl. Sci. IV **15** (1988), 411–445.
22. H. Hofer and E. Zehnder, *Periodic solutions on hypersurfaces and a result by C. Viterbo*, Invent. Math. **90** (1987), 1–9.
23. ———, *A new capacity for symplectic manifolds*, Analysis, et cetera, Academic Press, Boston, MA, 1990, pp. 405–427.
24. ———, *Symplectic Invariants and Hamiltonian Dynamics*, Birkhäuser, Basel, 1994.
25. F. Lalonde and D. McDuff, *The geometry of symplectic energy*, Ann. of Math. **141** (1995), 349–371.
26. G. Lu, *The Weinstein conjecture on some symplectic manifolds containing the holomorphic spheres*, Kyushu J. Math. **52** (1998), 331–351 and **54** (2000), 181–182.
27. L. Macarini and F. Schlenk, *A refinement of the Hofer–Zehnder theorem on the existence of closed trajectories near a hypersurface*, Bull. London Math. Soc. **37** (2005), 297–300.
28. D. McDuff and D. Salamon, *Introduction to Symplectic Topology*, Second ed., Oxford Mathematical Monographs, The Clarendon Press, Oxford University Press, New York, 1998.
29. ———, *J-holomorphic curves and symplectic topology*, AMS Colloquium Publications **52**, AMS, Providence, RI, 2004.
30. D. McDuff and L. Traynor, *The 4-dimensional symplectic camel and related results*, Symplectic Geometry (D. Salamon, ed.), LMS Lecture Note Series **192**, Cambridge Univ. Press, 1993.
31. Y.-G. Oh, *Symplectic topology as the geometry of action functional. I. Relative Floer theory on the cotangent bundle*, J. Differential Geom. **46** (1997), 499–577.
32. ———, *Symplectic topology as the geometry of action functional. II. Pants product and cohomological invariants*, Comm. Anal. Geom. **7** (1999), 1–54.
33. ———, *Chain level Floer theory and Hofer's geometry of the Hamiltonian diffeomorphism group*, Asian J. Math. **6** (2002), 579–624.
34. ———, *Mini-max theory, spectral invariants and geometry of the Hamiltonian diffeomorphism group*, math.SG/0206092.
35. ———, *Spectral invariants and length minimizing property of Hamiltonian paths*, math.SG/0212337.
36. L. Polterovich, *An obstacle to non-Lagrangian intersections*, The Floer Memorial Volume, Progr. Math. **133**, Birkhäuser, Basel, 1995, pp. 575–586.
37. ———, *Growth of maps, distortion in groups and symplectic geometry*, Invent. Math. **150** (2002), 655-686.
38. F. Schlenk, *Applications of Hofer's geometry to Hamiltonian dynamics*, to appear in Comment. Math. Helv.
39. M. Schwarz, *On the action spectrum for closed symplectically aspherical manifolds*, Pacific J. Math. **193** (2000), 419–461.
40. D. Théret, *A complete proof of Viterbo's uniqueness theorem on generating functions*, Topology Appl. **96** (1999), 249–266.
41. L. Traynor, *Symplectic embedding trees for generalized camel spaces*, Duke Math. J. **72** (1993), 573–594.
42. ———, *Symplectic homology via generating functions*, Geom. Funct. Anal. **4** (1994), 718–748.
43. C. Viterbo, *Symplectic topology as the geometry of generating functions*, Math. Ann. **292** (1992), 685–710.

44. ———, *Exact Lagrange submanifolds, periodic orbits and the cohomology of free loop spaces*, J. Differential Geom. **47** (1997), 420–468.
45. ———, *Functors and computations in Floer homology with applications. I*, Geom. Funct. Anal. **9** (1999), 985–1033.
46. ———, *Metric and isoperimetric problems in symplectic geometry*, J. Amer. Math. Soc. **13** (2000), 411–431.

Department of Mathematics, Hokkaido University, Sapporo 060-0810, Japan
E-mail address: urs@math.sci.hokudai.ac.jp

Department of Mathematics, UC Santa Cruz, CA 95064, U.S.A.
E-mail address: ginzburg@math.ucsc.edu

Mathematisches Institut, Universität Leipzig, 04109 Leipzig, Germany
E-mail address: schlenk@math.uni-leipzig.de

Contemporary Mathematics
Volume **387**, 2005

On Non-Bounded Generation of Discrete Subgroups in Rank-1 Lie Group

Koji Fujiwara

Dedicated to the memory of Robert Brooks

ABSTRACT. A discrete subgroup G in a rank-1 simple Lie group is not boundedly generated if G does not contain a nilpotent subgroup of finite index.

1. Introduction

The purpose of this note is to record some remarks on the non-bounded generation of some discrete groups. We point out that a theorem by Kotschick in [**K**] has more examples than mapping class groups.

A group G is said *boundedly generated* if there exist finitely many elements $g_1, \cdots, g_k \in G$ such that for any $g \in G$ there exist $n_i \in \mathbb{Z}$ with

$$g = g_1^{n_1} \cdots g_k^{n_k}.$$

One may say G is boundedly generated by $g_1, \cdots, g_k$.

Most non-uniform lattices in a Lie group of rank at least two are known to be boundedly generated (cf. [**T**]). For example $SL(n, \mathbb{Z}), n > 2$ and $SL(2, \mathbb{Z}[1/p])$ such that p is a prime number are boundedly generated.

On the other hand, a uniform lattice G in a rank-1 simple Lie group is not boundedly generated. It is because G is a non-virtually cyclic, word-hyperbolic group, which is known to have an infinite quotient which is a torsion group (cf. [**FLM**, Prop. 3.6]).

Margulis and Vinberg [**MV**] showed that many discrete subgroups in a rank-1 simple Lie group are virtually mapped by homomorphisms to non-abelian free groups, so that they are not boundedly generated. A group is said *virtually* to have some property if some subgroup of finite index in the group has this property.

It seems the following result is new in this generality.

2000 *Mathematics Subject Classification.* Primary 20F65, 22E40.

Key words and phrases. quasi-homomorphisms, bounded generation, discrete subgroup in Lie group.

THEOREM 1. *Let G be a discrete subgroup in a rank-1 simple Lie group. If G does not contain a nilpotent subgroup of finite index, then it is not boundedly generated.*

The author dedicates this paper to the memory of Robert Brooks. He was the first to construct many quasi-homomorphisms on non-abelian free groups, [**Br**]. The author thanks Alex Lubotzky for comments.

2. Proof and examples

Let G be a discrete group. A function $f : G \to \mathbb{R}$ is called a *quasi-homomorphism* if there exists constant C such that for any $g, h \in G$

$$|f(gh) - f(g) - f(h)| \leq C.$$

A quasi-homomorphism f is called *homogeneous* if for any $g \in G$ and $n \in \mathbb{Z}$

$$f(g^n) = nf(g).$$

If f is a quasi-homomorphism then the function, ϕ, defined by the following, is a homogeneous quasi-homomorphism:

$$\phi(g) = \lim_{n\to\infty} \frac{f(g^n)}{n}.$$

The set of all homogeneous quasi-homomorphisms on G is a vector space over $\mathbb{R}$, denoted by $HQH(G)$.

We quote a result by Kotschick relating the existence of many quasi-homomorphisms on G to the non-bounded generation of G.

PROPOSITION 2. [**K**, Prop. 5] *If G is boundedly generated by $g_1, \cdots, g_k$, then the dimension of $HQH(G)$ as a vector space is at most k.*

We remark that he introduced a weaker notion of bounded generation, called "weak bounded generation," and showed that it suffices to assume it in the proposition. It had been known [**BF**] that the dimension of HQH is infinite for mapping class groups (see Theorem 4 and Example 1), so that he applied Proposition 2 to mapping class groups, and gave a new proof to the following theorem by Farb-Lubotzky-Minsky.

THEOREM 3 ([**FLM**]). *The mapping class group MCG of a closed orientable surface of genus at least one is not boundedly generated.*

In fact, the following theorem implies that no subgroup G in MCG is boundedly generated if G is not virtually abelian (see Example 1).

THEOREM 4 ([**BF**]). *Suppose X is a Gromov-hyperbolic space. Assume a group G acts on X by isometries, weakly properly discontinuously. Then the dimension of $HQH(G)$ is infinite.*

We remark that they showed [**BF**] that the dimension of the vector space of quasi-homomorphisms on G divided by the subspace of bounded functions, $QH(G)$, is infinite by constructing an infinite sequence of quasi-homomorphisms $f_1, f_2, \cdots$ on G which are linearly independent in $QH(G)$. It is easy to see that the homogeneous quasi-homomorphisms, ϕ_i, for f_i are linearly independent in $HQH(G)$.

A geodesic space X is called *Gromov-hyperbolic* [**G**] if there is a constant $\delta \geq 0$ such that every geodesic triangle in X satisfies the following property: any side is in the δ-neighborhood of the union of the other two sides. In general, a simply connected complete Riemannian manifold is a Gromov-hyperbolic space if there is a negative constant such that the sectional curvature is bounded from above by the constant. A rank-1 symmetric space is a standard example.

An isometric action of G on a Gromov-hyperbolic space X is *weakly properly discontinuous* [**BF**] if

(1) G is not virtually cyclic;
(2) G contains an element which acts by a hyperbolic isometry on X;
(3) For every hyperbolic element $g \in G$, every $x \in X$, and every $C > 0$, there exists $N > 0$ such that the set

$$\{\gamma \in G | d(x, \gamma x) \leq C, d(g^N x, \gamma g^N x) \leq C)\}$$

is finite.

Clearly, the property 3 is satisfied by any proper discontinuous action.

As an example, let G be a discrete subgroup in a rank-1 simple Lie group, and X the rank-1 symmetric space associated to the Lie group. Then the isometric action of G on X is properly discontinuous. If G is not virtually nilpotent then the action is weakly properly discontinuous, so that $\dim HQH(G) = \infty$ by Theorem 4 (cf. Corollary 1.2 in [**F**]).

The following is immediate from Proposition 2.

COROLLARY 5. *Let G be as in Theorem 4. Then G is not boundedly generated.*

EXAMPLE. By Theorem 4, the following groups have infinite dimensional HQH, so that they are not boundedly generated:

(1) a non-virtually abelian subgroup in the mapping class group of a compact orientable surface, ([**BF**]);
(2) a non-virtually cyclic subgroup of a word-hyperbolic group, ([**EF**]);
(3) a discrete subgroup G of a rank-1 simple Lie group such that G is not virtually nilpotent, (Theorem 1.1, Corollary 1.2 in [**F**]).

The last example implies Theorem 1.

References

[BF] Mladen Bestvina and Koji Fujiwara, *Bounded cohomology of subgroups of mapping class groups*, Geom. Topol. **6** (2002), 69–89.

[Br] Robert Brooks, *Some remarks on bounded cohomology*, Riemann Surfaces and Related Topics, Ann. of Math. Stud., **97**, Princeton Univ. Press, 1981, pp. 53–63.

[EF] David B.A. Epstein and Koji Fujiwara, *The second bounded cohomology of word-hyperbolic groups*, Topology **36** (1997), no. 6, 1275–1289.

[FLM] Benson Farb, Alexander Lubotzky, and Yair Minsky, *Rank-1 phenomena for mapping class groups*, Duke Math. J. **106** (2001), no. 3, 581–597.

[F] Koji Fujiwara, *The second bounded cohomology of a group acting on a Gromov-hyperbolic space*, Proc. London Math. Soc. (3) **76** (1998), no. 1, 70–94.

[G] M. Gromov, *Hyperbolic groups*, Essays in Group Theory, MSRI Publ. **8**, Springer, 1987, pp. 75–263.

[K] D. Kotschick, *Quasi-homomorphisms and stable lengths in mapping class groups*, preprint, GR/0307362.

[MV] G.A. Margulis and E.B.Vinberg, *Some linear groups virtually having a free quotient*, J. Lie Theory **10** (2000), no. 1, 171–180.

[T] O. Tavgen, *Bounded generation of Chevalley groups over rings of algebraic S-integers*, Math. USSR-Izv. **36** (1991), 101–128.

Math Institute, Tohoku University, Sendai 980-8578, Japan
E-mail address: fujiwara@math.tohoku.ac.jp

Contemporary Mathematics
Volume **387**, 2005

Isospectral and Isoscattering Manifolds: A Survey of Techniques and Examples

Carolyn Gordon, Peter Perry, and Dorothee Schueth

We dedicate this paper to the memory of our friend and colleague Robert Brooks.

ABSTRACT. The method of torus actions developed by the first and third authors yields examples of isospectral, non-isometric metrics on compact manifolds and isophasal, non-isometric metrics on non-compact manifolds. In contrast to most examples constructed by the Sunada method, the resulting examples have different local geometry. In this review, we discuss insights into the inverse spectral problem gained through both of these approaches.

1. Introduction

One of Robert Brooks' ongoing research interests—and an area in which he made fundamental contributions—was the inverse spectral problem on Riemannian manifolds. Through ingenious constructions, he helped produce examples of Riemannian manifolds which were sufficiently symmetric to have the same spectral and scattering data, and yet were not isometric. Such examples illuminate the inverse spectral problem first of all by proving non-uniqueness, and secondly by helping to isolate geometric properties of Riemannian manifolds which are not determined by spectral data. In the present paper, we will review recent constructions of compact manifolds which are isospectral, and complete non-compact manifolds which are 'isoscattering' in a sense that we will make precise. We will emphasize recent progress in understanding 'isoscattering' manifolds and recent applications of the method of torus actions, due to the first and third authors, to 'isoscattering' problems.

We will consider the following inverse problems from the contrarian point of view of trying to construct counterexamples.

PROBLEM 1.1. (*Inverse spectral problem for compact manifolds*) *If* (M, g) *is a closed Riemannian manifold, the spectrum of the Laplacian consists of an infinite sequence* $\{\lambda_j\}_{j=0}^{\infty}$ *of nonnegative eigenvalues, and is described by the counting*

2000 *Mathematics Subject Classification.* Primary 58J53; Secondary 58J50.
Gordon supported in part by NSF grant DMS-0306752.
Perry supported in part by NSF grant DMS-0100829.
Schueth supported in part by the DFG Priority Programme 1154 and DFG SFB 647.

function

$$N(\lambda) = \#\{\lambda_j : \lambda_j \leq \lambda\}.$$

If (M,g) is a compact manifold with boundary and Dirichlet or Neumann conditions are imposed, the spectrum is again an infinite sequence of eigenvalues. Find manifolds (M_1,g_1) and (M_2,g_2) with the same spectrum.

We will call closed manifolds with the same spectrum (including multiplicities) *isospectral.* For compact manifolds with boundary, one may refer to *Dirichlet isospectral* or *Neumann isospectral* manifolds. For the constructions we will consider, the manifolds will be both Dirichlet and Neumann isospectral, so we will sometimes simply say "isospectral". (There is, however, one example known [**5**] of compact manifolds which are Neumann but not Dirichlet isospectral.)

PROBLEM 1.2. (*Inverse scattering problem, 'absolute' version*) *Let (M,g) be a non-compact, complete Riemannian manifold with Laplacian Δ. Let $R(z) = (\Delta - z)^{-1}$ be the resolvent operator, and suppose that the Laplacian has only continuous spectrum in $[c,\infty)$ for some $c \geq 0$. Thus $R(z)$, as a function from $L^2(M,dg)$ to itself, is analytic in $\mathbb{C}\backslash[c,\infty)$. Suppose that $R(z)$, viewed as a map from $\mathcal{C}_0^\infty(M)$ to $\mathcal{C}^\infty(M)$, admits a meromorphic continuation to a Riemann surface which covers the cut plane. Poles of the meromorphically continued operator are called* scattering resonances (*or sometimes* scattering poles). *Find complete, non-compact manifolds (M_1,g_1) and (M_2,g_2) with the same scattering resonances.*

We will call manifolds with the same scattering resonances (including multiplicities) *isopolar.*

PROBLEM 1.3. (*Inverse scattering problem, 'relative' version*) *Let (M,g_0) be a non-compact, complete Riemannian manifold of dimension n and suppose that g is a compactly supported perturbation of g_0 which is also a complete metric on M. Let H_0 be the Laplacian on (M,g_0) and let $H = \tau\Delta\tau^*$ where Δ is the Laplacian on (M,g) and $\tau : L^2(M,dg) \to L^2(M,dg_0)$ be the natural isometry. There is a real-valued, locally integrable function ξ on $\mathbb{R}$ with the property that*

$$\operatorname{Tr}\left(f(H) - f(H_0)\right) = -\int f'(\lambda)\xi(\lambda)\,d\lambda$$

for all smooth functions f which vanish sufficiently rapidly at infinity. The metric g_0 and operator H_0 are referred to as the background metric and the reference operator and remain fixed throughout the discussion. The function ξ is called the scattering phase for the pair (H,H_0), and is analogous to the counting function $N(\lambda)$ in Problem 1.1. Find metrics g_1 and g_2 on M so that the pairs (H_1,H_0) and (H_2,H_0) have the same scattering phase.

We will call such pairs of metrics *isophasal* and, when the common, fixed reference metric g_0 is understood, we will also refer to g_1 and g_2 as "isophasal metrics" and to (M,g_1) and (M,g_2) as "isophasal manifolds."

The scenario outlined in Problem 1.2 happens, among other examples, for metric perturbations of $\mathbb{R}^n$ and quotients of real hyperbolic space by geometrically finite discrete groups. We will discuss the mathematical and physical meaning of scattering resonances in §2 of what follows. Relative scattering as discussed in Problem 1.3 makes sense for *any* complete manifold, as was shown in a striking paper of Gilles Carron [**20**].

For Problem 1.1, there is a vast literature of examples of isospectral manifolds. See [**27**] for a survey of examples prior to 2000. We will not attempt to survey all the examples here but rather will emphasize the techniques and mention primarily recent examples. There are, roughly speaking, three methods for constructing examples of isospectral manifolds.

(a) *Explicit Construction*: Recent examples constructed by explicit computations include isospectral flat manifolds with surprising spectral properties ([**43**], [**44**], [**45**]), the first examples [**58**] of isospectral manifolds with boundary having different local geometry (these partially motivated and were later reinterpreted by the torus action method below) and the first examples of pairs of isospectral metrics on balls and spheres [**59**].
(b) *Representation-Theoretic Construction*: Representation theoretic methods, especially the celebrated Sunada technique [**56**], have provided the most systematic and widely used methods for constructing isospectral manifolds with the same local, but different global, geometry.
(c) *Torus Actions.* This method generally produces isospectral manifolds with different local geometry.

Among the many examples constructed by Sunada's method are Riemann surfaces of every genus greater than or equal to four [**17**], including huge families of mutually isospectral surfaces in high genus [**10**], and examples of isospectral plane domains [**34**]. As explained in §3 below, the Sunada technique (and other representation-theoretic techniques) produce isospectral quotients $H_1\backslash M$ and $H_2\backslash M$ of a given Riemannian manifold M by discrete groups H_i of isometries; thus the isospectral manifolds are locally isometric. Recently, however, Craig Sutton [**57**] modified Sunada's method to allow the subgroups H_i to be connected, and constructed isospectral simply-connected, normal homogenous spaces that are not locally isometric.

The method of torus actions ([**26**], [**28**], [**29**], [**36**], [**52**], [**53**], [**54**]) was developed to construct isospectral manifolds with different local geometry. The first author used this method to construct continuous families of isospectral metrics on the n-ball and $(n-1)$-sphere for all $n \geq 9$. The third author lowered n to 8 and also obtained pairs of isospectral metrics on the 6-ball and 5-sphere. She also showed that in all cases, one can arrange that the metrics on the balls are Euclidean except on an arbitrarily chosen smaller ball about the origin.

Both the Sunada technique and the method of torus actions have been extended to complete, noncompact manifolds in order to obtain non-isometric manifolds with the same scattering resonances (the isopolar manifolds of Problem 1.2) and, in some cases, the same scattering phase (the isophasal manifolds of Problem 1.3). As explained in §2, isophasality is a stronger condition than isopolarity in contexts where both notions are well-defined (see Remarks 2.2 and 2.3). The examples constructed by these two methods account for all known examples of complete, non-isometric manifolds with the same scattering data. The examples constructed by variants of Sunada's technique include finite-area Riemann surfaces (both isopolar and isophasal–see Bérard [**4**] and Zelditch [**66**]), Riemann surfaces of infinite area (isopolar and isophasal–see Guillopé-Zworski [**38**] and Brooks-Davidovich [**8**]), three-dimensional Schottky manifolds (isopolar–see Brooks-Gornet-Perry [**11**]), and surfaces that are isometric to Euclidean space outside a compact set (isopolar and isophasal–see Brooks-Perry [**12**]).

The generalization of the torus action method to noncompact manifolds is more recent ([**31**] and [**47**]). In [**31**], the first two authors showed the following: Let $\{g_t\}$ be any of the families of isospectral metrics on the unit ball in $\mathbb{R}^n$ constructed in [**28**] or [**54**], modified as in [**54**] so that the metrics are Euclidean outside of a ball of smaller radius about the origin. Extend the metrics to all of $\mathbb{R}^n$ so that they are Euclidean outside of the small ball. Then the resulting metrics are non-isometric, but are both isophasal and isopolar. In [**47**], the last two authors show how to use a similar construction to obtain non-isometric, isophasal and isopolar families of metrics on $\mathbb{R}^n$ which are hyperbolic off a small ball or more generally are perturbations of complete metrics which admit an $O(n)$ action by isometries.

In what follows we first review basic notions of spectral and scattering theory for the Laplacian on a Riemannian manifold (§2), recall the Sunada method (§3), and discuss the method of torus actions (§4). Finally, we pose several open problems (§5).

2. Spectral and Scattering Theory for the Laplacian

The Laplace-Beltrami operator on a Riemannian manifold is most easily defined via the method of quadratic forms. For a closed manifold, we denote by $\mathcal{H}^1(M,g)$ the completion of $\mathcal{C}^\infty(M)$ in the inner product

$$\langle\varphi,\psi\rangle = \int_M \nabla\varphi\cdot\overline{\nabla\psi}\,dg + \int_M \varphi\overline{\psi}\,dg. \tag{2.1}$$

For a complete, non-compact manifold, we denote by $\mathcal{H}_0^1(M,g)$ the completion of $\mathcal{C}_0^\infty(M)$ in the same inner product. If M is a compact manifold with boundary, we denote by $\mathcal{H}_D^1(M,g)$ the completion of $\mathcal{C}_0^\infty(M)$ in the inner product (2.1), and by $\mathcal{H}_N^1(M,g)$ the completion of $\mathcal{C}^\infty(M)$ [1] in the same inner product.

The Laplace-Beltrami operator on M is the positive operator Δ_M associated to the quadratic form

$$\mathfrak{q}(\varphi,\psi) = \int_M \nabla\varphi\cdot\overline{\nabla\psi}\,dg,$$

with form domain given by:

- $\mathcal{H}^1(M,g)$ if M is compact and without boundary,
- $\mathcal{H}_D^1(M,g)$ if M is compact, $\partial M\neq\emptyset$ and Dirichlet boundary conditions are imposed,
- $\mathcal{H}_N^1(M,g)$ if M is compact, $\partial M\neq\emptyset$ and Neumann boundary conditions are imposed, and
- $\mathcal{H}_0^1(M,g)$ if (M,g) is non-compact and complete.

The spectral theory of the operator Δ_M determines the behavior of solutions to the wave equation on (M,g). Consider the initial value problem for a function $u:\mathbb{R}\to L^2(M,g)$ and initial datum $\psi\in L^2(M,g)$:

$$\begin{aligned} u_{tt} &= -\Delta_M u \\ u(0) &= \psi \\ u_t(0) &= 0. \end{aligned} \tag{2.2}$$

[1] If M is a manifold with boundary, $\mathcal{C}^\infty(M)$ denotes the restrictions of $\mathcal{C}^\infty$ functions on the double of M, i.e., the manifold obtained by gluing two copies of M along ∂M.

Formally, the solution to this equation is

$$u(t) = \cos\left(t\sqrt{\Delta_M}\right)\psi$$

where the solution operator

$$E(t) = \cos\left(t\sqrt{\Delta_M}\right)$$

is defined by the functional calculus for the self-adjoint operator Δ_M.

On the level of functional analysis, the spectral theory of the Laplace operator determines the behavior of solutions in the following way. Recall that if ψ is a vector in $\mathcal{H}$, and A is a self-adjoint operator on $\mathcal{H}$, the linear functional $f \mapsto (f(A)\psi, \psi)$ on real-valued continuous functions f that vanish at infinity may be represented as integration with respect to a Borel measure μ_ψ on the real line:

$$(f(A)\psi, \psi) = \int_{\mathbb{R}} f(\lambda)\, d\mu_\psi(\lambda).$$

This measure is called the spectral measure for ψ with respect to the self-adjoint operator A. For any self-adjoint operator A on a Hilbert space $\mathcal{H}$, the orthogonal decomposition

$$\mathcal{H} = \mathcal{H}_{\text{p.p.}}(A) \oplus \mathcal{H}_{\text{a.c.}}(A) \oplus \mathcal{H}_{\text{s.c.}}(A)$$

holds, corresponding to vectors $\psi \in \mathcal{H}$ for which the spectral measure μ_ψ is pure point, absolutely continuous, or singularly continuous with respect to Lebesgue measure on the line (see [**49**], chapter VII and §VIII.3). Roughly and informally, initial data in $\mathcal{H}_{\text{p.p.}}(\Delta_M)$, $\mathcal{H}_{\text{p.p.}}(\Delta_M)$, and $\mathcal{H}_{\text{s.c.}}(\Delta_M)$ corresponds respectively to bound, escaping, and recurrent orbits for the wave equation. On compact manifolds $\mathcal{H} = \mathcal{H}_{\text{p.p.}}(\Delta_M)$ so that there are only bound orbits. On non-compact manifolds with simple geometry at infinity, we expect that $\mathcal{H} = \mathcal{H}_{\text{p.p.}}(\Delta_M) \oplus \mathcal{H}_{\text{a.c.}}(\Delta_M)$, i.e., all orbits are either bounded or escape to infinity, and there are no recurrent orbits.

2.1. Compact Manifolds. If M is a closed manifold or compact manifold with boundary, the spectrum of the Laplacian consists of discrete eigenvalues λ_j associated to normalized eigenfunctions φ_j. The solution is written in the familiar separation of variables form

$$u(x,t) = \sum_{j=0}^{\infty} (\varphi_j, \psi) \cos\left(t\sqrt{\lambda_j}\right) \varphi_j(x), \tag{2.3}$$

where $(\,\cdot\,,\,\cdot\,)$ is the $L^2(M,g)$-inner product. The eigenfunctions φ_j and the numbers $\sqrt{\lambda_j}$ determine standing wave patterns and frequencies of oscillation. These are determined by the geometry of the manifold and encode geometric data.

For purposes of comparison with the non-compact case, it will be useful to note that the numbers λ_j may be obtained as poles of the L^2-resolvent operator

$$R(z) = (\Delta_M - z)^{-1}$$

whose residues project onto the appropriate eigenspaces. The solution operator $E(t)$ is obtained from the resolvent $R(z)$ via the integral formula

$$E(t) = \frac{1}{2\pi i} \int_{\text{Re}(\lambda)=c} \lambda R(-\lambda^2) e^{t\lambda}\, d\lambda, \tag{2.4}$$

where $c < 0$, as follows from the inverse Laplace transform

$$\cos(tx) = \frac{1}{2\pi i} \int_{\mathrm{Re}(\lambda)=c} \frac{\lambda}{\lambda^2 + x^2} e^{t\lambda} \, d\lambda, \tag{2.5}$$

true for any $c < 0$, together with the spectral theorem for self-adjoint operators. One can recover the formula (2.3) from (2.5) and the meromorphy of the resolvent operator.

2.2. Non-Compact Manifolds: Scattering Resonances. If (M, g) is not compact but has "simple geometry at infinity,"[2] the Laplace operator may have *no* eigenvalues, corresponding to the fact that energy may "leak out" of any bounded region. If we examine the behavior of a solution to (2.2) on a non-compact manifold M with simple geometry at infinity, but restrict attention to a compact subset of M, we find an expansion analogous to (2.3) in which the cosines are replaced by complex exponentials and the eigenvalues are replaced by complex numbers ζ, the *scattering resonances* of the Laplacian, whose real parts determine a frequency of oscillation and whose imaginary parts determine a rate of energy decay for the associated normal mode.[3] For example, in the case of scattering by a compactly supported perturbation in $\mathbb{R}^n$ when n is odd, the expansion[4]

$$\chi \cos\left(t\sqrt{\Delta_M}\right) \chi\psi \sim \sum_j \sum_{k=0}^{N_j} c_{j,k} t^k \exp\left(i\zeta_j t\right) \varphi_j(x)$$

holds, where $\chi \in C_0^\infty(\mathbb{R}^n)$ is a cutoff function, the φ_j are resonance eigenfunctions, and the numbers $c_{j,k}$ depend on the initial data ψ. The secular terms (involving powers of t) may arise because the resonances are solutions of a non-self-adjoint eigenvalue problem. More precisely, the resonances are poles of the analytically continued operator

$$\widetilde{R}(k) = \chi \left(\Delta_M - k^2\right)^{-1} \chi,$$

initially defined on the half plane $\mathrm{Im}(k) < 0$ (corresponding to the cut plane $\mathbb{C} - [0, \infty)$ in the $\lambda = k^2$ variable) and extended to the complex k-plane. Like the resolvent of a non-symmetric matrix, the resolvent $\widetilde{R}(k)$ has a Laurent expansion near a given singularity ζ whose polar part takes the form $\sum_{j=1}^{N_\zeta} A_j \left(k - \zeta\right)^{-j}$. Here the A_j are finite-rank operators and the A_j for $j \geq 2$ are nilpotent. The resonance eigenfunctions φ_j are determined by the finite-rank residues of the resolvent. The *multiplicity* of a scattering resonance ζ is the dimension, m_ζ, of the space $\oplus_{j=1}^{N_\zeta} \mathrm{Ran}(A_j)$. The set of resonances ζ together with their multiplicities m_ζ forms the *resonance set* for Δ_M and constitutes a discrete set of 'scattering' data analogous to the eigenvalues. This resonance set is the subject of the 'absolute scattering' inverse problem, Problem 1.2.

---

[2]Examples include perturbations of the Euclidean metric and non-compact locally symmetric spaces.

[3]For an introduction to resonances and a review of earlier literature, see the surveys [**67**] and [**68**].

[4]So-called *resonance wave expansions* of this kind were first proved by Vainberg [**61**] for acoustical scattering by an obstacle in $\mathbb{R}^n$. Resonance wave expansions have also been obtained for certain hyperbolic surfaces by Christiansen and Zworski [**23**] and for scattering on $\mathbb{R}^n$ by a compactly supported perturbation by Tang and Zworski [**60**]. Although resonance wave expansions are expected to hold in "reasonable" scattering situations, the proof involves delicate estimates on the meromorphically continued resolvent and subtle remainder estimates.

In most cases of interest, the scattering poles for a complete, non-compact manifold M with geometric boundary $\partial_\infty M$ can also be characterized as poles of a scattering matrix $S(z) : \mathcal{C}^\infty(\partial_\infty M) \to \mathcal{C}^\infty(\partial_\infty M)$.

2.3. Non-Compact Manifolds: Scattering Phase. Relative scattering theory compares solutions of an evolution equation such as the wave equation (2.2) to solutions of the same equation for a simpler, 'unperturbed' system. For example, suppose that $(M, g) = (\mathbb{R}^n, g)$ where g is a metric on $\mathbb{R}^n$ which differs from the Euclidean metric g_0 only on a compact set. A natural comparison problem is then the wave equation for the Laplacian Δ_0 on $(\mathbb{R}^n, g_0)$.

Thus, comparison or relative scattering theory is very naturally a branch of perturbation theory for linear operators. One of the most fruitful versions of scattering theory at the level of operator theory is the trace-class scattering theory pioneered by Kato, Birman, Krein, and others; Yafaev's monograph [**64**] gives a very complete survey. For basics of scattering theory and a more concise review of the trace-class theory of scattering, see [**50**]. The trace-class theory concerns spectral and scattering theory for pairs of operators (A, B) for which $\varphi(A) - \varphi(B)$ belongs to the trace class for some monotone function φ. The following theorem of Carron [**20**] (actually proved in somewhat greater generality there) shows that we can apply trace-class scattering to many geometric situations.

THEOREM 2.1. *Suppose that (M, g_0) is a complete Riemannian manifold and that g is another complete metric on M with the property that $g - g_0$ is compactly supported. Let Δ_0 and Δ be the respective Laplacians on (M, g_0) and (M, g), let $\tau : L^2(M, g) \to L^2(M, g_0)$ be the natural isometry, let $H_0 = \Delta_0$, and let $H = \tau\Delta\tau^*$. Then for any integer $k > n/2$ and any $z \in \mathbb{C} - \mathbb{R}$, the operator*

$$(H - z)^{-k} - (H_0 - z)^{-k}$$

is a trace-class operator.

For such pairs (H, H_0), the trace-class theory guarantees that for every solution of the initial value problem

$$\begin{aligned} u_{tt} &= -Hu \\ u(0) &= \psi \\ u_t(0) &= 0 \end{aligned}$$

with initial data ψ in the absolutely continuous spectral subspace for H, there are corresponding initial data $\psi_\pm$ and solutions $u_\pm$ of the equation

$$\begin{aligned} u_{tt} &= -H_0 u \\ u(0) &= \psi_\pm \\ u_t(0) &= 0 \end{aligned}$$

with the property that $\lim_{t\to\pm\infty} \|u(t) - u_\pm(t)\| = 0$. That is, the solutions $u_\pm$ of the unperturbed equation are asymptotic to the solution u of the perturbed equation. The scattering operator is the map $S(H, H_0) : \psi_- \to \psi_+$ from the "past" to the "future" asymptote.

The operator $S(H, H_0)$ commutes with H_0. Thus, in a spectral representation for H_0, $S(H, H_0)$ acts by unitary operators $S(\lambda)$ on Hilbert spaces $\mathcal{H}_\lambda$ that arise in the spectral decomposition of H_0; here $\lambda \in [0, \infty)$ is a spectral parameter.[5]

In geometric situations, $S(\lambda)$ can be viewed as an operator from $\mathcal{C}^\infty(\partial_\infty M)$ to itself: elements of $\mathcal{C}^\infty(\partial_\infty M)$ should be thought of as 'radiation patterns' for a wave of energy λ, and $S(\lambda)$ as a map from incoming to outgoing radiation patterns. The trace-class theory guarantees that $S(\lambda) - I$ is a trace class operator, so that the operator determinant $\det S(\lambda)$ is well-defined. Since $S(\lambda)$ is unitary, it follows that

$$\det S(\lambda) = \exp(2\pi i \sigma(\lambda))$$

for a function σ on the real line. The function $\sigma(\lambda)$ is called the *scattering phase* and is determined by the *pair* (H, H_0).[6]

A fundamental result of Birman and Krein relates the scattering phase to the spectral shift function (SSF) for the pair (H, H_0). Under the trace-class condition in Theorem 2.1, there is a measurable, real-valued, and locally integrable function ξ on $\mathbb{R}$ with the property for all admissible functions f (including $\mathcal{C}_0^\infty(\mathbb{R})$ functions and the function $f(\lambda) = \exp(-t\lambda)$), the trace formula

$$\operatorname{Tr}\left(f(H) - f(H_0)\right) = -\int f'(\lambda)\xi(\lambda)\, d\lambda$$

holds. The celebrated Birman-Krein formula states that

$$\det S(\lambda) = \exp(2\pi i \xi(\lambda)). \tag{2.6}$$

Setting $f(\lambda) = \exp(-t\lambda)$ yields a 'relative heat trace'

$$\begin{aligned} H(t) &= \operatorname{Tr}\left(\exp(-tH) - \exp(-tH_0)\right) \\ &= -t \int \exp(-t\lambda)\, \xi(\lambda)\, d\lambda. \end{aligned}$$

which is thus determined by the scattering phase. It is not difficult to see that if pairs of operators (H_1, H_0) and (H_2, H_0) have the same relative heat trace, then the spectral shift functions are equal almost everywhere.

The scattering phase is the subject of the 'relative scattering inverse problem,' Problem 1.3.

---

[5]For example, if H_0 is the Euclidean Laplacian, $S(H, H_0)$ acts on $L^2(\mathbb{R}^n)$. Let $\mathcal{H}_\lambda = L^2(S^{n-1})$, and think of the space $\mathcal{H} = L^2((0,\infty) \times S^{n-1})$ as the 'constant fibre direct integral'

$$\int_{(0,\infty)}^{\oplus} \mathcal{H}_\lambda\, d\lambda.$$

If $\mathcal{F}$ is the Fourier transform, then the map

$$\mathcal{V}f(\lambda, \omega) = 2^{-1/2}\lambda^{(n-2)/4}\left(\mathcal{F}f\right)(\lambda^{1/2}\omega)$$

from $L^2(\mathbb{R}^n)$ to $\mathcal{H}$ (here $\lambda \in (0,\infty)$ and $\omega \in S^{n-1}$) gives the spectral representation for H_0. The operator $\mathcal{V}S(H, H_0)\mathcal{V}^{-1}$ acts on $\mathcal{H}$ as

$$\int_{(0,\infty)}^{\oplus} S(\lambda)\, d\lambda$$

where $S(\lambda)$ is a unitary operator on $\mathcal{H}_\lambda$ and $S(\lambda) - I$ is a trace-class operator on $\mathcal{H}_\lambda$.

[6]As we have defined it, σ is defined only modulo the integers. The Birman-Krein formula (see (2.6)) expresses the scattering phase in terms of the spectral shift function for the pair (H, H_0), which *is* uniquely determined.

REMARK 2.2. When (M, g_0) is Euclidean space and g is a compactly supported perturbation of g_0, the scattering phase $\sigma(\lambda)$ has a meromorphic extension to a double cover (n odd) or an infinite cover (n even) of the cut plane whose poles are exactly the scattering resonances. Thus the scattering phase determines the scattering resonances. On the other hand, the scattering resonances determine $\sigma(\lambda)$ only up to finitely many parameters, in analogy to the fact that the zeros of an entire function of finite order determine an entire function only up to an overall factor which is the exponential of a polynomial (if n is odd, the additional parameters are exactly the coefficients of a polynomial in such an exponential factor).

REMARK 2.3. When (M, g) is a Riemann surface with finite geometry and infinite area, there is also a natural 'comparison scattering theory' as explained in §3. In these cases, the corresponding scattering phase can be continued to the complex plane and has poles at the scattering resonances. One can show that the continued scattering phase is an entire function of finite order so that it is determined by its zeros up to finitely many parameters. For pairs of isopolar manifolds, the two scattering phases have the same poles when analytically continued; for pairs of isophasal manifolds, the scattering phases are actually the same, a stronger condition.

3. The Sunada Technique

Sunada's technique [**56**] reduces the problem of constructing isospectral or isoscattering manifolds to finding a geometric model for a triple of finite groups (G, H_1, H_2) (sometimes called a *Sunada triple*) that obeys a simple conjugacy condition.

DEFINITION 3.1. Let G be a finite group and let H_1 and H_2 be subgroups of G. We will say that H_1 is *almost conjugate* to H_2 in G if each G-conjugacy class $[g]_G$ intersects H_1 and H_2 in the same number of elements.

REMARK 3.2. The almost conjugacy condition is equivalent to a representation theoretic condition as follows. The left multiplication of G on the cosets in G/H_i gives rise to a natural action of G on the finite-dimensional vector space $L^2(G/H_i)$. The subgroups H_1 and H_2 of G are almost conjugate if and only if $L^2(G/H_1)$ and $L^2(G/H_2)$ are isomorphic as G-modules.

Recall that a group G acts *freely* on a manifold M if the only $g \in G$ with a non-trivial fixed point set is the identity element. A group action is called *effective* if no nontrivial group element acts as the identity.

THEOREM 3.3. *Let H_1 and H_2 be almost conjugate subgroups of a finite group G. Let (M, g) be a compact Riemannian manifold on which G acts on the left by isometries. Assume that H_1 and H_2 act freely. Then*

$$\operatorname{spec}(H_1 \backslash M, g) = \operatorname{spec}(H_2 \backslash M, g).$$

There are many proofs of this theorem, each one simple and elegant. (See the survey [**7**] for a full discussion and references.) Pierre Berard [**4**], motivated by an example of Peter Buser, developed a proof by "transplantation" in which eigenfunctions on one manifold can be explicitly transplanted to eigenfunctions with the same eigenvalue on the other manifold. See also Zelditch [**65**] for an independent construction of transplantation. In our presentation below, we give a simplified version of the transplantation argument by Robert Brooks and Orit

Davidovich [8] (see also [11] and [12] for similar constructions in different geometric contexts and [30] for an expanded treatment of transplantation).

The transplantation argument is based on the representation-theoretic version of the almost conjugacy condition given in Remark 3.2. Any G-module isomorphism τ between $L^2(G/H_1)$ and $L^2(G/H_2)$ is uniquely determined by the image $c = \tau(\chi_{H_1})$, where $\chi_{H_1} \in L^2(G/H_1)$ is the map that takes the value 1 on the coset H_1 and zero elsewhere. We may view c as a function on G satisfying $c(gh) = c(g)$ for all $g \in G$ and $h \in H_2$. Under the hypothesis of Theorem 3.3, identify $\mathcal{C}^\infty(H_i \backslash M_i)$ with $\mathcal{C}^\infty(M)^{H_i}$, the space of smooth functions on M invariant under the action of H_i. Then one shows that the map $T : \mathcal{C}^\infty(M) \to \mathcal{C}^\infty(M)$ given by

$$T(f)(x) = \sum_{g \in G} c(g) f(g \cdot x) \tag{3.1}$$

is a linear isomorphism from $\mathcal{C}^\infty(M)^{H_1}$ to $\mathcal{C}^\infty(M)^{H_2}$ intertwining the Laplacians.

Given a finite group G and almost conjugate subgroups H_1 and H_2, one can easily obtain examples of compact Riemannian manifolds M on which G acts by isometries in such a way that Theorem 3.3 can be applied. Indeed let M_0 be any compact Riemannian manifold whose fundamental group admits a surjection $\phi : \pi_1(M_0) \to G$, and let M be the Riemannian covering with fundamental group $\ker(\phi)$. Then G acts freely by isometries on M. (This condition is stronger than what is needed in Theorem 3.3. There we require that H_1 and H_2 act freely so that the quotients are manifolds but not that G act freely. In the more general case, $M_0 = G \backslash M$ is an orbifold.) In this way one obtains manifolds M_1 and M_2 which are isospectral and are covers of M_0, hence are locally isometric.

As noted in the introduction, Sunada's theorem has been used extensively to construct isospectral, non-isometric compact manifolds (see for example the first author's survey paper [27]), and there are a number of adaptations to the noncompact setting (see [4], [8], [11], [12], [38], [65]). Here we review one such adaptation, given by Brooks and Davidovich [8], to construct isopolar and isophasal Riemann surfaces with cusps and/or funnels.

Let M be a complete Riemann surface of infinite area and finite geometry, and let M^C be a conformal compactification. M^C has one boundary circle for each funnel and one boundary point for each cusp. Thus if M has N_c cusps and N_f funnels, a smooth function on ∂M^C is an element of $\mathbb{C}^{N_c} \oplus \left(\oplus_{j=1}^{N_f} \mathcal{C}^\infty(S^1) \right)$. Choose a *defining function* on M^C; this is a function ρ which is positive on M^C except at the boundary circles and cusp points, where it vanishes exactly to first order. Clearly, any two defining functions ρ and ρ' are related by $\rho = w\rho'$ for a strictly positive function $w \in \mathcal{C}^\infty(M^C)$.

Let Δ be the positive Laplacian on M with its natural hyperbolic metric. To define the scattering matrix, we first need a uniqueness result for generalized eigenfunctions (see, for example [41]). This result should be viewed as an analogue of uniqueness of solutions for the Dirichlet problem on a bounded domain.

PROPOSITION 3.4. *For s with* $\mathrm{Re}(s) = 1/2$ *and* $s \neq 1/2$, *let* $f_+ \in \mathcal{C}^\infty(\partial M^C)$. *Then there exists a unique solution* $u \in \mathcal{C}^\infty(M)$ *of the eigenvalue equation*

$$(\Delta - s(1-s))u = 0$$

with the property that

$$u \sim \rho^{1-s} f_+ + \rho^s f_- + \mathcal{O}(\rho)$$

as $\rho \downarrow 0$ *for a function* $f_- \in \mathcal{C}^\infty(M^C)$.

The uniqueness of u implies that the map $S(s) : f_+ \to f_-$ is well-defined; this map is called the scattering matrix, a map from $\mathcal{C}^\infty(\partial M^C)$ to itself. The functions $f_\pm$ can be thought of as incoming and outgoing radiation patterns. Although initially defined for $\mathrm{Re}(s) = 1/2$, the scattering matrix extends to a meromorphic operator-valued function of s. Roughly and informally, the poles of $S(s)$ coincide with the poles of the resolvent (see, for example, [**38**] for a detailed discussion). Two Riemann surfaces are called *isopolar* if the poles of their respective scattering matrices coincide. Although the scattering matrix appears to depend on the choice of defining function, this dependence is trivial and it can be shown that the poles of the scattering matrix are independent of the choice of defining function.

To obtain isopolar surfaces, Brooks and Davidovich began with a complete surface M_0 whose fundamental group surjects on a finite group G containing almost conjugate subgroups H_1 and H_2. They let M_0^C be a conformal compactification and chose a defining function ρ_0. Letting M be the complete Riemann surface that covers M_0 with covering group G, then a conformal compactification M^C of M covers M_0^C, and the defining function ρ_0 lifts to a G-invariant defining function ρ on M^C. Let $M_i = H_i \backslash M$. Then $M_i^C := H_i \backslash (M^C)$ is a conformal compactification of M_i and ρ descends to a defining function ρ_i on M_i^C. Identifying $\mathcal{C}^\infty(M_i^C)$ with $\mathcal{C}^\infty(M^C)^{H_i}$ and $\mathcal{C}^\infty(\partial M_i^C)$ with $\mathcal{C}^\infty(\partial M^C)^{H_i}$, then it is straightforward to check that the transplantation map T defined by equation 3.1 both on $\mathcal{C}^\infty(M^C)$ and on $\mathcal{C}^\infty(\partial M^C)$ intertwines the scattering matrices of M_1 and M_2. Thus these manifolds are isopolar. Brooks and Davidovich used this method (with very carefully chosen M_0) to construct isopolar surfaces of various genera with various numbers of ends and also to construct isopolar congruence surfaces.

As shown in [**38**], it is possible to define relative scattering from a Riemann surface of infinite area and finite geometry. For such a Riemann surface M,

$$M = Z \cup \left(\cup_{i=1}^{N_C} C_i\right) \cup \left(\cup_{j=1}^{N_f} F_j\right)$$

where C_i are cusps and F_j are funnels. The absolute scattering operator $S(s)$ acts on $\mathcal{C}^\infty(\partial M^C) = \mathbb{C}^{N_c} \oplus \left(\oplus_{j=1}^{N_f} \mathcal{C}^\infty(S^1)\right)$; if

$$S_0(s) = \mathbf{1} \oplus \left(\oplus_{j=1}^{N_f} S_{F_j}(s)\right)$$

where $\mathbf{1}$ is the identity on $\mathbb{C}^{N_C}$ and $S_{F_j}(s)$ is the scattering matrix for a hyperbolic half-funnel with Dirichlet conditions, then the relative scattering matrix is $S_{\mathrm{rel}}(s) = S(s)S_0(s)^{-1}$ and compares wave motion on M with wave motion on a disjoint union of funnels. The scattering phase is given by $\det(S_{\mathrm{rel}}(s)) = \exp(2\pi i \xi(s))$. The counterexamples constructed by Brooks and Davidovich (as well as earlier examples constructed by Guillopé and Zworski in [**38**]) are also isophasal.

4. The Method of Torus Actions

By a torus we will always mean a compact connected abelian Lie group. At the beginning of the development of the so-called method of torus actions lay the first author's observation of the general principle expressed in the theorem below. She used this idea to obtain the first examples of pairs of *locally nonisometric* isospectral closed manifolds (certain two-step nilmanifolds). (Earlier Z. Szabo had announced the first examples of locally nonisometric, isospectral manifolds with boundary, later published in [**58**].)

THEOREM 4.1. [**26**] *If a torus T acts on two closed Riemannian manifolds (M, g), (M', g') freely and isometrically with totally geodesic orbits, and if the quotients of the manifolds by any subtorus W of codimension at most one are isospectral when endowed with the submersion metrics g^W, ${g'}^W$, then (M, g) and (M', g') are isospectral.*

The proof is quite simple: We use the fact that if $M \to \overline{M}$ is a Riemannian submersion with totally geodesic fibers, then the spectrum of $\overline{M}$ coincides with the spectrum of the Laplacian on M restricted to functions constant on the fibers. Using Fourier decomposition with respect to the T-action and the fact that the T-orbits are totally geodesic (and hence that the W-orbits are also totally geodesic for each subtorus W), one thus shows that

$$\operatorname{spec}(M, g) = \operatorname{spec}(M/T, g^T) \cup \bigcup_W \left(\operatorname{spec}(M/W, g^W) \setminus \operatorname{spec}(M/T, g^T)\right),$$

where multiplicities are taken into account and W runs through the set of all subtori $W \subset T$ of codimension one. Since the analogous decomposition of the spectrum also holds for (M', g'), the theorem follows immediately.

In view of this, the theorem above seems almost tautological. Its usefulness, however, lies in the fact that there are lots of examples in which the submersion quotients $(M/W, g^W)$ and $(M'/W, {g'}^W)$ are actually *isometric* (and thus trivially isospectral), but still the "big" isospectral manifolds (M, g), (M', g') are nonisometric.

Particularly simple examples of this kind occur in the case where $M = M' = N \times T$ for some closed manifold N and the T-action is the canonical action on the second factor. Fix a Riemannian metric h on N and a translation invariant (i.e., flat) metric on T. The metric on T is specified by an inner product $\langle\, ,\rangle$ on the Lie algebra $\mathfrak{t}$. For each $(x, t) \in N \times T$ the tangent space $T_{(x,t)}(N \times T)$ is a direct sum $T_xN \oplus \mathfrak{t}$. Given a $\mathfrak{t}$-valued 1-form λ on N, we construct a metric g_λ on $N \times T$ so that the projection $\pi : (N \times T, g_\lambda) \to (N, h)$ is a Riemannian submersion with fibres isometric to T with its given metric. We specify g_λ at each $(x, t) \in N \times T$ by defining $\operatorname{Hor}_{(x,t)}$ (i.e., the g_λ-orthogonal complement of the tangent space to the T-orbit through (x, t)) to be the graph of $-\lambda_x : T_xN \to \mathfrak{t}$ and requiring:

- $g_\lambda(U, V) = \langle U, V\rangle$ for $U, V \in \mathfrak{t}$,
- $\operatorname{Hor}_{(x,t)} \perp \mathfrak{t}$, and
- $\left(\operatorname{Hor}_{(x,t)}, \langle\,\cdot\,,\,\cdot\,\rangle_g\right) \to (T_xN, \langle\,\cdot\,,\,\cdot\,\rangle_h)$ is an isometry.

REMARK 4.2. The pull-back of λ to $N \times T$ by the projection $N \times T \to N$ is a T-equivariant 1-form $\tilde{\lambda}$ that vanishes on vectors tangent to the T-orbits. Letting g denote the product metric on $N \times T$ defined by the metric h on N and the given flat metric on T, then the metric g_λ defined above may be expressed as $g_\lambda(X, Y) = g(X + \tilde{\lambda}(X), Y + \tilde{\lambda}(Y))$ for $X, Y \in T(N \times T)$. (This is one reason for the minus sign in the definition of $\operatorname{Hor}_{(x,t)}$. Also, the connection form on the principal T-bundle $N \times T$ which has the same horizontal distribution as g_λ is given by $(X, Z) \mapsto \lambda(X) + Z$ for all $X \in T_xN$, $Z \in \mathfrak{t}$.)

Suppose now that g_{λ_1} and g_{λ_2} are two such metrics and that there is an isometry F of N with the property that $\lambda_2 = F^*\lambda_1$. Then the map $(F, Id) : (N \times T, g_{\lambda_1}) \to (N \times T, g_{\lambda_2})$ is easily seen to be an isometry.

By the very construction of g_λ we have $g_\lambda^T = h$ for all λ. Let $W \subset T$ be a subtorus of codimension one. Then $T/W \cong S^1$ is a 1-dimensional torus with Lie algebra $\mathfrak{t}/\mathfrak{w}$, and $((N \times T)/W, g_\lambda^W)$ is isometric to $(N \times (T/W), g_{\lambda^W})$, where λ^W is the $\mathfrak{t}/\mathfrak{w}$-valued 1-form on N given by λ, followed by the canonical projection. Here, $\mathfrak{w}$ denotes the Lie algebra of W. We may view λ^W as a real-valued 1-form on N by choosing $\mu \in \mathfrak{t}^*$ with $\ker \mu = \mathfrak{w}$ and identifying λ^W with $\mu \circ \lambda$.

Now let λ' be a second $\mathfrak{t}$-valued 1-form on N. By our observations above, the two submersion quotients $((N \times T)/W, g_\lambda^W)$ and $((N \times T)/W, g_{\lambda'}^W)$ will be isometric if the two $\mathfrak{t}/\mathfrak{w}$-valued 1-forms λ^W and λ'^W are intertwined by an isometry of (N, h). We have thus proved:

THEOREM 4.3. [**53**] *Let λ, λ' be $\mathfrak{t}$-valued 1-forms on N. Assume that for each $\mu \in \mathfrak{t}^*$ there exists $F_\mu \in \mathrm{Isom}(N, h)$ such that $\mu \circ \lambda = F_\mu^*(\mu \circ \lambda')$. Then $(N \times T, g_\lambda)$ and $(N \times T, g_{\lambda'})$ are isospectral.*

In all the applications thus far of Theorem 4.3, mild genericity conditions on the choices of λ have sufficed to guarantee that the metrics g_λ and $g_{\lambda'}$ are not isometric, provided of course that $\lambda' \neq F^*\lambda$ for any isometry F of (N, h).

The dependence of the isometry F_μ on μ is crucial here: If there were an isometry F of (N, h) satisfying the condition above for all μ, then λ and λ' themselves would be intertwined by F and hence give rise to isometric metrics g_λ and $g_{\lambda'}$.

There are several generalizations of this theorem. In its most general version (see Theorem 4.10 below), the torus action is not even required to be free anymore. However, already Theorem 4.3 has many nice applications. In most of them, the key to constructing suitable 1-forms λ, λ' are pairs or families of so-called isospectral j-maps, defined as follows.

Let H be a compact connected semisimple Lie group with Lie algebra $\mathfrak{h}$, and let the Lie algebra $\mathfrak{t}$ of T be endowed with a fixed euclidean inner product.

DEFINITION 4.4.

(i) Two linear maps $j, j' : \mathfrak{t} \to \mathfrak{h}$ are called *isospectral* if for each $Z \in \mathfrak{t}$ there is $a_Z \in H$ such that $j'_Z = \mathrm{Ad}_{a_Z}(j_Z)$.
(ii) j and j' are called *equivalent* if there is $\Phi \in \mathrm{Aut}(\mathfrak{h})$ and $C \in \mathrm{O}(\mathfrak{t})$ such that $j'_Z = \Phi(j_{C(Z)})$ for all $Z \in \mathfrak{t}$.

REMARK 4.5. Let $\mathfrak{t} = \mathbb{R}^2$, equipped with the standard metric, and denote by $\mathcal{J}$ the vector space of all linear maps from $\mathfrak{t}$ to $\mathfrak{h}$.

(i) [**36**] If $\mathfrak{h} = \mathfrak{so}(m)$, where m is any positive integer other than $1, 2, 3, 4$, or 6, then there is a Zariski open subset $\mathcal{O}$ of $\mathcal{J}$ such that each $j \in \mathcal{O}$ belongs to a d-parameter family of isospectral, inequivalent elements of $\mathcal{J}$. Here $d \geq m(m-1)/2 - [m/2]([m/2]+2) > 1$. For $m = 6$, there exist at least 1-parameter families in $\mathcal{J}$ with these properties.
(ii) [**53**] If $\mathfrak{h} = \mathfrak{su}(m)$, where $m \geq 3$, then there is a Zariski open subset $\mathcal{O}$ of $\mathcal{J}$ such that each $j \in \mathcal{O}$ belongs to a continuous family of isospectral, inequivalent elements of $\mathcal{J}$.
(iii) [**48**] Recently, Emily Proctor established results analogous to those in (i) (multiparameter families) for $\mathfrak{su}(m \geq 5)$ and $\mathfrak{sp}(m \geq 2)$.

EXAMPLE 4.6. [**29**] Let $(N, h) := S^{m-1 \geq 4}$, endowed with the standard metric. Let T be two-dimensional. For each linear map $j : \mathfrak{t} \to \mathfrak{so}(m)$ define a T-valued 1-form λ^j on S^{m-1} by

$$\langle \lambda^j_q(X), Z \rangle := \langle j_Z q, X \rangle$$

for all $X \in T_q S^{m-1}$. If two such maps $j, j' : \mathfrak{t} \to \mathfrak{so}(m)$ are isospectral, then the associated forms $\lambda^j, \lambda^{j'}$ satisfy the condition of Theorem 4.3. In fact, if $\mu \in \mathfrak{t}^*$ and $Z \in \mathfrak{t}$ is the dual vector with respect to the inner product on $\mathfrak{t}$, and a_Z is chosen as in Definition 4.4(i), then the isometry $F_\mu := a_Z \in \mathrm{SO}(m)$ of S^{m-1} satisfies $\mu \circ \lambda^j = F_\mu(\mu \circ \lambda^{j'})$. Theorem 4.3 thus yields pairs of isospectral metrics on $S^{m-1} \times T$.

Actually the construction of these manifolds $(S^{m-1} \times T, g_{\lambda^j})$ in [**29**] had not been done using the approach above of associating metrics to certain 1-forms λ; rather, the manifolds there occurred as submanifolds of certain two-step nilpotent Lie groups with a left invariant metric. These submanifolds, in turn, were the boundaries of certain Dirichlet- and Neumann-isospectral subdomains diffeomorphic to the product $B^m \times T$ of a ball with a torus, which had been given in [**36**]. The latter had been the first examples of *continuous* families of isospectral metrics which were not locally isometric.

The isospectral metrics constructed above on $S^{m-1} \times T$ are in general not locally isometric when j and j' are inequivalent. For example, the metrics can in general be distinguished by the maximum of the associated scalar curvature function on the manifold.

By using multiparameter families of isospectral j-maps, one obtains multiparameter families of isospectral metrics on $S^{m-1} \times T$ which, again, can be shown to be nontrivial in most cases.

Independently of [**29**], Szabo had constructed pairs of isospectral metrics on certain products of spheres (or balls) with tori [**58**]. Excitingly, these examples include a pair of manifolds where one is homogeneous and the other is not even locally so.

EXAMPLE 4.7. Although in the example above the dimension of the sphere factor was required to be at least four, pairs of $\mathfrak{t} \cong \mathbb{R}^2$-valued 1-forms λ, λ' (not arising from j-maps) which satisfy the condition of Theorem 4.3 nontrivially can be found even on S^2. Using such 1-forms, the third author constructed in [**53**] pairs of isospectral metrics on $S^2 \times T$ (with T two-dimensional) which can be distinguished by the dimension of the locus of the maximal scalar curvature. No examples of locally nonisometric isospectral manifolds in dimension lower than four are known so far.

EXAMPLE 4.8. [**53**] Let again $\dim(\mathfrak{t}) = 2$, let $\mathfrak{h}$ be any of the Lie algebras from Remark 4.5, and let H be a Lie group with Lie algebra $\mathfrak{h}$, endowed with a bi-invariant metric h. We will let H play the role of the manifold N in the discussion above, so that each $\mathfrak{t}$-valued 1-form λ gives rise to a Riemannian metric g_λ on $H \times T$. If we choose λ to be *left-invariant* (i.,e., λ is defined by a linear map $\mathfrak{h} \to \mathfrak{t}$), then g_λ will be a left-invariant metric on $H \times T$.

For each linear map $j : \mathfrak{t} \to \mathfrak{h}$, define a left invariant $\mathfrak{t}$-valued 1-form λ^j on H by

$$\langle \lambda^j(X), Z \rangle = \langle j_Z, X \rangle$$

for all $Z \in \mathfrak{t}$ and all $X \in \mathfrak{h}$. If $j, j' : \mathfrak{t} \to \mathfrak{h}$ are isospectral, then $\lambda^j, \lambda^{j'}$ again satisfy the conditions of Theorem 4.3: For $\mu \in \mathfrak{t}$, the isometry F_μ of (H, h) which satisfies $\mu \circ \lambda^j = F_\mu^*(\mu \circ \lambda^{j'})$ is now conjugation $I_{a_Z} = L_{a_Z} \circ R_{a_z}^{-1} \in \mathrm{Isom}(H, h)$, where again Z is the vector dual to μ, and $a_Z \in H$ is chosen as in Definition 4.4(i). Theorem 4.3 thus gives us isospectral left-invariant metrics on $H \times T$.

The metrics in this example are homogeneous and can in general be distinguished, as shown in [**53**], by the norm of the associated Ricci tensor.

REMARK 4.9. The construction of the 1-forms $\langle \lambda^j(.), Z\rangle$ on S^{m-1} in Example 4.6 can be interpreted as taking duals to the Killing vectorfields corresponding to the $j_Z \in \mathfrak{so}(m)$, induced on S^{m-1} by the action of $\mathrm{SO}(m)$. Viewed in this way, the construction immediately generalizes to any other base manifold (N, h) admitting an effective isometric H-action, where H is a Lie group whose Lie algebra is one of those from Remark 4.5. Then one canonically obtains isospectral metrics of the type g_{λ^j} on $N \times T$, using pairs or families of isospectral j-maps from $\mathfrak{t}$ to $\mathfrak{h}$.

As an illustration, if we endow $N := H$ with a bi-invariant metric and consider the *left* action of H on itself, then we obtain *right* invariant isospectral metrics g_{λ^j} on $H \times T$. The left invariant isospectral metrics g_{λ^j} from Example 4.8 correspond, in the same sense, to the right action of H on itself (up to the sign of the 1-forms λ^j).

The two cases above, namely, that N is chosen to be either S^{m-1} (in case $H = \mathrm{SO}(m)$) or to be H itself, are in a sense extreme; see [**55**] for a discussion of the case of other homogeneous spaces $N = H/K$.

Below we will, as promised, present the current state of the art—as formulated in Theorem 4.10—concerning the method of torus actions. There had been several intermediate steps:

- The case of nontrivial T-bundles with totally geodesic fibers: [**26**] (certain two-step nilmanifolds, constituting the first examples of locally nonisometric isospectral manifolds), [**52**] (the first examples of isospectral metrics on simply connected manifolds; namely, certain products of spheres); [**53**] (including continuous isospectral families of left invariant metrics on irreducible compact Lie groups);
- the case without the assumption of totally geodesic fibers, but still with a free T-action: [**33**] (including examples of continuous isospectral families of negatively curved manifolds with boundary), [**53**] (including the first examples of pairs of conformally equivalent, locally nonisometric manifolds);
- general T-actions, but still for compact manifolds [**28**] (the first examples of continuous families of isospectral metrics on spheres and balls; namely, continuous families on $B^{n\geq 9}$ and $S^{n-1\geq 8}$), [**54**] (continuous families of isospectral metrics on B^8 and S^7, and pairs on B^6 and S^5); [**32**] (the first examples of isospectral potentials and conformally equivalent isospectral metrics on simply connected manifolds);
- finally, general T-actions on noncompact manifolds [**31**] (isophasal scattering metrics which are compact perturbations of the euclidean metric on $\mathbb{R}^n$), [**47**] (isophasal scattering metrics which are compact perturbations of any rotational metric on $\mathbb{R}^n$).

We remark that Z. Szabo [**59**] constructed the first examples of pairs of isospectral pairs of metrics on balls and spheres using a different technique involving explicit computations. His construction slightly preceded the construction cited above of continuous families of isospectral metrics on balls and spheres.

Before presenting the method we will use for constructing isophasal metrics, we review basic properties of group actions, in particular, torus actions. Given an action of a compact Lie group G on a manifold M, the *principal orbits* are the orbits with minimal isotropy. The union of the principal orbits is an open dense subset $\widehat{M}$ of M. There exists a subgroup H of G such the isotropy group of every element of $\widehat{M}$ is conjugate to H. Moreover, the isotropy group of an arbitrary element of M contains a subgroup conjugate to H. In case G is a torus, it follows that the isotropy group of every element contains H itself. In particular, if a torus action is effective, then H is trivial and so the action on the principal orbits is free. Thus $\widehat{M}$ is a principal G-bundle.

THEOREM 4.10. *Let T be a torus which acts effectively on two complete Riemannian manifolds (M,g) and (M',g') by isometries. For each subtorus $W \subset T$ of codimension one, suppose that there exists a T-equivariant diffeomorphism $F_W : M \longrightarrow M'$ which satisfies $F_W^* \,\mathrm{dvol}_{g'} = \mathrm{dvol}_g$ and induces an isometry $\bar{F}_W$ between the quotient manifolds $(\widehat{M}/W, g^W)$ and $(\widehat{M}'/W, g'^W)$, where $\widehat{M}$ (resp. $\widehat{M}'$) denote the union of the principal orbits in M (resp. M').*

(i) *Suppose (M,g) and (M',g') are compact. Then (M,g) and (M',g') are isospectral; if the manifolds have boundary then they are Dirichlet and Neumann isospectral.*

(ii) *Suppose $M = M'$, M is noncompact, and that g and g' are compact perturbations of a complete T-invariant Riemannian metric g_0 on M with $\mathrm{dvol}_{g_0} = \mathrm{dvol}_g = \mathrm{dvol}_{g'}$. Furthermore, assume that the maps F_W can be chosen such that they commute with Δ_{g_0}. Then (Δ_g, Δ_{g_0}) and $(\Delta_{g'}, \Delta_{g_0})$ have the same scattering phase.*

Part (i) of this theorem was first formulated in a slightly different version by the first author in [**28**]. For the proof of (i) (in the version above) and (ii) see [**54**] and [**47**], respectively. In each case, the heart of the proof consists in showing that there exists an L^2-norm preserving isometry from $\mathcal{H}^1(M,g')$ to $\mathcal{H}^1(M,g)$, where $\mathcal{H}^1$ actually means either $\mathcal{H}^1$ (closed case) or $\mathcal{H}^1_D$ resp. $\mathcal{H}^1_N$ (case with boundary) in (i), and $\mathcal{H}^1_0$ in (ii), as explained in section 2.

To construct this isometry, we decompose $\mathcal{H} := \mathcal{H}^1(M,g)$ and $\mathcal{H}' := \mathcal{H}^1(M',g')$ using Fourier decomposition with respect to the T-action and obtain

$$\mathcal{H} = \mathcal{H}_T \oplus \bigoplus_W (\mathcal{H}_W \ominus \mathcal{H}_T), \qquad \mathcal{H}' = \mathcal{H}'_T \oplus \bigoplus_W (\mathcal{H}'_W \ominus \mathcal{H}'_T),$$

where the sum runs over all subtori $W \subset T$ of codimension one, and $\mathcal{H}_W$ denotes the subspace of W-invariant functions (similarly for T and for $\mathcal{H}'$). Therefore, it suffices to find an L^2-norm preserving isometry from $\mathcal{H}'_W$ to $\mathcal{H}_W$ for each W. It turns out that the pullbacks F_W^* by the maps F_W chosen as in the assumptions do map $\mathcal{H}'_W$ to $\mathcal{H}_W$ isometrically. Preservation of L^2-norms is trivial here by the assumption $F_W^*(\mathrm{dvol}_{g'}) = \mathrm{dvol}_g$; preservation of $\mathcal{H}^1$-norms then follows from preservation of norms of gradients (of smooth W-invariant functions), which in turn follows from the assumption that F_W induces an isometry from $(M/W, g^W)$ to $(M'/W, g'^W)$.

There is a useful specialization of Theorem 4.10 in which $M = M'$, and g, g' arise from g_0 by changing the horizontal distribution on $\widehat{M}$ using a pair λ, λ' of $\mathfrak{t}$-valued 1-forms on M; this specialization is actually a generalization of Theorem 4.3. (See Remark 4.2 to clarify the relationship between Theorem 4.3 and the construction below.)

Begin with a complete Riemannian manifold (M, g_0) on which a torus T acts by isometries. For $Z \in \mathfrak{t}$, the action of T on M gives rise to a vector field Z^* on M given by $Z^*_p = \frac{d}{dt}\big|_{t=0} e^{tZ} \cdot p$. Given a $\mathfrak{t}$-valued 1-form λ on M which is T-invariant and vanishes on vectors tangent to the T-orbits, we define a new metric g_λ on M by

$$g_\lambda(X, Y) := g_0(X + \lambda(X)^*, Y + \lambda(Y)^*)$$

for all $X, Y \in T_pM$, $p \in M$. Note that by the presupposed properties of λ, the metric g_λ is again a T-invariant Riemannian metric, coincides with g_0 on vectors tangent to the T-orbits, and satisfies $g_0^T = g_\lambda^T$. The metrics g_0 and g_λ differ only by the associated horizontal distributions on $\widehat{M}$, which are related by

$$\mathrm{Hor}_p(g_\lambda) = \{X - \lambda(X)^* \mid X \in \mathrm{Hor}_p(g_0)\}$$

where $\mathrm{Hor}_p(g)$ denotes the g-orthogonal complement of $\mathfrak{t}$ in $T_p(M)$. In particular, the volume elements of g_0 and g coincide. In this situation we have:

THEOREM 4.11. [**47**], [**54**] *Let λ, λ' be T-invariant, $\mathfrak{t}$-valued 1-forms on M which vanish on vectors tangent to the T-orbits. Assume that for each $\mu \in \mathfrak{t}^*$ there exists a T-equivariant isometry F_μ of (M, g_0) such that $\mu \circ \lambda = F^*_\mu(\mu \circ \lambda')$. Moreover, assume λ, λ' to be compactly supported (in case M is noncompact). Then (M, g_λ) and $(M, g_{\lambda'})$ satisfy the conditions of Theorem* 4.10 (*with respect to g_0 in part* (ii)).

A typical situation in which Theorem 4.11 can be applied occurs when (M, g_0) admits an effective isometric $H \times T$-action, where H is a compact Lie group whose Lie algebra $\mathfrak{h}$ is one of the Lie algebras from Remark 4.5, and T is two-dimensional. Again one considers pairs or families of isospectral maps $j : \mathfrak{t} \to \mathfrak{h}$, and for each j defines a $\mathfrak{t}$-valued 1-form λ^j on M by letting $\langle \lambda^j(.), Z\rangle$ be dual to the Killing field corresponding to $j_Z \in \mathfrak{h}$ on M, induced by the action of H (compare Remark 4.9). Here, $\langle\, ,\, \rangle$ denotes some fixed auxiliary scalar product on $\mathfrak{t}$ (not to be confused with the metrics on any of the T-orbits, which now anyway are, in general, no longer isometric to each other).

Since the actions of H and T commute, it is clear that these λ^j will be T-invariant. Also, if j and j' are isospectral, then for each $Z \in \mathfrak{t}$ the 1-form $\langle \lambda^j(.), Z\rangle$ is the pullback of $\langle \lambda^{j'}(.), Z\rangle$ by the element $a_Z \in H$ from the isospectrality condition on j, j'; so these $a_Z \in H$ can serve as the T-equivariant g_0-isometries F_μ required in the assumption of Theorem 4.11. There is one difficulty, namely, the condition that the λ^j vanish on vectors tangent to the T-orbits. This can be ensured by assuming the action of $H \times T$ on M to be such that T-orbits and H-orbits meet g_0-orthogonally in every point. Even if this is not the case, one can achieve this condition by modifying the λ^j a bit: Namely, by first multiplying them with the squared norm of the volume form of the T-orbits, and then projecting the form thus obtained to its horizontal part (which will then still be smooth); see [**55**].

In the case that M is noncompact, the 1-forms constructed above will have noncompact support. This, however, can easily be mended using an idea from [**54**]: If one already has *some* pair of $\mathfrak{t}$-valued 1-forms λ, λ' on M which satisfy the conditions of the theorem, then so do $\psi\lambda, \psi\lambda'$, where ψ is any smooth function on M which is invariant under T and under all the F_μ from the assumption (more precisely, with the property that the F_μ can be *chosen* such that they preserve ψ). In our case, where the F_μ are actually elements of the compact group H, we can choose ψ to be any nontrivial smooth $H \times T$-invariant function on M; e.g., with support in an invariant neighborhood of any of the $H \times T$-orbits. This idea can be used in the compact case, too: It shows that the 1-forms can be chosen to have support in arbitrary small subsets of (M, g_0), outside which the associated metrics will be equal to g_0.

In the following examples, the isospectral (resp. isophasal) metrics either turned out to be, or were constructed to be, of the type just described; that is, they are associated with $\mathfrak{t}$-valued 1-forms on M which are of the above form λ^j—possibly modified as mentioned to ensure horizontality and/or compact support. Here we only list examples with a *nonfree* T-action:

- [**28**] Continuous multiparameter families of isospectral metrics on $S^{m+3\geq 8}$ and $B^{m+4\geq 9}$, associated with the standard action of $\mathrm{SO}(m)\times T \subset \mathrm{SO}(m)\times \mathrm{SO}(4) \subset \mathrm{SO}(m+4)$ on $\mathbb{R}^{m+4}$;
- [**54**] Continuous families of isospectral metrics on S^7 and B^8 associated with the action of $\mathrm{SU}(3) \times T \subset \mathrm{U}(3) \times \mathrm{U}(1) \subset \mathrm{U}(4)$ on $\mathbb{C}^4 \cong \mathbb{R}^8$;
- [**31**] Continuous multiparameter families of isophasal scattering metrics on $\mathbb{R}^{m+4\geq 9}$ which are compact perturbations of the Euclidean metric;
- [**47**] Continuous multiparameter families of isophasal scattering metrics on $\mathbb{R}^{m+4\geq 9}$ which are compact perturbations of *any* rotational (that is, $\mathrm{O}(m+4)$-invariant) metric on $\mathbb{R}^{m+4}$; for example, compact perturbations of the hyperbolic metric.

The same $\mathrm{SO}(m) \times T$-action on $\mathbb{R}^{m+4}$ as in the first item is used in the third and fourth one. Using the $\mathrm{SU}(3) \times T$-action from the second item, one also obtains isophasal scattering metrics on $\mathbb{R}^8$.

Finally, we mention that using certain suitable pairs of $\mathfrak{t} \cong \mathbb{R}^2$-valued 1-forms λ, λ' related to those used in Example 4.7, it is possible to apply Theorem 4.11 to obtain pairs of isospectral, resp. isophasal, metrics also on $S^2 \times S^3$ [**2**], S^5, B^6 [**54**] and $\mathbb{R}^6$.

5. Summary and Open Problems

5.1. Structure of isospectral or isopolar sets of metrics. We list below, in various contexts, the lowest dimensions in which examples of isospectral or isopolar manifolds are known.

- Pairs of isospectral manifolds with different global, though the same local, geometry: dimension 2 [**63**]. (As discussed in Section 3, there are large families of isospectral Riemann surfaces constructed by Sunada's technique [**10**].)
- Continuous isospectral deformations with different global, though the same local, geometry: dimension 5 [**35**].
- Pairs of isospectral metrics with different local geometry: dimension 4. (Metrics on $S^2 \times T^2$ [**53**]; see Section 4.)
- Continuous isospectral deformations with different local geometry: dimension 6. (Metrics on $S^4 \times T^2$ [**29**]; see Section 4.)

- Pairs of isospectral simply-connected manifolds: dimension 5. (Metrics on $S^2 \times S^3$ [**2**] and on S^5 [**54**]; see Section 4.)
- Continuous isospectral deformations of metrics on a simply-connected manifold: dimension 7. (Metrics on S^7 [**54**], see section 4.)
- Pairs of isophasal manifolds with same global, different local geometry: dimension 2. (Riemann surfaces [**8**], [**38**]; see Section 3.)
- Pairs of isophasal metrics with different local geometry: dimension 6. (Compactly supported perturbations of the Euclidean metric on $\mathbf{R}^6$; see Section 4.)
- Continuous isophasal deformations of metrics with different local geometry: dimension 8. (Compactly supported perturbations of the Euclidean metric on $\mathbf{R}^8$; see Section 4.)

PROBLEM 5.1. *In all cases above for which the lowest known dimension is greater than two, it remains open whether lower-dimensional examples exist. What is the critical dimension in each case?*

Given a closed Riemannian manifold (M, g), the set of all Riemannian metrics on M isospectral to g will be called the *isospectral set* of g. A more general problem than the one above is to describe the structure of the isospectral set of a metric. One can define a $\mathcal{C}^\infty$ topology on the set of all isometry classes of metrics on M.

PROBLEM 5.2. *Is the isospectral set of a metric always compact in the $\mathcal{C}^\infty$ topology?*

In dimension 2, the answer is yes, as proven by Osgood, Phillips, and Sarnak [**46**]. In higher dimensions, the problem remains open, although a number of partial results are known (in conformal classes on 3-manifolds [**16**], [**21**], [**22**]; on negatively curved 3-manifolds [**1**], [**13**]; for 3-manifolds with metrics in a spectrally determined neighborhood of a constant curvature metric [**15**]).

Negatively curved metrics on closed manifolds are spectrally rigid; i.e., they do not admit nontrivial continuous isospectral deformations; this fact was proven by Guillemin and Kazhdan [**37**] in dimension 2 and by Croke and Sharafutdinov [**24**], [**25**] in arbitrary dimension. In two dimensions, this local rigidity along with the compactness result cited above leads one to ask the following question:

PROBLEM 5.3. *Is the isospectral set of a negatively curved metric on a surface always finite?*

For Riemann surfaces with the hyperbolic metric, McKean proved that isospectral sets are always finite and P. Buser [**19**] obtained an explicit, though huge, upper bound depending only on the genus. However, the problem remains open for surfaces of variable negative curvature.

5.2. Geometry of the Spectrum. The various examples of isospectral manifolds reveal a number of geometric and topological properties that are not spectrally determined. In the following list, we indicate the first example which revealed that the specified property is not spectrally determined. (This list is not complete.)

(a) Fundamental group [**62**].
(b) Diameter [**18**].
(c) Maximal scalar curvature [**29**].

(d) $\int_M \mathrm{Scal}_g^2\, dv_g$, $\int_M \|\mathrm{Ric}_g\|^2 dv_g$, and $\int_M \|R_g\|^2 dv_g$. [**52**], [**53**]. (Note that a specific linear combination of these three integrals gives the second heat invariant, a spectrally determined quantity.)

(e) Whether a closed manifold is homogeneous and whether it is locally homogeneous [**58**], [**59**].

(f) Whether a closed manifold has constant scalar curvature; whether a manifold with boundary is Einstein and whether its curvature tensor is parallel [**33**].

Of course there is still a large gap between the known spectral invariants and the geometric invariants that the counterexamples tell us are not spectrally determined. Some questions suggested by the list above are:

PROBLEM 5.4. *The heat invariants are integrals of (in general very complicated) polynomial expressions in the curvature and its covariants. Are the heat invariants the only spectral invariants of that form? (See* (d) *above.)*

PROBLEM 5.5. *Does the spectrum of a closed Riemannian manifold determine whether the manifold is Einstein and whether it is locally symmetric? (See* (f) *above.)*

5.3. Obstacle Scattering. The most interesting open problem in inverse scattering involves obstacle scattering. Let $\mathcal{O}$ be a compact connected subset of $\mathbb{R}^n$ with smooth boundary, let H_0 be the Euclidean Laplacian on $\mathbb{R}^n$, let $\Omega = \mathbb{R}^n \setminus \mathcal{O}$, and let H be the Laplacian on Ω with Dirichlet boundary conditions on $\partial\mathcal{O}$. Although the pair (H, H_0) act on different spaces, one can define relative scattering theory and a spectral shift for the pair (H, H_0). Given the scattering phase for an obstacle $\mathcal{O}$, its *isophasal set* is the set of all obstacles in $\mathbb{R}^n$ modulo isometries of $\mathbb{R}^n$ which have the same scattering phase. Hassell and Zelditch [**39**] showed that the isophasal set of a given obstacle in $\mathbb{R}^2$ is compact in a suitable topology on obstacles. Hassell and Zworski [**40**] showed that the sphere S^2 is uniquely determined by its scattering poles (when viewed as the boundary of an obstacle $\mathcal{O} \subset \mathbb{R}^3$).

PROBLEM 5.6. *Does the isophasal set of an obstacle contain more than one element?*

PROBLEM 5.7. *What is the 'critical dimension' for polar and phasal rigidity for obstacle scattering?*

References

1. M. T. Anderson, *Remarks on the compactness of isospectral sets in low dimensions*, Duke Math. J. **63** (1991), no. 3, 699–711.
2. W. Ballmann, *On the construction of isospectral manifolds*, preprint (2001).
3. P. Bérard, *Transplantation et isospectralité.* I, Math. Ann. **292** (1992), no. 3, 547–559.
4. ———, *Transplantation et isospectralité.* II, J. London Math. Soc. (2) **48** (1993), no. 3, 565–576.
5. P. Bérard and D. L. Webb, *One cannot hear the orientability of surfaces*, C. R. Acad. Sci. Paris **320**, no. 1 (1995), 533–536.
6. R. Brooks, *Constructing isospectral manifolds*, Amer. Math. Monthly **95** (1988), no. 9, 823–839.
7. ———, *The Sunada method*, Tel Aviv Topology Conference: Rothenberg Festschrift (1998), Contemp. Math., **231**, Amer. Math. Soc., Providence, RI, 1999, pp. 25–35.
8. R. Brooks and O. Davidovich, *Isoscattering on surfaces*, J. Geom. Anal. **13** (2003), no. 1, 39–53.

9. R. Brooks and C. Gordon, *Isospectral families of conformally equivalent Riemannian metrics*, Bull. Amer. Math. Soc. (N.S.) **23** (1990), no. 2, 433–436.
10. R. Brooks, R. Gornet, and W. H. Gustafson, *Mutually isospectral Riemann surfaces*, Adv. Math. **138** (1998), no. 2, 306–322.
11. R. Brooks, R. Gornet, and P. Perry, *Isoscattering Schottky manifolds*, Geom. Funct. Anal. **10** (2000), no. 2, 307–326.
12. R. Brooks and P. A. Perry, *Isophasal scattering manifolds in two dimensions*, Comm. Math. Phys. **223** (2001), no. 3, 465–474.
13. R. Brooks, Peter A. Perry, and P. Petersen, *Compactness and finiteness theorems for isospectral manifolds*, J. Reine Angew. Math. **426** (1992), 67–89.
14. ———, *Finiteness of diffeomorphism types of isospectral manifolds*, Differential Geometry: Riemannian Geometry (Los Angeles, CA, 1990), Proc. Sympos. Pure Math., **54**, Part 3, Amer. Math. Soc., Providence, RI, 1993, pp. 89–94.
15. ———, *Spectral geometry in dimension* 3, Acta Math. **173** (1994), no. 2, 283–305.
16. R. Brooks, P. Perry, and P. Yang, *Isospectral sets of conformally equivalent metrics*, Duke Math. J. **58** (1989), no. 1, 131–150.
17. R. Brooks and R. Tse, *Isospectral surfaces of small genus*, Nagoya Math. J. **107** (1987), 13–24.
18. P. Buser, *Isospectral Riemann surfaces*, Ann. Inst. Fourier **36** (1986), 167–192.
19. ———, *Geometry and Spectra of Compact Riemann Surfaces*, Birkhäuser, Boston, 1992.
20. G. Carron, *Déterminant relatif et la fonction Xi*. Amer. J. Math. **124** (2002), no. 2, 307–352.
21. S.-Y. A. Chang and P. C. Yang, *Compactness of isospectral conformal metrics on S^3*, Comment. Math. Helv. **64** (1989), no. 3, 363–374.
22. ———, *Isospectral conformal metrics on 3-manifolds* J. Amer. Math. Soc. **3** (1990), no. 1, 117–145.
23. T. Christiansen and M. Zworski, *Resonance wave expansions: two hyperbolic examples*, Comm. Math. Phys. **212** (2000), no. 2, 323–336.
24. C. Croke and V. A. Sharafutdinov, *Spectral rigidity of a compact negatively curved manifold*, Topology **37** (1998), no. 6, 1265–1273.
25. ———, *Spectral rigidity of a compact negatively curved manifold*, Topology **37** (1998), no. 6, 1265–1273.
26. C. S. Gordon, *Isospectral closed Riemannian manifolds which are not locally isometric*. II, Geometry of the Spectrum (Seattle, WA, 1993), Contemp. Math., **173**, Amer. Math. Soc., Providence, RI, 1994, pp. 121–131.
27. ———, *Survey of isospectral manifolds*, Handbook of Differential Geometry, Vol. I, North-Holland, Amsterdam, 2000, pp. 747–778.
28. ———, *Isospectral deformations of metrics on spheres*, Invent. Math. **145** (2001), no. 2, 317–331.
29. C. S. Gordon, R. Gornet, D. Schueth, D. L. Webb, and E. N. Wilson, *Isospectral deformations of closed Riemannian manifolds with different scalar curvature*, Ann. Inst. Fourier (Grenoble) **48** (1998), no. 2, 593–607.
30. C. S. Gordon, E. Makover, and D. L. Webb, *Transplantation and Jacobians of isospectral Riemann surfaces*, to appear in Adv. Math.
31. C. S. Gordon and P. A. Perry, *Continuous families of isophasal scattering manifolds*, preprint.
32. C. S. Gordon and D. Schueth, *Isospectral potentials and conformally equivalent isospectral metrics on spheres, balls and Lie groups*, J. Geom. Anal. **13** (2003), no. 2, 279–306.
33. C. S. Gordon and Z. I. Szabo, *Isospectral deformations of negatively curved Riemannian manifolds with boundary which are not locally isometric*, Duke Math. J. **113** (2002), no. 2, 355–383.
34. C. S. Gordon, D. L. Webb, and S. Wolpert, *Isospectral plane domains and surfaces via Riemannian orbifolds*, Invent. Math. **110** (1992), no. 1, 1–22.
35. C. S. Gordon and E. N. Wilson, *Isospectral deformations of compact solvmanifolds*, J. Differential. Geom. **19** (1984), 241–256.
36. ———, *Continuous families of isospectral Riemannian metrics which are not locally isometric*, J. Differential Geom. **47** (1997), no. 3, 504–529.
37. V. Guillemin and D. Kazhdan, *Some inverse spectral results for negatively curved 2-manifolds*, Topology **19** (1980), no. 3, 301–312.
38. L. Guillopé and M. Zworski, *Scattering asymptotics for Riemann surfaces*, Ann. Math. **145** (1997), 597–660.

39. A. Hassell and S. Zelditch, *Determinants of Laplacians in exterior domains*, Internat. Math. Res. Notices **1999**, no. 18, 971–1004.
40. A. Hassell and M. Zworski, *Resonant rigidity of* S^2, J. Funct. Anal. **169** (1999), no. 2, 604–609.
41. M. Joshi and A. Sá Barreto, *Inverse scattering on asymptotically hyperbolic manifolds*, Acta Math. **184** (2000), 41–86.
42. P. D. Lax and R. S. Phillips, *Scattering theory*, second edition. With appendices by C. S. Morawetz and G. Schmidt. Pure and Appl. Math., **26**. Academic Press, Inc., Boston, MA, 1989.
43. R. Miatello and J. P. Rossetti, *Flat manifolds isospectral on p-forms*, J. Geom. Anal. **11** (2001), 649–667.
44. ———, *Comparison of twisted p-form spectra for flat manifolds with diagonal holonomy*, Ann. Global Anal. Geom. **21** (2002), no. 4, 341–376.
45. ———, *Length spectra and p-spectra of compact flat manifolds*, J. Geom. Anal. **13** (2003), no. 4, 631–657.
46. B. Osgood, R. Phillips, and P. Sarnak, *Compact isospectral sets of metrics*, J. Funct. Anal. **180** (1988), 212-234.
47. P. A. Perry and D. Schueth, *Continuous families of isophasal scattering manifolds which are asymptotically hyperbolic*, preprint (2004).
48. E. Proctor, *Isospectral metrics on classical compact simple Lie groups*, to appear in Mich. Math J.
49. M. Reed and B. Simon, *Methods of Modern Mathematical Physics, I. Functional Analysis*, New York, Academic Press, 1972.
50. ———, *Methods of Modern Mathematical Physics, III. Scattering Theory*, New York, Academic Press, 1979.
51. ———, *Methods of Modern Mathematical Physics, IV. Analysis of Operators*, New York, Academic Press, 1978.
52. D. Schueth, *Continuous families of isospectral metrics on simply connected manifolds*, Ann. of Math. (2) **149** (1999), no. 1, 287–308.
53. ———, *Isospectral manifolds with different local geometries*, J. reine angew. Math. **534** (2001), 41–94.
54. ———, *Isospectral metrics on five-dimensional spheres*, J. Differential Geom. **58** (2001), no. 1, 87–111.
55. ———, *Constructing isospectral metrics via principal connections*, Geometric Analysis and Nonlinear Partial Differential Equations (S. Hildebrandt, H. Karcher eds.), Springer, 2003, pp. 69-79.
56. T. Sunada, *Riemannian coverings and isospectral manifolds*, Ann. of Math. (2) **121** (1985), no. 1, 169–186.
57. C. J. Sutton, *Isospectral simply-connected homogeneous spaces and the spectral rigidity of group actions*, Comment. Math. Helv. **77** (2002), no. 4, 701–717.
58. Z. I. Szabo, *Locally non-isometric yet super isospectral spaces*, Geom. Funct. Anal. **9** (1999), no. 1, 185–214.
59. ———, *Isospectral pairs of metrics on balls, spheres, and other manifolds with different local geometries*, Ann. of Math. (2) **154** (2001), no. 2, 437–475.
60. S.-H. Tang and M. Zworski, *Resonance expansions of scattered waves*, Comm. Pure Appl. Math. **53** (2000), no. 10, 1305–1334.
61. B. R. Vainberg, *Exterior elliptic problems that depend polynomially on the spectral parameter, and the asymptotic behavior for large values of the time of the solutions of nonstationary problems* (Russian), Mat. Sb. (N.S.) **92** (134) (1973), 224–241, 343.
62. M.-F. Vignéras, *Exemples de sous-groupes discrets non conjugués de* $PSL(2, R)$ *qui ont même fonction zéta de Selberg*, C. R. Acad. Sci. Paris Sér. A-B **287** (1978), no. 2, A47–A49.
63. ———, *Variétés riemanniennes isospectrales et non isométriques*, Ann. of Math. (2) **112** (1980), no. 1, 21–32.
64. D. R. Yafaev, *Mathematical Scattering Theory. General Theory*, translated from the Russian by J. R. Schulenberger. Translations of Mathematical Monographs, **105**. American Mathematical Society, Providence, RI, 1992.
65. S. Zelditch, *Isospectrality in the FIO category*, J. Differential Geom. **35** (1992), no. 3, 689–710.
66. ———, *Kuznecov sum formulae and Szegö limit formulae on manifolds*, Comm. Partial Differential Equations **17** (1992), no. 1-2, 221–260.

67. M. Zworski, *Counting scattering poles*, Spectral and scattering theory (Sanda, 1992), 301–331, Lecture Notes in Pure and Appl. Math., **161**, Dekker, New York, 1994.
68. ———, *Resonances in physics and geometry*, Notices Amer. Math. Soc. **46** (1999), no. 3, 319–328.

Department of Mathematics, Dartmouth College, Hanover, New Hampshire 03755, U.S.A.
E-mail address: Carolyn.S.Gordon@dartmouth.edu

Department of Mathematics, University of Kentucky, Lexington, Kentucky 40506–0027, U.S.A.
E-mail address: perry@ms.uky.edu

Inst. f. Math., HU Berlin, 10099 Berlin, Germany
E-mail address: schueth@math.hu-berlin.de

Contemporary Mathematics
Volume **387**, 2005

Filling Area Conjecture, Optimal Systolic Inequalities, and the Fiber Class in Abelian Covers

Mikhail G. Katz* and Christine Lescop†

Dedicated to the memory of Robert Brooks, a colleague and friend.

ABSTRACT. We investigate the filling area conjecture, optimal systolic inequalities, and the related problem of the nonvanishing of certain linking numbers in 3-manifolds.

1. Filling radius and systole

The filling radius of a simple loop C in the plane is defined as the largest radius, $R > 0$, of a circle that fits inside C, see Figure 1.1:

$$\operatorname{FillRad}(C \subset \mathbb{R}^2) = R.$$

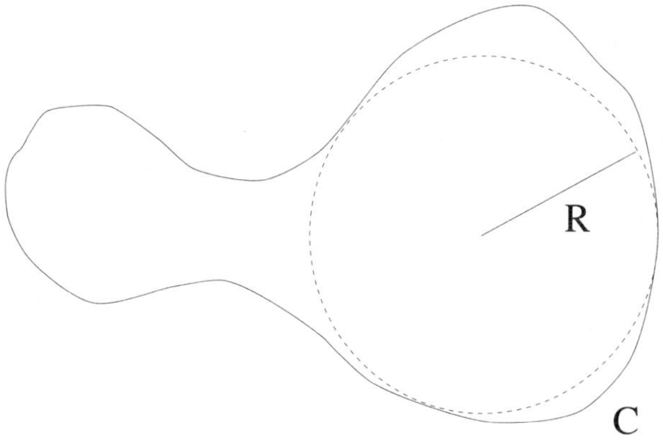

FIGURE 1.1. Largest inscribed circle has radius R

2000 *Mathematics Subject Classification.* Primary 53C23; Secondary 57N65, 57M27, 52C07.
Key words and phrases. Abel-Jacobi map, Casson invariant, filling radius, filling area conjecture, Hermite constant, linking number, Loewner inequality, Pu's inequality, stable norm, systole.
*Supported by the Israel Science Foundation (grants no. 620/00-10.0 and 84/03).
†Supported by CNRS (UMR 5582).

There is a kind of a dual point of view that allows one to generalize this notion in an extremely fruitful way, as shown in [**14**]. Namely, we consider the ϵ-neighborhoods of the loop C, denoted $U_\epsilon C \subset \mathbb{R}^2$, see Figure 1.2.

FIGURE 1.2. Neighborhoods of loop C

As $\epsilon > 0$ increases, the ϵ-neighborhood $U_\epsilon C$ swallows up more and more of the interior of the loop. The "last" point to be swallowed up is precisely the center of a largest inscribed circle. Therefore we can reformulate the above definition by setting

$$\operatorname{FillRad}(C \subset \mathbb{R}^2) = \inf \{\epsilon > 0 \,|\, \text{loop } C \text{ contracts to a point in } U_\epsilon C\}.$$

Given a compact manifold X imbedded in, say, Euclidean space E, we could define the filling radius *relative* to the imbedding, by minimizing the size of the neighborhood $U_\epsilon X \subset E$ in which X could be homotoped to something smaller dimensional, e.g., to a lower dimensional polyhedron. Technically it is more convenient to work with a homological definition.

Denote by A the coefficient ring $\mathbb{Z}$ or $\mathbb{Z}_2$, depending on whether or not X is oriented. Then the fundamental class, denoted $[X]$, of a compact n-dimensional manifold X, is a generator of $H_n(X; A) = A$, and we set

$$\operatorname{FillRad}(X \subset E) = \inf \left\{\epsilon > 0 \,\middle|\, \iota_\epsilon([X]) = 0 \in H_n(U_\epsilon X)\right\}, \tag{1.1}$$

where ι_ϵ is the inclusion homomorphism.

To define an "absolute" filling radius in a situation where X is equipped with a Riemannian metric g, M. Gromov proceeds as follows, cf. (1.3). One exploits an imbedding due to C. Kuratowski [**31**] (while the spelling of the author's first name used in [**31**] starts with a "C", the correct spelling seems to be Kazimierz). One imbeds X in the Banach space $L^\infty(X)$ of bounded Borel functions on X, equipped with the sup norm $\| \ \|$. Namely, we map a point $x \in X$ to the function $f_x \in L^\infty(X)$ defined by the formula $f_x(y) = d(x, y)$ for all $y \in X$, where d is the distance function defined by the metric. By the triangle inequality we have

$$d(x, y) = \|f_x - f_y\|, \tag{1.2}$$

and therefore the imbedding is strongly isometric, in the precise sense of formula (1.2) which says that internal distance and ambient distance coincide. Such a strongly isometric imbedding is impossible if the ambient space is a Hilbert space, even when X is the Riemannian circle (the distance between opposite points must be π, not 2!). We then set $E = L^\infty(X)$ in formula (1.1) and define

$$\text{FillRad}(X) = \text{FillRad}\left(X \subset L^\infty(X)\right). \tag{1.3}$$

Exact calculation of the filling radius is discussed in Section 5. The related invariant called filling volume is defined in Section 3. For the relation between filling radius and systole, see (5.1). The defining text for this material is [**16**], with more details in [**14, 15**]. See also the recent survey [**11**].

The systole, $\text{sys}\pi_1(g)$, of a compact non simply connected Riemannian manifold (X, g) is the least length of a noncontractible loop $\gamma \subset X$:

$$\text{sys}\pi_1(g) = \min_{[\gamma]\neq 0 \in \pi_1(X)} \text{length}(\gamma). \tag{1.4}$$

This notion of systole is apparently unrelated to the systolic arrays of [**30**]. We will be concerned with comparing this Riemannian invariant to the total volume of the metric, as in Loewner's inequality (2.2).

In Section 2, we recall the classical Loewner inequality, as well as Pu's inequality. In Section 3, we define an invariant called the filling volume, and state Gromov's filling volume (e.g., area) conjecture. The notion of the successive minima of a lattice in Banach space is recalled in Section 4. The results on the precise calculation of filling invariants (both the filling radius and the filling area) are reviewed in Section 5. Stable and conformal systoles are defined in Section 6.

In Section 7, we recall the definition of the Abel-Jacobi map to the torus, and present M. Gromov's optimal stable systolic inequality for n-tori. Higher dimensional optimal generalisations of Loewner's inequality are discussed in Section 8 and Section 9. Thus, in Section 8, we recall J. Hebda's inequality combining systoles of dimension and codimension 1 in the case of unit Betti number, and present a generalisation to arbitrary Betti number.

Optimal inequalities combining varieties of 1-systoles appear in Section 9. Here a necessary topological condition is the nonvanishing of the homology class of the lift to the maximal free abelian cover, of the typical fiber of the Abel-Jacobi map. The relation of this condition to the generalized Casson invariant λ of [**33**] is examined in Section 10. The example of nilmanifolds is discussed in Section 11, and the general case treated in Section 12.

2. Inequalities of C. Loewner and P. Pu

The Hermite constant, denoted γ_n, can be defined as the optimal constant in the inequality

$$\text{sys}\pi_1(\mathbb{T}^n)^2 \leq \gamma_n \, \text{vol}(\mathbb{T}^n)^{2/n}, \tag{2.1}$$

over the class of all *flat* tori $\mathbb{T}^n$, cf. Section 4. Here γ_n is asymptotically linear in n, cf. [**32**, pp. 334, 337]. The precise value is known for small n, e.g., $\gamma_2 = \frac{2}{\sqrt{3}}$ (see Lemma 4.5), $\gamma_3 = 2^{\frac{1}{3}}, \ldots$. It can be shown via the filling invariants that an inequality of type (2.1) remains valid in the class of *all* metrics, but with a nonsharp constant on the order of n^{4n} [**14**]. M. Gromov's proof uses inequality (5.1) as a starting point.

Around 1949, Charles Loewner proved the first systolic inequality, cf. [**38**]. He showed that every Riemannian metric g on the torus $\mathbb{T}^2$ satisfies the inequality

$$\text{sys}\pi_1(g)^2 \leq \gamma_2 \, \text{area}(g), \tag{2.2}$$

while a metric satisfying the boundary case of equality in (2.2) is necessarily flat, and is homothetic to the quotient of $\mathbb{C}$ by the lattice spanned by the cube roots of unity.

The following estimate is found in [**14**, Corollary 5.2.B]. Namely, every aspherical compact surface (Σ, g) admits a metric ball $B = B_p\left(\frac{1}{2}\operatorname{sys}\pi_1(g)\right) \subset \Sigma$ of radius $\frac{1}{2}\operatorname{sys}\pi_1(g)$, which satisfies

$$\operatorname{sys}\pi_1(g)^2 \leq \frac{4}{3}\operatorname{area}(B). \tag{2.3}$$

P. Pu's inequality [**38**] admits an immediate generalisation via Gromov's inequality (2.3). Namely, every surface (X, g) which is not a 2-sphere satisfies the inequality

$$\operatorname{sys}\pi_1(g)^2 \leq \frac{\pi}{2}\operatorname{area}(g), \tag{2.4}$$

where the boundary case of equality in (2.4) is attained precisely when, on the one hand, the surface X is a real projective plane, and on the other, the metric g is of constant Gaussian curvature.

3. Filling area conjecture

Consider a compact manifold N of dimension $n \geq 1$ with a distance function d. Here d could be induced by a Riemannian metric g, but could also be more general. The notion of the Filling Volume, $\operatorname{FillVol}(N^n, d)$, of N was introduced in [**14**], where it is shown that when $n \geq 2$,

$$\operatorname{FillVol}(N^n, d) = \inf_g \operatorname{vol}_{n+1}(X^{n+1}, g) \tag{3.1}$$

where X is any fixed manifold such that $\partial X = N$. Here one can even take a cylinder $X = N \times [0, \infty)$. The infimum is taken over all complete Riemannian metrics g on X for which the boundary distance function is $\geq d$, i.e., the length of the shortest path in X between boundary points p and q is $\geq d(p, q)$. In the case $n = 1$, the topology of the 2-dimensional filling can affect the infimum, as is shown by example in [**14**, Counterexamples 2.2.B].

The precise value of the filling volume is not known for any Riemannian metric. However, the following values for the canonical metrics on the spheres (of sectional curvature +1) is conjectured in [**14**], immediately after Proposition 2.2.A.

CONJECTURE 3.1. $\operatorname{FillVol}(S^n, can) = \frac{1}{2}\omega_{n+1}$, *where* ω_{n+1} *denotes the volume of the unit* $(n+1)$*-sphere.*

This conjecture is still open in all dimensions. The case $n = 1$ can be broken into separate problems depending on the genus s of the filling, cf. (5.3).

4. Lattices and successive minima

Let $b > 0$. By a *lattice* L in Euclidean space $\mathbb{R}^b$, we mean a discrete subgroup isomorphic to $\mathbb{Z}^b$, i.e., the integer span of a linearly independent set of b vectors.

More generally, let B be a finite-dimensional Banach space, i.e., a vector space together with a norm $\| \ \|$. Let $L \subset (B, \| \ \|)$ be a lattice of maximal rank $\operatorname{rank}(L) = \dim(B)$.

DEFINITION 4.1. For each $k = 1, 2, \ldots, \mathrm{rank}(L)$, define the *k-th successive minimum* of the lattice L by

$$\lambda_k(L, \|\ \|) = \inf\left\{ \lambda \in \mathbb{R} \;\middle|\; \begin{array}{l} \exists \text{ lin. indep. } v_1, \ldots, v_k \in L \\ \text{with } \|v_i\| \leq \lambda \end{array} \right\}. \tag{4.1}$$

Thus, the first successive minimum, $\lambda_1(L, \|\ \|)$ is the least length of a nonzero vector in L. If $\mathrm{rank}(L) \geq 2$, then given a pair of vectors $S = \{v, w\}$ in L, define the length $|S|$ of S by setting $|S| = \max(\|v\|, \|w\|)$. Then the second successive minimum, $\lambda_2(L, \|\ \|)$ is the least length of a pair of non-proportional vectors in L: $\lambda_2(L) = \inf_S |S|$, where S runs over all linearly independent pairs $\{v, w\} \subset L$.

EXAMPLE 4.2. The standard lattice $\mathbb{Z}^b \subset \mathbb{R}^b$ satisfies

$$\lambda_1(\mathbb{Z}^b) = \ldots = \lambda_b(\mathbb{Z}^b) = 1.$$

Note that the torus $\mathbb{T}^b = \mathbb{R}^b/\mathbb{Z}^b$ satisfies $\mathrm{vol}(\mathbb{T}^b) = 1$ as it has the unit cube as a fundamental domain.

Here the volume of the torus $\mathbb{R}^b/L$ (also called the covolume of the lattice L) is by definition the volume of a fundamental domain for L, e.g., a parallelepiped spanned by a $\mathbb{Z}$-basis for L. The volume can be calculated as the square root of the determinant of the Gram matrix of such a basis. Given a finite set $S = \{v_i\}$ in $\mathbb{R}^b$, we define its *Gram matrix* as the matrix of inner products $Gram(S) = (\langle v_i, v_j \rangle)$. Then the parallelepiped P spanned by the vectors $\{v_i\}$ satisfies $\mathrm{vol}(P) = \det(Gram(S))^{1/2}$.

EXAMPLE 4.3. Consider the lattice $L_\zeta \subset \mathbb{R}^2 = \mathbb{C}$ spanned by $1 \in \mathbb{C}$ and the sixth root of unity $\zeta = e^{\frac{2\pi i}{6}} \in \mathbb{C}$. Then $L_\zeta = \mathbb{Z}\zeta + \mathbb{Z}1$ satisfies $\lambda_1(L_\zeta) = \lambda_2(L_\zeta) = 1$. Meanwhile, the torus $\mathbb{T}^2 = \mathbb{R}^2/L_\zeta$ satisfies $\mathrm{area}(\mathbb{T}^2) = \frac{\sqrt{3}}{2}$.

DEFINITION 4.4. The Hermite constant γ_b is defined in one of the following two equivalent ways:

(a) γ_b is the *square* of the maximal first successive minimum, among all lattices of unit covolume;
(b) γ_b is defined by the formula

$$\sqrt{\gamma_b} = \sup\left\{ \frac{\lambda_1(L)}{\mathrm{vol}(\mathbb{R}^b/L)^{\frac{1}{b}}} \;\middle|\; L \subseteq (\mathbb{R}^b, \|\ \|) \right\}, \tag{4.2}$$

where the supremum is extended over all lattices L in $\mathbb{R}^b$ with a Euclidean norm $\|\ \|$.

A lattice realizing the supremum is called *critical.* Such a lattice may be thought of as realizing a densest packing in $\mathbb{R}^b$, when we place balls of radius $\frac{1}{2}\lambda_1(L)$ at the points of L.

LEMMA 4.5. *Let $b = 2$. Then we have the following value for the Hermite constant: $\gamma_2 = \frac{2}{\sqrt{3}} = 1.1547\ldots$. The corresponding optimal lattice is homothetic to the $\mathbb{Z}$-span of cube roots of unity in $\mathbb{C}$.*

PROOF. Consider a lattice $L \subset \mathbb{C} = \mathbb{R}^2$. Clearly, multiplying L by nonzero complex numbers does not change the value of the quotient $\frac{\lambda_1(L)^2}{\mathrm{area}(\mathbb{C}/L)}$. Choose a "shortest" vector $z \in L$, i.e., we have $|z| = \lambda_1(L)$. By replacing L by the lattice $z^{-1}L$, we may assume that the complex number $+1 \in \mathbb{C}$ is a shortest element in the lattice. We will denote the new lattice by the same letter L, so that now $\lambda_1(L) = 1$. Now complete $+1$ to a $\mathbb{Z}$-basis $\{\tau, 1\}$, so that

$$L = \mathbb{Z}\tau + \mathbb{Z}1.$$

Thus $|\tau| \geq \lambda_1(L) = 1$. Consider the real part $\Re(\tau)$. We can adjust the basis by adding or subtracting a suitable integer to τ, so as to satisfy the condition $-\frac{1}{2} \leq \Re(\tau) \leq \frac{1}{2}$. Thus, the second basis vector τ may be assumed to lie in the closure of the standard fundamental domain, cf. [**42**, p. 78]

$$D = \left\{ z \in \mathbb{C} \,\middle|\, |z| > 1,\ |\Re(z)| < \frac{1}{2}, \Im(z) > 0 \right\}$$

(for the action of $PSL(2, \mathbb{Z})$ in the upper half-plane of $\mathbb{C}$). Then the imaginary part satisfies $\Im(\tau) \geq \frac{\sqrt{3}}{2}$, with equality possible precisely for $\tau = e^{i\frac{\pi}{3}}$ or $\tau = e^{i\frac{2\pi}{3}}$. The proof is concluded by calculating the area of the parallelogram in $\mathbb{C}$ spanned by τ and $+1$ as follows: $\frac{\lambda_1(L)^2}{\text{area}(\mathbb{C}/L)} = \frac{1}{\Im(\tau)} \leq \frac{2}{\sqrt{3}}$. □

EXAMPLE 4.6. In dimensions $b \geq 3$, the Hermite constants are harder to compute, but explicit values (as well as the associated critical lattices) are known for small dimensions, e.g., $\gamma_3 = 2^{\frac{1}{3}} = 1.2599...$, while in dimension 4, one has $\gamma_4 = \sqrt{2} = 1.4142...$ [**10**].

The lattice L^* dual to L may be described as follows. If L is the $\mathbb{Z}$-span of vectors (x_i), then L^* is the $\mathbb{Z}$-span of a dual basis (y_j) satisfying

$$\langle x_i, y_j \rangle = \delta_{ij}. \tag{4.3}$$

DEFINITION 4.7. The Bergé-Martinet constant, denoted γ_b', is defined as follows:

$$\gamma_b' = \sup \left\{ \lambda_1(L)\lambda_1(L^*) \,\middle|\, L \subseteq \mathbb{R}^b \right\}, \tag{4.4}$$

where the supremum is extended over all lattices L in $\mathbb{R}^b$ with its Euclidean norm. A lattice attaining the supremum is called *dual-critical.*

Like the Hermite constant γ_b, the Bergé-Martinet constant γ_b' is asymptotically linear in b. Its value is known in dimensions up to 4.

EXAMPLE 4.8. In dimension 3, the value of the Bergé-Martinet constant, $\gamma_3' = \sqrt{\frac{3}{2}} = 1.2247\ldots$, is slightly below the Hermite constant $\gamma_3 = 2^{\frac{1}{3}} = 1.2599\ldots$. It is attained by the face-centered cubic lattice, which is not isodual [**35**, p. 31], [**5**, Proposition 2.13(iii)], [**10**].

5. Precise calculation of filling invariants

Unlike the filling area and volume, the filling radius lends itself somewhat easier to precise calculation. The following result was proved in [**19**].

THEOREM 5.1. *Let X be a length space with distance function d. Let $Y \subset X$ be a subset. Let $R > 0$. Assume that the following two conditions are satisfied:*

(a) $\text{diam}(Y) \leq 2R$;
(b) $d(x, Y) \leq 2R$ *for all* $x \in X$.

Then $\text{FillRad}(X) \leq R$.

COROLLARY 5.2. *Let S^1 be the Riemannian circle. Then*

$$\mathrm{FillRad}(S^1) = \frac{1}{6}\,\mathrm{length}(S^1).$$

More precisely, the neighborhood $U_\epsilon S^1 \subset L^\infty$ has the homotopy type of S^1 for $\epsilon < \frac{1}{2}d_1$, and the homotopy type of the sphere S^3 when $\frac{1}{2}d_1 < \epsilon < \frac{1}{2}d_2$, where $d_i = \frac{i}{2i+1}\,\mathrm{length}(S^1)$.

PROOF. Using the set of vertices of a regular inscribed triangle as our subset $Y \subset S^1$ in Theorem 5.1 proves the upper bound. The lower bound follows from the general inequality

$$\mathrm{sys}\pi_1(X) \le 6\,\mathrm{FillRad}(X) \tag{5.1}$$

valid for an arbitrary essential manifold X [**14**]. The more precise result on the homotopy type of the neighborhood was proved in [**21**] using a kind of Morse theory on the space of subsets of S^1, for which regular odd polygons are the extrema, as discussed below. □

Let $d_i(X)$ be the i-th extremal value of the diameter functional on the space of finite subsets of X, with the convention that $d_0 = 0$. Here a subset $Y \subset X$ is extremal if small perturbations of size $\epsilon > 0$ can decrease the diameter of Y at most quadratically in ϵ. For the Riemannian circle we have a complete classification of the extrema, namely sets of vertices of odd regular polygons inscribed in the circle. Thus, we obtain the following extremal values:

$$d_i(S^1) = \frac{i}{2i+1}\,\mathrm{length}(S^1). \tag{5.2}$$

For the n-sphere, the first extremum is the set of vertices of a regular inscribed $n+1$-simplex, but the higher ones are difficult to classify even on S^2 [**20**].

The precise calculation of the filling radius is possible in the following cases [**19, 21, 22, 23**].

THEOREM 5.3. *The filling radius of the following two-point homogeneous spaces X equals half the first extremal value $d_1(X)$:*

(a) *$X = S^n$, where $d_1(S^n) = \arccos\left(-\frac{1}{n+1}\right) K^{-1/2}$, where K denotes the (constant) value of sectional curvature;*
(b) *$X = \mathbb{R}P^n$, where $d_1(\mathbb{R}P^n) = \frac{\pi}{3}K^{-1/2}$;*
(c) *$X = \mathbb{C}P^2$, where $d_1(\mathbb{C}P^2) = \arccos\left(-\frac{1}{3}\right) K^{-1/2}$, where K denotes the maximal value of sectional curvatures.*

Meanwhile, for $\mathbb{C}P^3$ we have a strict inequality

$$\mathrm{FillRad}(\mathbb{C}P^3) > \frac{1}{2}\, d_1(\mathbb{C}P^3) = \frac{1}{2}\arccos\left(-\tfrac{1}{3}\right) K^{-1/2}.$$

REMARK 5.4. In general, understanding the homotopy type of neighborhoods of X in L^∞ depends on the study of the extremal values of the diameter functional, which is an interesting open question. For example, the value of $d_2(\mathbb{C}P^n)$ is unknown for $n \ge 2$, but progress was made in [**23**]. Successful calculation of these extrema depends on analyzing certain semialgebraic sets, defined in terms of the distance functions among points, in the Cartesian (or symmetric) powers of the space.

We now turn to the related invariant of the filling area. We report on the recent progress on calculating the optimal ratio $\sigma_{2,s}$ of (5.3), in the case of hyperelliptic fillings, in the following sense. We introduce the constant

$$\sigma_{2,s} = \sup_{\Sigma_s} \frac{\text{length}(S^1)^2}{\text{area}(\Sigma_s)}, \tag{5.3}$$

where Σ_s denotes an orientable filling, of genus s, of a Riemannian circle S^1, meaning that the inclusion of the boundary $S^1 \subset \Sigma_s$ is strongly isometric, cf. formula (1.2). A related invariant appears in the systolic inequality (9.4).

For simply connected fillings (i.e., by a disk D), we have $\sigma_{2,0} = 2\pi$, by Pu's inequality (2.4) applied to the real projective plane obtained from $\Sigma_0 = D$ by identifying pairs of opposite points of the boundary circle. The case of genus 1 fillings has recently been solved, as well [**1**]. Thus,

$$\sigma_{2,s} = 2\pi \text{ for } s = 0 \text{ or } 1. \tag{5.4}$$

Given a filling by an orientable surface X_1 of genus s, one identifies pairs of antipodal points of the boundary, so to form a nonorientable closed surface X_2. We construct the orientable double cover $X_3 \to X_2$ with deck transformation τ. The pair (X, τ) is thus a real Riemann surface of genus $2s$, which can be called ovalless since τ is fixed point free, cf. [**1**]. Assuming that X_3 is in addition hyperelliptic [**36**], one can prove that it satisfies a bound equivalent to the formula $\sigma_{2,s} = 2\pi$. The result follows from a combination of two ingredients. On the one hand, one exploits integral geometric comparison with orbifold metrics of constant positive curvature on real surfaces of even positive genus, where the conical singularities correspond to Weierstrass points on the hyperelliptic surface. On the other hand, one exploits an analysis of the combinatorics on unions of closed curves, arising as geodesics of such orbifold metrics.

6. Stable and conformal systoles

In a Riemannian manifold (X, g), we define the volume $\text{vol}_k(\sigma)$ of a Lipschitz k-simplex $\sigma : \Delta^k \to X$ to be the integral over the k-simplex Δ^k of the "volume form" of the pullback $\sigma^*(g)$. The stable norm $\|h\|$ of an element $h \in H_k(X;\mathbb{R})$ is the infimum of the volumes $\text{vol}_k(c) = \Sigma_i |r_i| \text{vol}_k(\sigma_i)$ over all real Lipschitz cycles $c = \Sigma_i r_i \sigma_i$ representing h. We define the stable k-systole of the metric g by setting

$$\text{stsys}_k(g) = \lambda_1 \left(H_k(X;\mathbb{Z})_{\mathbb{R}}, \| \ \| \right), \tag{6.1}$$

where $\| \ \|$ is the stable norm in homology associated with the metric g, while $H_k(X;\mathbb{Z})_{\mathbb{R}}$ denotes the integral lattice in real homology, and λ_1 is the first successive minimum, cf. Definition 4.1.

The conformally invariant norm $\| \ \|_n$ in $H_1(X^n;\mathbb{R})$ is by definition dual to the conformally invariant L^n-norm in de Rham cohomology. The latter norm on $H^1(X,\mathbb{R})$ is the quotient norm of the corresponding norm on closed one-forms. Thus, given $\alpha \in H^1(X,\mathbb{R})$, we set

$$\|\alpha\|_n = \inf_{\omega} \left\{ \left(\int_X |\omega_x|^n dvol(x) \right)^{\frac{1}{n}} \middle| \omega \in \alpha \right\},$$

where ω runs over closed one-forms representing α. The conformal 1-systole of (X^n, g) is the quantity

$$\text{confsys}_1(g) = \lambda_1\left(H_1(X;\mathbb{Z})_{\mathbb{R}}, \|\ \|_n\right), \tag{6.2}$$

satisfying $\text{stsys}_1(g) \leq \text{confsys}_1(g)\,\text{vol}(g)^{\frac{1}{n}}$ on an n-manifold (X, g).

7. Gromov's optimal inequality for n-tori

Let X be a smooth compact manifold with positive first Betti number. We define the Jacobi variety $J_1(X)$ of X by setting

$$J_1(X) = H_1(X;\mathbb{R})/H_1(X;\mathbb{Z})_{\mathbb{R}}. \tag{7.1}$$

The homotopy class of the *Abel-Jacobi map* can be defined as follows. Consider the abelianisation homomorphism on the fundamental group of X, followed by quotienting by torsion. Consider a map $K(\pi_1(X),1) \to \mathbb{T}^{b_1(X)} = J_1(X)$ associated with the resulting homomorphism. The composition $X \to K(\pi_1(X),1) \to \mathbb{T}^{b_1(X)}$ represents the desired homotopy class. When X is equipped with a metric, the Abel-Jacobi map

$$\mathcal{A}_X : X \to J_1(X), \tag{7.2}$$

can be induced by the harmonic one-forms on X, originally introduced by A. Lichnerowicz [**34**], cf. [**16**, 4.21]. More precisely, let E be the space of harmonic 1-forms on X, with dual E^* canonically identified with $H_1(X, \mathbb{R})$. By integrating an integral harmonic 1-form along paths from a basepoint $x_0 \in X$, we obtain a map to $\mathbb{R}/\mathbb{Z} = S^1$. In order to define a map $X \to J_1(X)$ without choosing a basis for cohomology, we argue as follows. Let x be a point in the universal cover $\widetilde{X}$ of X. Thus x is represented by a point of X together with a path c from x_0 to it. By integrating along the path c, we obtain a linear form, $h \to \int_c h$, on E. We thus obtain a map $\widetilde{X} \to E^* = H_1(X, \mathbb{R})$, which, furthermore, descends to a map

$$\overline{\mathcal{A}}_X : \overline{X} \to E^*, \quad c \mapsto \left(h \mapsto \int_c h\right), \tag{7.3}$$

where $\overline{X}$ is the universal free abelian cover. By passing to quotients, this map descends to the Abel-Jacobi map (7.2). The Abel-Jacobi map defined using harmonic forms is unique up to translations of the Jacobi torus $J_1(X)$.

An optimal higher-dimensional generalisation of Loewner's inequality (2.2) is due to M. Gromov [**16**, pp. 259-260] (cf. [**11**, inequality (5.14)]) based on the techniques of D. Burago and S. Ivanov [**7, 8**]. We give below a slight generalisation of Gromov's statement.

DEFINITION 7.1. Given a map $f : X \to Y$ between closed manifolds of the same dimension, we denote by $\deg(f)$, either the algebraic degree of f when both manifolds are orientable, or the absolute degree [**6**], otherwise.

We denote by $\mathcal{A}_X$ the Abel-Jacobi map of X, cf. (7.2); by γ_n, the Hermite constant, cf. Definition 4.4; and by $\text{stsys}_1(g)$, the stable 1-systole of a metric g, cf. formula (6.1).

THEOREM 7.2 (M. Gromov). *Let X^n be a compact manifold of equal dimension and first Betti number:* $\dim(X) = b_1(X) = n$. *Then every metric g on X satisfies the following optimal inequality:*

$$\deg(\mathcal{A}_X)\,\mathrm{stsys}_1(g)^n \le (\gamma_n)^{\frac{n}{2}}\,\mathrm{vol}_n(g). \tag{7.4}$$

The boundary case of equality in inequality (7.4) *is attained by flat tori whose group of deck transformations is a critical lattice in $\mathbb{R}^n$.*

Note that the inequality is nonvacuous in the orientable case, only if the real cuplength of X is n, i.e., the Abel-Jacobi map $\mathcal{A}_X$ is of nonzero algebraic degree.

The method of proof is to construct a suitable map A from X to its Jacobi torus,

$$A : X \to J_1(X). \tag{7.5}$$

Here the torus is endowed with the flat Riemannian metric of least volume which dominates the stable norm. The map A has the following properties:

(a) A belongs to the homotopy class of the Abel-Jacobi map $\mathcal{A}_X$;
(b) the derivative of A is volume-decreasing at every point.

The existence of such a map A follows from the techniques of [**7, 8**], as explained in [**16**, pp. 259-260]. Inequality (7.4) now follows from the definition of the Hermite constant. This approach was systematized and generalized in [**18, 2**], as explained in Section 9.

8. Inequalities combining dimension and codimension 1

The material treated here is in [**3, 4**]. An optimal stable systolic inequality, for n-manifolds X with $b_1(X,\mathbb{R}) = 1$, is due to J. Hebda [**17**]:

$$\mathrm{stsys}_1(g)\,\mathrm{sys}_{n-1}(g) \le \mathrm{vol}_n(g), \tag{8.1}$$

with equality if and only if (X, g) admits a Riemannian submersion with connected minimal fibers onto a circle.

Here the systole, sys_k, and the stable systole, stsys_k, coincide in codimension 1 (i.e., when $k = n-1$) for orientable manifolds. Thus $\mathrm{sys}_{n-1} = \mathrm{stsys}_{n-1}$. In general, the k-systole is defined by minimizing the k-dimensional area of *integral* k-cycles which are not nullhomologous.

PROOF OF (8.1). Let $\omega \in H^1(X,\mathbb{Z})_\mathbb{R}$ be a primitive element in the integer lattice in cohomology, and similarly $\alpha \in H_1(X,\mathbb{Z})_\mathbb{R}$ in homology. Let $\eta \in \omega$ be the harmonic 1-form for the metric g. Then there exists a map f to the circle such that

$$f : X \to S^1 = \mathbb{R}/\mathbb{Z}, \quad df = \eta.$$

Using the Cauchy-Schwartz inequality, we obtain

$$\begin{aligned} \|\omega\|_2^*(vol_n(g))^{1/2} &= |\eta|_2(vol_n(g))^{1/2} \\ &\ge \int_X |df|dvol_n. \end{aligned} \tag{8.2}$$

Using the coarea formula, cf. [**13**, 3.2.11], [**9**, p. 267], we obtain

$$\begin{aligned} \|\omega\|_2^*(vol_n(g))^{1/2} &\geq \int_{S^1} vol_{n-1}\left(f^{-1}(t)\right) dt \\ &\geq \| \operatorname{PD}(\omega)\|_\infty \\ &\geq \operatorname{sys}_{n-1}(g), \end{aligned} \tag{8.3}$$

since the hypersurface $f^{-1}(t) \subset X$ is Poincaré dual to ω for every regular value t of f. Let us normalize the metric to unit total volume to simplify the calculations. In the case $b_1(X) = 1$, we have

$$\|\alpha\|_2 \, \|\omega\|_2^* = 1. \tag{8.4}$$

Therefore

$$\begin{aligned} \operatorname{stsys}_1(g)\operatorname{sys}_{n-1}(g) &= \|\alpha\|_\infty \operatorname{sys}_{n-1}(g) \\ &\leq \|\alpha\|_\infty \, \|\omega\|_2^* \\ &\leq \|\alpha\|_2 \, \|\omega\|_2^* \\ &= 1, \end{aligned}$$

proving Hebda's inequality (8.1). □

An optimal inequality, involving the conformal 1-systole (6.2), is proved in [**4**], namely inequality (8.5) below. This inequality generalizes simultaneously Loewner's inequality (2.2), Hebda's inequality (8.1), the inequality [**3**, Corollary 2.3], as well as certain results of G. Paternain [**37**]. Let X be a compact, oriented, n-dimensional manifold with positive first Betti number $b_1(X) \geq 1$. Then every metric g on X satisfies

$$\operatorname{confsys}_1(g)\operatorname{sys}_{n-1}(g) \leq \gamma'_{b_1(X)} \operatorname{vol}_n(g)^{\frac{n-1}{n}}, \tag{8.5}$$

where equality occurs if and only if the following three conditions are satisfied:

(a) the stable norm in $H_1(X;\mathbb{R})$ is Euclidean;
(b) the deck transformations of the Jacobi torus (7.1) form a dual-critical lattice (4.4);
(c) the Abel-Jacobi map (7.2) is a Riemannian submersion with connected minimal fibers.

The proof of inequality (8.5) is similar to the calculation (8.3). To conclude the argument, identity (8.4) is replaced by the inequality

$$\|\alpha\|_2 \, \|\omega\|_2^* \leq \gamma'_{b_1(X)} \tag{8.6}$$

for a pair of minimizing elements

$$\alpha \in H_1(X,\mathbb{Z})_\mathbb{R} \setminus \{0\}, \quad \omega \in H^1(X,\mathbb{Z})_\mathbb{R} \setminus \{0\}.$$

Note that inequality (8.6) results from the definition of the Bergé-Martinet's constant γ'_b combined with the fact that the L^2 norm in cohomology is actually a Euclidean norm since harmonic forms form a vector space.

9. Inequalities combining varieties of 1-systoles

A generalisation of Gromov's inequality (7.4) was obtained in [**18**], with a number of further generalisations in [**2**]. The following two theorems were proved in [**18**] (the second one is an immediate corollary of the first one). Denote by $[\overline{\mathrm{F}_X}] \in H_{n-b}(\overline{\mathrm{X}}; A)$ the class of a typical fiber of the map $\overline{\mathcal{A}}$ of (7.3), where $A = \mathbb{Z}$ or $\mathbb{Z}_2$ according as $\overline{\mathrm{X}}$ is orientable or not. Following Gromov [**14**], denote by $\deg(\mathcal{A}_X)$ the infimum of $(n-b)$-volumes of cycles representing the class $[\overline{\mathrm{F}_X}]$. Note that this quantity is a topological invariant only if $n = b$.

THEOREM 9.1 ([**18**]). *Let X be a closed Riemannian manifold of dimension $n = \dim(X)$. Let $b = b_1(X)$, and assume $n \geq b \geq 1$. Then every metric g on X satisfies the inequality*

$$\deg(\mathcal{A}_X)\,\mathrm{stsys}_1(g)^b \leq (\gamma_b)^{\frac{b}{2}} \mathrm{vol}_n(g). \tag{9.1}$$

THEOREM 9.2 ([**18**]). *Let X be a closed orientable manifold. Let $b = b_1(X)$. Assume that* $\dim(X) = b+1$, *and* $[\overline{\mathrm{F}_X}] \neq 0$. *Then every metric g on X satisfies the following optimal inequality:*

$$\mathrm{stsys}_1(g)^b\,\mathrm{sys}\pi_1(g) \leq (\gamma_b)^{\frac{b}{2}}\,\mathrm{vol}_{b+1}(g). \tag{9.2}$$

The main tool in the proof of Theorems 9.1 and 9.2 is a generalisation of the construction of an area-decreasing map (7.5) in a situation where the dimension is bigger than the Betti number. A suitable version of the coarea formula is then applied to this map to conclude the proof.

The paper [**2**] contains a number of generalisations of such stable inequalities as well as the study of the boundary case of equality, with conclusions along the closing lines of Section 8. In particular, the Abel-Jacobi map is a submersion.

Now let X be a closed 3-manifold. Let λ be the extension of the Casson-Walker invariant defined in [**33**]. The general definition is a bit involved, but in our case the invariant can be expressed in terms of the self-linking number of a typical fiber of the Abel-Jacobi map, cf. (10.2). We show in Section 10 that the nonvanishing of the λ invariant is a sufficient condition for the nonvanishing of the class $[\overline{\mathrm{F}_X}]$.

The argument runs roughly as follows. Consider the map $\overline{\mathcal{A}} : \overline{\mathrm{X}} \to \mathbb{R}^2$ of (7.3). Let $a, b \in \mathbb{R}^2$. Let $\bar{F}_a = \overline{\mathcal{A}}^{-1}(a)$ and $\bar{F}_b = \overline{\mathcal{A}}^{-1}(b)$ be lifts of the corresponding fibers $F_a, F_b \subset X$. Choose a properly imbedded ray $r_b \subset \mathbb{R}^2$ joining b to infinity while avoiding a (as well as its translates), and consider the complete surface $S = \overline{\mathcal{A}}^{-1}(r_b) \subset \overline{\mathrm{X}}$ with $\partial S = \bar{F}_b$. Furthermore, $S \cap g.\bar{F}_a = \emptyset$ for all $g \in G$, where G denotes the group of deck transformations of the cover $\overline{\mathrm{X}} \to X$. If $\bar{F}_a$ is zero-homologous in $\overline{\mathrm{X}}$, choose a compact surface $M \subset \overline{\mathrm{X}}$ with

$$\partial M = \bar{F}_a.$$

The linking number $\ell_X(F_a, F_b)$ in X can therefore be computed as the *algebraic* intersection

$$\begin{aligned}
\ell_X(F_a, F_b) &= \overline{\mathcal{A}}(M) \cap F_b \\
&= \sum_{g \in G} g.M \cap \bar{F}_b \\
&= \sum_{g \in G} \partial\,(g.M \cap S) \\
&= 0,
\end{aligned}$$

where all sums and intersections are algebraic, and the curve $M \cap S$ is compact by construction. The details of a rigorous proof are found in Section 12.

Therefore we obtain the following immediate corollary of Theorem 9.2.

COROLLARY 9.3. *Suppose X is a three-manifold such that $b_1(X) = 2$, with nonvanishing λ-invariant. Then every metric g on X satisfies the inequality*

$$\operatorname{stsys}_1(g)^2 \operatorname{sys}\pi_1(g) \leq \gamma_2 \operatorname{vol}(g). \tag{9.3}$$

In the boundary case of equality, the manifold must be one of the bundles of Example 11.1.

Analyzing the condition $[\mathrm{F}_X] \neq 0$ leads to interesting questions about 3-manifolds, see Section 10.

REMARK 9.4 (Relation to LS). In dimension 3, there is a connection between systoles and the Lusternik-Schnirelmann category, denoted $\operatorname{cat}_{\mathrm{LS}}$. A new invariant $\operatorname{cat}_{\mathrm{sys}}$, called *systolic category*, is defined in terms of the existence of systolic inequalities of a suitable type [**24**]. It turns out that the two categories coincide in dimension three: $\operatorname{cat}_{\mathrm{LS}} = \operatorname{cat}_{\mathrm{sys}}$. This and other phenomena relating the two categories are discussed in [**24, 25**].

REMARK 9.5 (Codimension 2). In the case $n - b = 2$, inequality (9.1) is related to the filling area conjecture. For simplicity, let $X = \mathbb{R}P^2 \times \mathbb{T}^2$. It was proved in [**18**] that every metric g on $\mathbb{T}^2 \times \mathbb{R}P^2$ satisfies a "Pu times Loewner" inequality

$$\operatorname{sys}\pi_1(g)^2 \operatorname{stsys}_1(g)^2 \leq \frac{\tilde{\sigma}_2}{4} \gamma_2 \operatorname{vol}_4(g), \tag{9.4}$$

where $\tilde{\sigma}_2$ is defined similarly to (5.3) by allowing all possible topology of fillings.

All the inequalities discussed so far depend on the rather restrictive hypothesis $\dim(X) \geq b_1(X)$. In the case $\dim(X) < b_1(X)$ it is harder to obtain optimal inequalities, and typically the extremal metrics are not smooth, see [**11**]. Recently, progress was achieved toward removing the topological assumption in Loewner's inequality (2.2) (similarly to the generalisation (2.4) of Pu's inequality), which involves the study of surfaces of genus $s \geq 2$. Namely, all hyperelliptic surfaces satisfy Loewner's inequality

$$\operatorname{sys}\pi_1(X)^2 \leq \gamma_2 \operatorname{area}(X) \quad \text{if } X \text{ hyperelliptic}, \tag{9.5}$$

and in particular [**12**] all metrics in genus 2 [**26**]. Moreover, all surfaces of genus at least 20 satisfy Loewner's inequality, as well [**27**].

10. Fiber class, linking number, and the generalized Casson invariant λ

The results of this section tie in with Corollary 9.3. Namely, Proposition 10.1 provides a necessary condition so that the invariant λ does not vanish. The condition is the nonvanishing of the fiber class in the maximal free abelian cover. Thus the nonvanishing of λ is a sufficient condition for the inequality (9.3) to be satisfied.

From now on, unless otherwise mentioned, all the manifolds considered are smooth, compact, and oriented. Boundaries are oriented according to the outward normal first convention.

In any (oriented) 3-manifold X, the *linking number* of two disjoint closed curves γ and d with null homology classes in $H_1(X; \mathbb{Q})$ is defined as follows. Let $N(\gamma)$ be a tubular neighborhood of γ disjoint from d. There exists a curve $C(k\gamma)$ in $N(\gamma)$

that is homologous to $k\gamma$ in $N(\gamma)$ for some integer $k > 0$, and that bounds a surface $\Sigma(C(k\gamma))$ transverse to d. Then the *linking number* of γ and d in X is defined as

$$\ell_X(\gamma, d) = \frac{\langle \Sigma(C(k\gamma)), d\rangle_X}{k}$$

where $\langle \; , \; \rangle_X$ denotes the algebraic intersection number in X.

PROPOSITION 10.1. *Let X be a connected closed three-manifold whose first Betti number is two. Let S_1 and S_2 be two surfaces imbedded in X whose homology classes generate $H_2(X;\mathbb{Z})$. The intersection of S_1 and S_2 is denoted by γ. Let γ' be a parallel of γ in S_1. Let $\overline{X}$ be the maximal free abelian cover of X, and let $\overline{\gamma}$ be a lift of γ in $\overline{X}$. If the homology class of $\overline{\gamma}$ vanishes in $H_1(\overline{X};\mathbb{Q})$, then the linking number*

$$\ell_X(\gamma, \gamma') \tag{10.1}$$

of γ and γ' in X vanishes, as well.

This proposition will be proved in Section 12 as a particular case of a more general result, Proposition 12.1, as suggested by Alexis Marin.

REMARK 10.2. The two parallels of γ in S_1 are homotopic to each other in the complement of γ in X and they are also homotopic to the two parallels of γ in S_2. We could equivalently define γ' as the parallel of γ with respect to the framing of γ induced by S_1 or S_2.

REMARK 10.3. By Poincaré duality, under the assumptions of the proposition, there exists a basis (a_1, a_2) of $H_1(X;\mathbb{Q})$ such that the homology class $[d]$ of any closed curve d of X reads

$$[d] = \langle d, S_1\rangle_X a_1 + \langle d, S_2\rangle_X a_2,$$

cf. (4.3). In particular, we have $[\gamma] = 0$ in $H_1(X;\mathbb{Q})$. Therefore, the number $\ell_X(\gamma, \gamma')$ is well-defined. In fact, if λ denotes the extension of the Casson invariant defined in [**33**], and if $|\mathrm{Torsion}(H_1(X))|$ is the cardinality of the torsion part of $H_1(X;\mathbb{Z})$, then

$$\ell_X(\gamma, \gamma') = -\frac{\lambda(X)}{|\mathrm{Torsion}(H_1(X))|}, \tag{10.2}$$

cf. (11.2). Thus the linking number does not depend on the choice of the transverse surfaces S_1 and S_2 whose homology classes generate $H_2(X;\mathbb{Z})$. An easy direct proof of this fact is given in [**33**, pp. 93-94].

DEFINITION 10.4. Let X be a connected topological space, and let

$$g : \pi_1(X) \to G$$

be a homomorphism. The *connected cover*

$$\hat{X}(Ker(g)) \tag{10.3}$$

of X associated to g is the connected cover of X where a loop of X lifts homeomorphically if and only if its homotopy class is in the kernel of g.

In particular, we have $\pi_1(\hat{X}(\mathrm{Ker}(g))) = \mathrm{Ker}(g)$. The covering group is the image of g. For example, the maximal free abelian cover $\overline{X}$ of X is associated with the natural composition

$$\pi_1(X) \longrightarrow H_1(X) \longrightarrow H_1(X)/\mathrm{Torsion}(H_1(X)).$$

Since $[\gamma] = 0$ in $H_1(X;\mathbb{Q})$, the covering map ρ from $\overline{X}$ to X maps $\overline{\gamma}$ to γ, homeomorphically. In particular, $\overline{\gamma}$ is a closed curve.

11. Fiber class in NIL geometry

Proposition 10.1 may be illustrated by the following example of the non-trivial circle bundles N_e over the torus. For them, we shall directly see that

- the roles of γ and γ' can be played by two disjoint fibers F and F',
- $\ell_{N_e}(F, F')$ does not vanish, and
- $H_1(\overline{N_e};\mathbb{Z}) = \mathbb{Z}\left[\overline{F}\right]$.

EXAMPLE 11.1. Let $p_e : N_e \longrightarrow S^1 \times S^1$ be the circle bundle over the torus with Euler number $e \neq 0$. Denote by $x = S^1 \times \{1\}$ and $y = \{1\} \times S^1$ the two S^1-factors of the base $S^1 \times S^1 = x \times y$ of N_e.

Let $D \subset x \times y$ be a disk disjoint from the union $x \cup y$. The bundle is trivial both over D and over its complement. Thus we have

$$N_e = S^1 \times s\left(\overline{(x \times y) \setminus D}\right) \cup_{S^1 \times \partial D} S^1 \times D,$$

where $s : \overline{(x \times y) \setminus D} \longrightarrow N_e$ is a section, and the gluing map reads

$$(u, s(v)) \in S^1 \times (\partial D \cong S^1) \mapsto (uv^{\pm e}, v) \tag{11.1}$$

(up to signs). Let $F = p_e^{-1}(x \cap y)$ be the fiber. Then

$$H_1(N_e) = \mathbb{Z}[s(x)] \oplus \mathbb{Z}[s(y)] \oplus \frac{\mathbb{Z}}{|e|\mathbb{Z}}[F].$$

The surfaces S_1 and S_2 of Proposition 10.1 may be chosen to be $S_1 = p_e^{-1}(x)$ and $S_2 = p_e^{-1}(y)$, since these surfaces are dual to the basis

$$([s(y)], [s(x)])$$

of $H_1(N_e)/\mathrm{Torsion}$ with respect to the algebraic intersection. Then the intersection $S_1 \cap S_2$ is precisely the fiber F, and its parallel induced by the surfaces is another fiber F'.

In this case, if the loop $s(\partial D)$ bounds a section of the bundle over $s\left((x \times y) \setminus D\right)$, then the loop $(s(\partial D) \pm eF)$ bounds a section of the bundle over D. This allows us to see that $|e|F$ bounds a surface that is pierced once by F', so that the linking number satisfies

$$\ell_{N_e}(F, F') = \pm \frac{1}{e}, \tag{11.2}$$

cf. (10.2). Now, let us study the maximal free abelian covering

$$\rho : \overline{N_e} \longrightarrow N_e \tag{11.3}$$

associated with the map

$$\pi_1(N_e) \longrightarrow (H_1(N_e)/\mathrm{Torsion}) = \mathbb{Z}[s(x)] \oplus \mathbb{Z}[s(y)] = G.$$

The free abelian group G acts on $\overline{N_e}$ freely as the covering group. Therefore, $H_1(\overline{N_e})$ becomes a $\mathbb{Z}[G]$-module. The ring $\mathbb{Z}[G]$ is denoted by R and is identified with $\mathbb{Z}[t_x^{\pm 1}, t_y^{\pm 1}]$ by mapping $[s(x)]$ and $[s(y)]$ to t_x and t_y, respectively. Consider the standard product covering

$$\begin{array}{rccc} \rho_T : & \mathbb{R}^2 & \longrightarrow & (S^1)^2 \\ & (u,v) & \mapsto & (\exp(2i\pi u), \exp(2i\pi v)). \end{array}$$

Then $\overline{N_e}$ is obtained from the trivialisation of ρ of (11.3) over the complement of D,

$$\rho^{-1}\left(p_e^{-1}\left((S^1)^2 \setminus D\right)\right) = F \times \left(\mathbb{R}^2 \setminus \rho_T^{-1}(D)\right),$$

by filling in the $\mathbb{Z}^2$ holes (with boundaries the lifts of $F \times \partial D$) by $\mathbb{Z}^2$ copies of $S^1 \times D$. Here we use the same gluing map (11.1) as before, $\mathbb{Z}^2$ times, equivariantly. Then

$$\begin{aligned} H_1(\overline{N_e}) &= \frac{R[\overline{F}] \oplus R[\overline{\partial D}]}{(t_x - 1)R[\overline{F}] + (t_y - 1)R[\overline{F}] + R\left([\overline{\partial D}] \pm e[\overline{F}]\right)} \\ &= \frac{R}{(t_x - 1)R + (t_y - 1)R}[\overline{F}] \\ &= \mathbb{Z}\left[\overline{F}\right]. \end{aligned}$$

12. A nonvanishing result (following A. Marin)

As it will be shown in Remark 12.2 below, Proposition 10.1 is a particular case of Proposition 12.1 that has been suggested and proved by Alexis Marin.

PROPOSITION 12.1. *Let X be a connected closed three-manifold. Let S_1 and S_2 be two surfaces imbedded in X that intersect transversely in X along a curve γ. Let γ' be a parallel of γ in S_1. Let $\hat{X}_2$ be the connected cover of X associated with the composition*

$$\begin{array}{rccc} f_{2*} : \pi_1(X) \longrightarrow & H_1(X) & \longrightarrow & \mathbb{Z} \\ & [d] & \mapsto & \langle d, S_2 \rangle_X \end{array} \tag{12.1}$$

where the first map is the Hurewicz homomorphism, cf. (10.3). *Let $\hat{\gamma}$ be a lift of γ in $\hat{X}_2$. If the homology class of $\hat{\gamma}$ vanishes in $H_1(\hat{X}_2;\mathbb{Q})$, then the linking number $\ell_X(\gamma,\gamma')$ of γ and γ' in X is well-defined and vanishes.*

REMARK 12.2. Let $\hat{X}$ be the connected cover of X associated with the composition

$$\begin{array}{rccc} A_* : \pi_1(X) \longrightarrow & H_1(X) & \longrightarrow & \mathbb{Z} \oplus \mathbb{Z} \\ & [d] & \mapsto & (\langle d, S_1 \rangle_X, \langle d, S_2 \rangle_X) \end{array} \tag{12.2}$$

where the first map is the Hurewicz homomorphism. Then $\hat{X}$ is a cover of $\hat{X}_2$. Suppose the homology class of a lift of γ in $\hat{X}$ vanishes in $H_1(\hat{X};\mathbb{Q})$. Then its image in $\hat{X}_2$ under the covering map, that is a lift of γ in $\hat{X}_2$, vanishes, as well. Therefore, assuming that Proposition 12.1 is true, replacing $\hat{X}_2$ by $\hat{X}$ in its statement yields another true proposition. In particular, Proposition 12.1 implies Proposition 10.1 because in this case the above $\hat{X}$ is precisely the maximal free abelian cover $\overline{\text{X}}$.

REMARK 12.3. In Proposition 12.1 that applies to 3-manifolds with arbitrarily large Betti numbers, the covering group is the subgroup $f_{2*}(\pi_1(X))$ of $\mathbb{Z}$ that is isomorphic to $\{0\}$ or $\mathbb{Z}$.

PROOF OF PROPOSITION 12.1. Let $\rho_2 : \hat{X}_2 \longrightarrow X$ be the covering map. Assume that the homology class of $\hat{\gamma}$ vanishes in $H_1(\hat{X}_2; \mathbb{Q})$. Note that this implies that γ is rationally null-homologous in X and that $\ell_X(\gamma, \gamma')$ is well-defined.

There exists a surface Σ in $\hat{X}_2$ whose boundary $\partial\Sigma$ lies in a small tubular neighborhood $N(\hat{\gamma})$ of $\hat{\gamma}$ such that

- $\partial\Sigma$ is homologous to $k[\gamma]$ in $N(\hat{\gamma})$, for some $k > 0$,
- $N(\hat{\gamma})$ does not meet $\rho_2^{-1}(\gamma')$,
- the restriction of ρ_2 to $N(\hat{\gamma})$ is injective.

Then

$$k\ell_X(\gamma, \gamma') = \langle \rho_2(\Sigma), \gamma' \rangle_X = \langle \Sigma, \rho_2^{-1}(\gamma') \rangle_{\hat{X}_2}$$

and $k\ell_X(\gamma, \gamma')$ is the sum over the lifts of γ' of their algebraic intersections with Σ.

We now apply the following Lemma 12.4 to complete the proof of Proposition 12.1. □

LEMMA 12.4. *For any lift $\hat{\gamma}'$ of γ', the algebraic intersection of Σ and $\hat{\gamma}'$ vanishes.*

PROOF OF LEMMA 12.4. The proof consists in seeing $\langle \Sigma, \hat{\gamma}' \rangle_{\hat{X}_2}$ as the algebraic boundary of the compact oriented intersection of Σ with a noncompact surface $\hat{S}_1^*$ that is bounded by $\hat{\gamma}'$ and that does not meet $N(\hat{\gamma})$.

Let us first construct $\hat{S}_1^*$.

Consider a tubular neighborhood $S_2 \times [-1, 1]$ of $(S_2 = S_2 \times 0)$ such that $S_1 \cap (S_2 \times [-1, 1])$ reads $\gamma \times [-1, 1]$ in $S_2 \times [-1, 1]$ (up to orientations).

Let $f_2 : X \to S^1$ be the map to the unit circle in $\mathbb{C}$, defined as follows on the subset $S_2 \times [-1, 1] \subset X$:

$$\begin{array}{llll} f_2 : & S_2 \times [-1,1] & \longrightarrow & S^1 \\ & (x,t) & \mapsto & \exp(i\pi t), \end{array}$$

while f_2 maps the complement of $S_2 \times [-1, 1]$ in X to the point (-1).

Note that the map f_2 induces the morphism f_{2*} of (12.1). Thus the composition $f_2 \circ \rho_2$ induces the trivial map from $\pi_1(\hat{X}_2)$ to $\pi_1(S^1)$. Therefore, it factors through the covering

$$\begin{array}{llll} \mathrm{Exp} : & \mathbb{R} & \longrightarrow & S^1 \\ & u & \mapsto & \exp(2i\pi u) \end{array}$$

yielding a map $\hat{f}_2 : \hat{X}_2 \longrightarrow \mathbb{R}$ such that $\mathrm{Exp} \circ \hat{f}_2 = f_2 \circ \rho_2$.

Without loss assume that

$$\gamma' = S_1 \cap f_2^{-1}(i)$$

and that

$$N(\hat{\gamma}) \subset \hat{f}_2^{-1}\left[-\tfrac{1}{8}, \tfrac{1}{8}\right].$$

Now, any lift $\hat{\gamma}'$ of γ' reads

$$\rho_2^{-1}(S_1) \cap \hat{f}_2^{-1}(\{n + 1/4\}) \tag{12.3}$$

for some $n \in \mathbb{Z}$, and splits $\rho_2^{-1}(S_1)$ as the union of

$$\hat{S}_1^+ = \rho_2^{-1}(S_1) \cap \hat{f}_2^{-1}\left(n + \tfrac{1}{4} + [0, +\infty[\right)$$

and

$$\hat{S}_1^- = \rho_2^{-1}(S_1) \cap \hat{f}_2^{-1}\left(n + \tfrac{1}{4} +]-\infty, 0]\right).$$

One of the non-compact subsurfaces $\hat{S}_1^+$ and $\hat{S}_1^-$ does not meet $N(\hat{\gamma})$. This will be our surface $\hat{S}_1^*$.

We may assume that the surface Σ is transverse to $\hat{S}_1^*$. Then these two surfaces intersect along an oriented curve $\Sigma \cap \hat{S}_1^*$ that is oriented so that the triple

(a) tangent vector $\vec{t}(\Sigma \cap \hat{S}_1^*)$ to the intersection curve,
(b) positive normal vector $\vec{n}(\Sigma)$ to Σ,
(c) positive normal vector $\vec{n}(\hat{S}_1^*)$ to $\hat{S}_1^*$

is direct. The curve $\Sigma \cap \hat{S}_1^*$ is a collection of arcs and closed curves properly imbedded in $\hat{S}_1^*$ since $\hat{S}_1^*$ does not meet $\partial\Sigma$. Since the oriented boundary of $\Sigma \cap \hat{S}_1^*$ (that obviously vanishes in $H_0(X)$) represents the algebraic intersection of $(\partial \hat{S}_1^* = \pm\hat{\gamma}')$ with Σ, the lemma is proved.

Let us be slightly more explicit about this last argument. A point x of the oriented boundary of $\Sigma \cap \hat{S}_1^*$ gets the sign $\varepsilon = \pm 1$ such that the tangent vector $\vec{t}_x(\Sigma \cap \hat{S}_1^*)$ of $\Sigma \cap \hat{S}_1^*$ at x is oriented as $\varepsilon \vec{n}_x(\Sigma \cap \hat{S}_1^*)$ where $\vec{n}_x(\Sigma \cap \hat{S}_1^*)$ is the outward normal vector to the curve at x. (In other words, final points of arcs get a positive sign while initial points get a minus sign.) On the other hand, $\vec{n}_x(\Sigma \cap \hat{S}_1^*)$ may be identified to the outward normal of $\hat{S}_1^*$ and the triple

$$\left(\vec{n}_x\left(\Sigma \cap \hat{S}_1^*\right), \vec{t}_x\left(\partial \hat{S}_1^*\right), \vec{n}_x\left(\hat{S}_1^*\right)\right)$$

is direct. Furthermore, $\vec{n}_x(\Sigma)$ may be identified with $\operatorname{sign}(x)\vec{t}_x(\partial \hat{S}_1^*)$ where $\operatorname{sign}(x)$ is the sign of the intersection of $\partial \hat{S}_1^*$ and Σ at x. Thus, the triple

$$\left(\varepsilon \vec{n}_x\left(\Sigma \cap \hat{S}_1^*\right),\ \operatorname{sign}(x)\vec{t}_x\left(\partial \hat{S}_1^*\right),\ \vec{n}_x\left(\hat{S}_1^*\right)\right)$$

is direct, too. This shows that $\varepsilon = \operatorname{sign}(x)$. Then the sum of the ε vanishes because there are as many initial points of arcs in $\Sigma \cap \hat{S}_1^*$ as there are final points, and that makes the algebraic intersection of $\hat{\gamma}'$ and Σ that is the sum of the $\operatorname{sign}(x)$ vanish, too. This concludes the proof of Lemma 12.4 and the proof of the proposition. □

Note that Lemma 12.4 yields the following corollary.

COROLLARY 12.5. *Let X be a 3-manifold with Betti number 2. Suppose the lift L of a typical fiber of the Abel-Jacobi map, to an infinite cyclic cover C of X is rationally zero-homologous. Equip L with the framing induced by the Abel-Jacobi map. Then the self-linking of $L \subset C$ is zero.*

This corollary is obviously not true when C is replaced by the compact manifold X itself, or by any *finite* cyclic cover of X.

Acknowledgments

We are grateful to A. Marin for suggesting and proving Proposition 12.1.

References

1. V. Bangert, C. Croke, S. Ivanov, and M. Katz, *Filling area conjecture and ovalless real hyperelliptic surfaces*, Geometric and Functional Analysis (GAFA) **56** (2005), no. 3. See `arXiv:math.DG/0405583`
2. ———, *Boundary case of equality in optimal Loewner-type inequalities*, Trans. Amer. Math. Soc., to appear. See `arXiv:math.DG/0406008`
3. V. Bangert and M. Katz, *Stable systolic inequalities and cohomology products*, Comm. Pure Appl. Math. **56** (2003), 979-997. `math.DG/0204181`

4. ———, *An optimal Loewner-type systolic inequality and harmonic one-forms of constant norm*, Comm. Anal. Geom. **12** (2004), number 3, 703-732. See arXiv:math.DG/0304494
5. A.-M. Bergé and J. Martinet, *Sur un problème de dualité lié aux sphères en géométrie des nombres*, J. Number Theory **32** (1989), 14-42.
6. R. B. S. Brooks, R. F. Brown, and H. Schirmer, *The absolute degree and the Nielsen root number of compositions and Cartesian products of maps*, Theory of Fixed Points and its Applications (São Paulo, 1999), Topology Appl. **116** (2001), no. 1, 5-27.
7. D. Burago and S. Ivanov, *Riemannian tori without conjugate points are flat*, Geom. Funct. Anal. **4** (1994), no. 3, 259-269.
8. ———, *On asymptotic volume of tori*, Geom. Funct. Anal. **5** (1995), no. 5, 800-808.
9. I. Chavel, *Riemannian Geometry–A Modern Introduction*, Cambridge Tracts in Mathematics, **108**, Cambridge University Press, Cambridge, 1993.
10. J. H. Conway, and N. J. A. Sloane, *On lattices equivalent to their duals*, J. Number Theory **48** (1994), no. 3, 373-382.
11. C. Croke and M. Katz, *Universal volume bounds in Riemannian manifolds*, Surveys in Differential Geometry **VIII** (2003), 109-137. Available at arXiv:math.DG/0302248
12. H. M. Farkas and I. Kra, *Riemann Surfaces, Second edition*, Graduate Texts in Mathematics **71**, Springer-Verlag, New York, 1992.
13. H. Federer, *Geometric Measure Theory*, Springer, 1969.
14. M. Gromov, *Filling Riemannian manifolds*, J. Diff. Geom. **18** (1983), 1-147.
15. ———, *Systoles and intersystolic inequalities*, Actes de la Table Ronde de Géométrie Différentielle (Luminy, 1992), Sémin. Congr., vol. 1, Soc. Math. France, Paris, 1996, pp. 291-362. www.emis.de/journals/SC/1996/1/ps/smf_sem-cong_1_291-362.ps.gz
16. ———, *Metric structures for Riemannian and non-Riemannian spaces*, Progr. in Mathematics **152**, Birkhäuser, Boston, 1999.
17. J. Hebda, *The collars of a Riemannian manifold and stable isosystolic inequalities*, Pacific J. Math. **121** (1986), 339-356.
18. S. Ivanov and M. Katz, *Generalized degree and optimal Loewner-type inequalities*, Israel J. Math. **141** (2004), 221-233. arXiv:math.DG/0405019
19. M. Katz, *The filling radius of two-point homogeneous spaces*, J. Diff. Geom. **18** (1983), 505-511.
20. ———, *Diameter-extremal subsets of spheres*, Discrete Comput. Geom. **4** (1989), 117-137.
21. ———, *On neighborhoods of the Kuratowski imbedding beyond the first extremum of the diameter functional*, Fundamenta Math. **137** (1991), 15-29.
22. ———, *The rational filling radius of complex projective space*, Topology and its Appl. **42** (1991), 201-215.
23. ———, *Pyramids in the complex projective plane*, Geom. Dedicata **40** (1991), 171-190.
24. M. Katz and Y. Rudyak, *Lusternik-Schnirelmann category and systolic category of low dimensional manifolds*, Comm. Pure Appl. Math., to appear. See arXiv:math.DG/0410456
25. ———, *Bounding volume by systoles of 3-manifolds.* Available at the site arXiv:math.DG/0504008
26. M. Katz and S. Sabourau, *Hyperelliptic surfaces are Loewner*, Proc. Amer. Math. Soc., to appear. See arXiv:math.DG/0407009
27. ———, *Entropy of systolically extremal surfaces and asymptotic bounds*, Ergodic Theory and Dynamical Systems, to appear. See arXiv:math.DG/0410312
28. ———, *An optimal systolic inequality for CAT(0) metrics in genus two*, preprint. See arXiv:math.DG/0501017
29. S. Kodani, *On two-dimensional isosystolic inequalities*, Kodai Math. J. **10** (1987), no. 3, 314-327.
30. H. T. Kung and C. E. Leiserson, *Systolic arrays (for VLSI)*, Sparse Matrix Proceedings 1978 (Sympos. Sparse Matrix Comput., Knoxville, Tenn., 1978), SIAM, Philadelphia, Pa., 1979, pp. 256-282.
31. C. Kuratowski, *Quelques problèmes concernant les espaces métriques non-séparables*, Fund. Math. **25** (1935), 534-545.
32. J. C. Lagarias, H. W. Lenstra, Jr. and C.P. Schnorr, *Bounds for Korkin-Zolotarev reduced bases and successive minima of a lattice and its reciprocal lattice*, Combinatorica **10** (1990), 343-358.

33. C. Lescop, *Global surgery formula for the Casson-Walker invariant*, Annals of Mathematics Studies **140**, Princeton University Press, Princeton, 1996.
34. A. Lichnerowicz, *Applications harmoniques dans un tore*, C.R. Acad. Sci., Sér. A, **269** (1969), 912-916.
35. J. Milnor and D. Husemoller, *Symmetric Bilinear Forms*, Springer, 1973.
36. R. Miranda, *Algebraic Curves and Riemann Surfaces*, Graduate Studies in Mathematics **5**, American Mathematical Society, Providence, RI, 1995.
37. G. P. Paternain, *Schrödinger operators with magnetic fields and minimal action functionals*, Israel J. Math. **123** (2001), 1-27.
38. P.M. Pu, *Some inequalities in certain nonorientable Riemannian manifolds*, Pacific J. Math. **2** (1952), 55-71.
39. S. Sabourau, *Systoles des surfaces plates singulières de genre deux*, Math. Zeitschrift **247** (2004), no. 4, 693–709.
40. ———, *Entropy and systoles on surfaces*, preprint.
41. ———, *Systolic volume and minimal entropy of aspherical manifolds*, preprint.
42. J.-P. Serre, *A course in Arithmetic*, Graduate Texts in Mathematics **7**, Springer-Verlag, New York-Heidelberg, 1973.

Department of Mathematics, Bar-Ilan University, Ramat-Gan 52900, Israel
E-mail address: katzmik@math.biu.ac.il

Institut Fourier, BP 74, Université de Grenoble I, 38402 Saint Martin d'Hères cedex, France
E-mail address: Christine.Lescop@ujf-grenoble.fr

Contemporary Mathematics
Volume **387**, 2005

An Invitation to Deninger's Work on Arithmetic Zeta Functions

Eric Leichtnam

ABSTRACT. Deninger has elaborated an infinite dimensional cohomological formalism in order to study general arithmetic zeta functions and to prove Lefschetz trace formulas, cohomological formulas, functional equations, "generalized" Riemann hypothesis, etc. For most zeta functions (in particular for the Riemann zeta function), the existence of such a cohomology is still open.

In the case of an elliptic curve E_0 over a finite field, admitting Hasse's theorem, Deninger has constructed a flow ϕ^t acting on an explicit laminated Riemannian foliated space $(S(E_0), \mathcal{F}, g, \phi^t)$. He has proved that the Lefschetz trace formula for ϕ^t acting on the corresponding L^2-leafwise cohomology groups is identical to the usual explicit formula which connects the closed points of E_0 to the zeroes of the zeta function of E_0. In the case of a smooth proper absolutely irreducible variety Y over $\mathbb{F}_q$, Deninger has constructed, in a *non natural way*, an infinite dimensional cohomology allowing to prove *only* a cohomological formula and the functional equation for the zeta function ζ_Y of Y. The underlying foliated space $(S_Y, \mathcal{F}, g, \phi^t)$ which should allow to construct in a natural way this cohomology is not known to exist, unless Y is an elliptic curve. We state a precise geometric hypothesis that such a $(S_Y, \mathcal{F}, g, \phi^t)$ should satisfy in order to provide a *natural construction* of this cohomology and we recall how it leads to a new proof *all* the known properties for ζ_Y. Moreover, we extract from Deninger's work on the Riemann zeta function $\widehat{\zeta}(s)$ a precise geometric conjecture stating the existence of a laminated Riemannian foliated space $(S_{\mathbb{Q}}, \mathcal{F}, g, \phi^t)$ endowed with an action of a flow ϕ^t and with a certain transverse measure. This conjecture assumes that the corresponding Lefschetz trace formula for the L^2-leafwise cohomology should coincide with the known explicit formula for $\widehat{\zeta}(s)$. We explain how it should imply all the expected properties for $\widehat{\zeta}(s)$.

We explain the compatibility of Deninger's approach and that of A. Connes, as well as that of N. Katz–P. Sarnak. We came to be interested in Deninger' s program because of its extraordinary internal coherence and its connections with Non-Commutative Geometry. In this direction, we describe for instance the connection between the algebraic Riemann Roch theorem for E_0 and its measured foliated analogue. We also show that the same cross-product von Neumann algebra appears in both approaches of Connes and Deninger, respectively.

2000 *Mathematics Subject Classification.* Primary 11M26, 14F20, 46L10; Secondary 11R56, 14H05, 37C27.

1. Introduction

The Riemann zeta function is given by

$$\widehat{\zeta}(s) = 2^{-1/2}\pi^{-s/2}\Gamma(s/2)\prod_{p\in\mathcal{P}}\frac{1}{1-p^{-s}},$$

where $\mathcal{P} = \{2, 3, 5, \ldots\}$ denotes the usual set of prime numbers. The notion of zeta function has been widely generalized. For instance, there is a class of zeta functions, arising in algebraic geometry over a finite field $\mathbb{F}_q$, that we describe briefly. Let Y be a smooth absolutely irreducible projective variety Y over $\mathbb{F}_q$ of dimension d, its zeta function is defined by

$$\zeta_Y(s) = \prod_{y\in|Y|}\frac{1}{1-(Ny)^{-s}},$$

where $|Y|$ denotes the set of closed points of Y and Ny is the cardinal of the residue field of y. Fix an algebraic closure $\overline{\mathbb{F}_q}$ of $\mathbb{F}_q$. Recall that the (geometric) $\overline{\mathbb{F}_q}$-linear Frobenius morphism Fr acts on the set $\overline{Y} = Y \otimes_{\mathbb{F}_q} \overline{\mathbb{F}_q}$ of geometric points of Y. It sends the point P with coordinates $(a_i), a_i \in \overline{\mathbb{F}_q}$, to the point $\mathrm{Fr}(P)$ with coordinates (a_i^q) (see [**15**, p. 454]). A closed point of Y is then by definition an orbit of Fr acting on the set of geometric points $\overline{Y}$. We shall denote for any $r \in \mathbb{N}$ by N_r the number of points of $\overline{Y}$ which are fixed by Fr^r or in other words, whose coordinates belong to $\mathbb{F}_{q^r}$. Recall then that

$$\zeta_Y(s) = \exp\left(\sum_{r=1}^{+\infty}\frac{N_r}{r}q^{-rs}\right).$$

The table below compares several properties of the zeta functions $\widehat{\zeta}$ and ζ_Y.

Arithmetic properties	$\widehat{\zeta}(s)$	$\zeta_Y(s)$
Underlying geometry	?	Y, the Frobenius Fr acts on $\overline{Y}$
Cohomology	?	l-adic cohomology groups $H^j(\overline{Y};\mathbb{Q}_l)$ on which Fr acts, $0 \le j \le 2d$.
Explicit formulas	See formulas (5.1).	Lefschetz trace formula: $N_r = \sum_{j=0}^{2d}(-1)^j\,\mathrm{TR}\,(\mathrm{Fr}^r : H^j(\overline{Y};\mathbb{Q}_l))$ For an example of a continuous version, see (2.2).
Cohomological formula for zeta	?	$\zeta_Y(s) = \prod_{j=0}^{2d}\left(\det(\mathrm{Id} - q^{-s}\,\mathrm{Fr} : H^j(\overline{Y};\mathbb{Q}_l)\right)^{(-1)^{1+j}}$ (Consequence of Lefschetz trace formula)

Arithmetic properties	$\widehat{\zeta}(s)$	$\zeta_Y(s)$
Functional equation	$\widehat{\zeta}(s) = \widehat{\zeta}(1-s)$	$\zeta_Y(d-s) = \pm q^{(\frac{d}{2}-s)E}\zeta_Y(s)$, results from the cohomological formula and Fr-equivariant Poincaré duality.
Riemann Hypothesis	?	proved by Deligne using sophisticated cohomological tools.

Motivated by such a table, C. Deninger [**7**] has elaborated a certain infinite dimensional cohomological formalism which should provide a unifying approach to numerous zeta functions, such as ζ_Y, $\widehat{\zeta}$, $\widehat{\zeta}_K$ (K a number field), the zeta function of an arithmetic variety, the Dirichlet L-functions, and more generally the motivic L-functions $L(M, s)$ where M is a motive over a number field.

Thus, to $\overline{\operatorname{Spec}\mathbb{Z}}$ (resp. M a motive over $\mathbb{Q}$) Deninger wishes to associate real (resp. complex) infinite dimensional cohomology groups $H^j(\overline{\operatorname{Spec}\mathbb{Z}}; \mathbb{R})$ (resp. $H^j(\overline{\operatorname{Spec}\mathbb{Z}}; M; \mathbb{C})$), $0 \le j \le 2$, endowed with an action of a one parameter group $((\phi^t)^*)_{t\in\mathbb{R}}$ (i.e., a continuous version of the Frobenius Fr), satisfying the following five properties:

(a) A Lefschetz trace formula holds for $(\phi^t)^*_{t\in\mathbb{R}}$ acting on $H^j(\overline{\operatorname{Spec}\mathbb{Z}}; \mathbb{R})$ and coincides with the explicit formula (5.1) (see Section 5) for $\widehat{\zeta}$.

(b) A cohomological formula holds for the various zeta (or L) functions.

(c) These cohomology groups are endowed with a cup product $\cup$ and a $(\phi^t)^*_{t\in\mathbb{R}}$-equivariant Poincaré duality holds:

$$H^j(\overline{\operatorname{Spec}\mathbb{Z}}; \mathbb{R}) \times H^{2-j}(\overline{\operatorname{Spec}\mathbb{Z}}; \mathbb{R}) \to \mathbb{R} \tag{1.1}$$

$$(\omega_1, \omega_2) \to \operatorname{Tr}(\omega_1 \cup \omega_2) \tag{1.2}$$

which implies the expected functional equation for $\widehat{\zeta}$.

(d) The set of zeroes of $\widehat{\zeta}$ coincide with the spectrum of Θ acting on $H^1(\overline{\operatorname{Spec}\mathbb{Z}}; \mathbb{R})$ where $e^{t\Theta} = (\phi^t)^*$.

(e) There is a Hodge star operator $*$ acting on $H^1(\overline{\operatorname{Spec}\mathbb{Z}}; \mathbb{R})$ commuting with $(\phi^t)^*$ such that $\omega \to \operatorname{Tr}(\omega \cup *\omega)$ induces a real Hilbert space structure on $H^1(\overline{\operatorname{Spec}\mathbb{Z}}; \mathbb{R})$. This would imply the Riemann hypothesis for $\widehat{\zeta}$.

Of course, quite similar properties are expected to hold more generally for the L-functions $L(M, s)$.

Remark. Similarly, one should prove the Riemann hypothesis for more general zeta functions. In the case of the smooth projective curves (resp. varieties) over a finite field, one should obtain a new proof of the theorem of Weil (resp. Deligne) completely different from that of Weil (resp. Deligne).

At the moment, Deninger's cohomological formalism is proved to exist only in very special cases. In Section 2 we consider the case of an elliptic curve E_0 over a finite field. Admitting Hasse's theorem, we recall Deninger's construction of the foliated laminated Riemannian space $(S(E_0), \mathcal{F}, g, \phi^t)$ endowed with an action of a flow ϕ^t. The cohomology groups appear here as L^2-leafwise cohomology groups and we show in detail how this cohomological formalism works in this case. We recall Deninger's proof of the fact that the Lefschetz trace formula for ϕ^t acting on these cohomology groups is identical with the explicit formula for ζ_{E_0}. We view the foliated space $(S(E_0), \mathcal{F}, \phi^t)$ as a noncommutative analogue of the set of geometric points $(\overline{E_0}, \mathrm{Fr})$, (ϕ^t) playing the role of the continuous version of the Frobenius.

In Section 3, we establish a connection with Connes' result [**4**] applied to E_0. Recall that the set $|E_0|$ of closed points of E_0 is the orbit space of the Frobenius acting on $\overline{E_0}$. In Deninger's [resp. Connes'] approach, the orbit space of the continuous [resp. discrete] Frobenius represents the noncommutative space of closed points of E_0. We show that in both cases the von Neumann algebra playing the role of the "bounded measurable functions" on this noncommutative space is the hyperfinite type $\mathrm{III}_{\frac{1}{q}}$-factor. Thus, the spaces of closed points in Deninger's and Connes' approach coincide when we viewed them as noncommutative measured spaces. Moreover, we compare the algebraic Riemann Roch theorem with its measured foliated analogue due to Connes.

In Section 4, we consider a smooth proper absolutely irreducible variety Y over $\mathbb{F}_q$ and we review Deninger's (non natural) construction of an infinite dimensional cohomology satisfying Poincaré Duality and allowing to prove only the expected cohomological formula and the functional equation for ζ_Y. The underlying foliated space $(S_Y, \mathcal{F}, g, \phi^t)$ which should allow to construct in a natural way this cohomology is not known to exist. We state a precise geometric hypothesis that such a foliated space $(S_Y, \mathcal{F}, g, \phi^t)$ (and the flow ϕ^t acting on S_Y) should satisfy in order to provide a natural construction of the expected cohomology. We show, following Deninger, how this geometric hypothesis provides a new proof all the expected properties for ζ_Y including the Riemann hypothesis (i.e., Deligne's Theorem).

In Section 5, we extract from several papers of Deninger about the Riemann zeta function $\widehat{\zeta}$, a conjecture stating the existence of a Riemannian laminated foliated space $(\overline{S_{\mathbb{Q}}}, \mathcal{F})$ endowed with a flow ϕ^t and satisfying certain geometric properties; $(\overline{S_{\mathbb{Q}}}, \mathcal{F})$ should fill the case labeled "underlying geometry" in the previous table. We establish a brief comparison between this conjectural foliated space and the noncommutative space constructed by Connes' [**4**] for $\widehat{\zeta}$. We explain how this conjecture implies the cohomological formula for $\widehat{\zeta}$ and the Riemann hypothesis. Lastly, we insert three appendices in which we explain various standard notions that are used throughout the paper.

The two geometric hypotheses and the conjecture stated here should be viewed as a working programme; we shall explain how they raise open questions and should generate interesting research both in number theory and in noncommutative geometry. We have tried to explain the arguments in their favour and ideas from Deninger which should allow to work in this subject. The author took great pleasure in learning C. Deninger's beautiful theory. Hopefully, the present paper will serve as an invitation to this fascinating circle of ideas.

2. The case of an elliptic curve E_0 over a finite field

2.1. The zeta function $\zeta_{E_0}(s)$ and the explicit formula. Let E_0 be an elliptic curve over a finite field $\mathbb{F}_q$ where $q = p^f$ and p is a prime number. Recall that the zeta function $\zeta_{E_0}(s)$ of E_0 is given by

$$\zeta_{E_0}(s) = \prod_{w \in |E_0|} \frac{1}{1 - (Nw)^{-s}} = \frac{(1 - \xi q^{-s})(1 - \overline{\xi} q^{-s})}{(1 - q^{-s})(1 - q^{1-s})} \tag{2.1}$$

where $|E_0|$ denotes the set of closed points of E_0 and ξ is a complex number which by Hasse's theorem satisfies $|\xi| = \sqrt{q}$. The explicit formula for $\zeta_{E_0}(s)$ takes the following form. Let $\alpha \in C_c^\infty(\mathbb{R},\mathbb{R})$ and set for any real s, $\Phi(s) = \int_{\mathbb{R}} e^{st}\alpha(t)\,dt$. Then one has

$$\begin{aligned}(2.2)\qquad &\sum_{\nu\in\mathbb{Z}}\Phi\left(\frac{2\pi\nu i}{\log q}\right) - \sum_{\rho\in\zeta_{E_0}^{-1}\{0\}}\Phi(\rho) + \sum_{\nu\in\mathbb{Z}}\Phi\left(1+\frac{2\pi\nu i}{\log q}\right) = \\ &= \sum_{w\in|E_0|}\log Nw\left(\sum_{k\geq 1}\alpha(k\log Nw) + \sum_{k\leq -1}(Nw)^k\alpha(k\log Nw)\right).\end{aligned}$$

The idea of the proof is to apply the residue theorem to

$$s \to \left(\int_0^{+\infty}\sqrt{t}\,\alpha(\log t)\,t^s\frac{dt}{t}\right)\frac{\zeta'_{E_0}}{\zeta_{E_0}}(s)$$

and to use the functional equation $\zeta_{E_0}(s) = \zeta_{E_0}(1-s)$. We are going to explain briefly how Deninger managed (see [**13**] for the details) to interpret this explicit formula (2.2) as a Lefschetz trace formula.

2.2. The Riemannian laminated foliated space $(S(E_0), \mathcal{F}, g, \phi^t)$. Let $\phi_0 : E_0 \to E_0$ be the q-th power Frobenius endomorphism of E_0 over $\mathbb{F}_q$. Deninger has used (see [**13**]) the following result due to Oort [**21**]:

LEMMA 1. *There exists*

(1) *a complete local integral domain R with field of fractions L a finite extension of $\mathbb{Q}_p$ ($q = p^r$) such that $R/\mathcal{M} = \mathbb{F}_q$ where $\mathcal{M}$ is the maximal ideal of R.*
(2) *an elliptic curve $\mathcal{E}$ over spec R together with an endomorphism $\phi : \mathcal{E} \to \mathcal{E}$ such that*

$$(\mathcal{E},\phi)\otimes\mathbb{F}_q = (E_0,\phi_0).$$

REMARKS.

(1) If the elliptic curve E_0 is ordinary, then one may take for R the ring of Witt vectors of $\mathbb{F}_q$, $W(\mathbb{F}_q)$, and then there is a canonical choice of the lifting $(\mathcal{E},\phi)$. On the contrary, if E_0 is supersingular ([**26**, p. 137]), then there is no canonical choice of $(\mathcal{E},\phi)$.
(2) It is possible to lift both a curve of genus ≥ 2 (over $\mathbb{F}_q$) but Hurwitz's formula ([**26**, p. 41]) shows that one cannot lift its Frobenius morphism in characteristic zero.

Now (still following [**13**]), we denote by $E = \mathcal{E}\otimes_R L$ the generic fibre. Then $\mathrm{End}_L(E)\otimes\mathbb{Q} = K$ is a field K which is either $\mathbb{Q}$ or an imaginary quadratic extension of $\mathbb{Q}$. We fix an embedding $\mathcal{I} : L \subset \mathbb{C}$ and consider the complex analytic elliptic curve $E(\mathbb{C})$. Let ω be a non zero holomorphic one form on $E(\mathbb{C})$ and let Γ be its period lattice. Then the Abel-Jacobi map

$$E(\mathbb{C}) \to \mathbb{C}/\Gamma,\ p \to \int_0^p \omega \bmod \Gamma$$

induces an isomorphism. Having fixed the embedding $\mathcal{I}$, there is a natural homomorphism

$$\Theta : K = \mathrm{End}_L(E) \to \mathrm{End}(\mathbb{C}/\Gamma).$$

Identify the Lie algebra $\mathrm{T}_0(\mathbb{C}/\Gamma)$ of $\mathbb{C}/\Gamma$ with $\mathbb{C}$. Then the map $\alpha \mapsto \mathrm{D}_0\Theta(\alpha)$ (where $\mathrm{D}_0\Theta(\alpha)$ denotes the tangent map at 0) defines two embeddings $\mathrm{End}_L(E) \to \mathbb{C}$ and $\mathcal{J} : K \to \mathbb{C}$. Now, set $\xi = \phi \otimes \mathrm{Id}_L \in \mathrm{End}_L(E) \subset K$ and identify ξ with $\mathcal{J}(\xi) \in \mathrm{T}_0(\mathbb{C}/\Gamma) = \mathbb{C}$. Then ξ defines by multiplication an element of $\mathrm{End}(\mathbb{C}/\Gamma)$ and $\xi\Gamma \subset \Gamma$.

Thus, the complex elliptic curve $\mathbb{C}/\Gamma$ endowed with the multiplication by ξ represents a lift of the elliptic curve E_0 and of its Frobenius morphism. Now, we set

$$V = \cup_{n\in\mathbb{N}}\xi^{-n}\Gamma, \quad \mathrm{T}\Gamma = \lim_{+\infty\leftarrow n} \frac{\Gamma}{\xi^n\Gamma}, \quad \text{and } V_\xi\Gamma = \mathrm{T}\Gamma \otimes_{\mathbb{Z}} \mathbb{Q}.$$

The set $\mathrm{T}\Gamma$ is a Tate module defined by a projective limit and $V_\xi\Gamma$ is a $\mathbb{Q}_p$-vector space of dimension belonging to $\{1, 2\}$. Any element v of V acts on $\mathbb{C} \times V_\xi\Gamma$ by $v.(z, \hat{v}) = (z + v, \hat{v} - v)$, we denote by $\frac{\mathbb{C}\times V_\xi\Gamma}{V}$ the quotient space.

LEMMA 2.

(1) *Let F be a finite set of representatives in Γ of the quotient group $\frac{\Gamma}{\xi\Gamma}$. Then any element of $V_\xi\Gamma$ is of the form $\sum_{l\geq -k} a_l\xi^l$ where $k \in \mathbb{N}$ and the $a_l \in F$. Moreover, the multiplication by ξ defines an automorphism of $V_\xi\Gamma$.*

(2) *The natural homomorphism*

$$\frac{\mathbb{C} \times \mathrm{T}\Gamma}{\Gamma} \to \frac{\mathbb{C} \times V_\xi\Gamma}{V}$$

defines a $\{\xi^l,\ l \in \mathbb{Z}\}$-equivariant isomorphism where the action of ξ is induced by the diagonal action on $\mathbb{C} \times \mathrm{T}\Gamma$ and $\mathbb{C} \times V_\xi\Gamma$, respectively.

PROOF. (1) Observe that any element of $\mathrm{T}\Gamma$ is of the form $\sum_{l\in N} a_l\xi^l$ where the $a_l \in F$. Using Bezout theorem one checks that a prime number $\widehat{p}$ not dividing q (i.e., $\neq p$) induces by multiplication an automorphism of $\mathrm{T}\Gamma$. If $\xi \notin \mathbb{Z}$ then the elliptic curve $\mathbb{C}/\Gamma$ has complex multiplication and its endomorphism ring is invariant under complex conjugation (see [**15**, p. 330]). So, in all cases, we have $\bar{\xi}\Gamma \subset \Gamma$. Then using the equality $\bar{\xi}\xi = q$, one gets the results of (1). Part (2) is left to the reader. $\square$

Now, any element $q^\nu \in q^{\mathbb{Z}}$ acts on $\frac{\mathbb{C}\times V_\xi\Gamma}{V} \times \mathbb{R}^{+*}$ by

$$q^\nu.([z, \hat{v}], x) = ([\xi^{-\nu}z, \xi^{-\nu}\hat{v}], xq^\nu).$$

In [**13**], Deninger has introduced the (compact) laminated Riemannian foliated space $(S(E_0), \mathcal{F})$ where

$$S(E_0) = \frac{\mathbb{C} \times V_\xi\Gamma}{V} \times_{q^{\mathbb{Z}}} \mathbb{R}^{+*},$$

and the leaves of $\mathcal{F}$ are the images of the sets $\mathbb{C} \times \{\hat{v}\} \times \{x\}$ by the natural map $\pi : \mathbb{C} \times V_\xi\Gamma \times \mathbb{R}^{+*} \to S(E_0)$. Observe that the domain of a typical foliation chart is locally isomorphic to $D \times \Omega \times]1, 2[$ where D is an open disk of $\mathbb{C}$, Ω is an open subset of $\mathrm{T}\Gamma$ so that the leaves are given by $D \times \{\omega\} \times \{x\}$ for $(\omega, x) \in \Omega \times]1, 2[$; the term "laminated" refers to the fact that the local transversal to the foliation $\mathcal{F}$ is the disconnected space $\Omega \times]1, 2[$.

REMARK. Using the fact that V (resp. $q^{\mathbb{Z}}$) acts freely on $V_\xi\Gamma$ (resp. $\mathbb{R}^{+*}$), the reader can check that $(S(E_0), \mathcal{F})$ has trivial holonomy.

One defines a flow ϕ^t acting on $(S(E_0), \mathcal{F})$ and sending each leaf into another leaf by $\phi^t(z, \hat{v}, x) = (z, \hat{v}, xe^t)$. Let μ_ξ denote a Haar measure on the group $V_\xi\Gamma$. Then one has the following

LEMMA 3.

(1) *The measure*

$$dx_1 dx_2 \otimes \mu_\xi \otimes \frac{dx}{x}$$

on $\mathbb{C} \times V_\xi\Gamma \times \mathbb{R}^{+*}$ *induces a measure* μ *on* $S(E_0)$.

(2) *The measure* μ *is invariant under the action of* ϕ^t.

PROOF.

(1) We just have to check that for any $\nu \in \mathbb{N}^*$ and any Borel subset A of $V_\xi\Gamma$, one has

$$\mu_\xi(\xi^\nu A) = |\xi|^{-2\nu}\mu_\xi(A) = q^{-\nu}\mu_\xi(A).$$

Since $(\xi^{-\nu})_*\mu_\xi$ is also a Haar measure on $V_\xi\Gamma$ it suffices to check this equality for $A = \mathrm{T}\Gamma$. But this is an immediate consequence of the fact that

$$\mathrm{T}\Gamma/(\xi^\nu \mathrm{T}\Gamma) \simeq \Gamma/(\xi^\nu\Gamma)$$

has $|\xi|^{2\nu} = q^\nu$ elements.

(2) This is obvious. □

Using the fact that $|\xi| = \sqrt{q}$, one checks that the Riemannian metric on the bundle $\mathrm{T}\mathbb{C} \times V_\xi\Gamma \times \mathbb{R}^{+*}$ given by

$$\tilde{g}_{z,\hat{v},x}(\eta_1, \eta_2) = x\mathrm{Re}\,(\eta_1\overline{\eta_2})$$

induces a Riemannian metric g along the leaves of $(S(E_0), \mathcal{F})$ so that the following property is satisfied:

$$\forall \eta \in \mathrm{T}_{[z,\hat{v},x]}\mathcal{F},\ g(\mathrm{T}_{[z,\hat{v},x]}\phi^t(\eta), \mathrm{T}_{[z,\hat{v},x]}\phi^t(\eta)) = e^t g(\eta, \eta). \tag{2.3}$$

On each *path* connected component $(Y, \mathcal{F})$ of $(S(E_0), \mathcal{F})$ one defines a metric $\widehat{g}$ by

$$\widehat{g}_{|\mathrm{T}\mathcal{F}} = g,\ \left(\frac{d}{dt}\phi^t(x)\right)_{t=0} \perp \mathrm{T}\mathcal{F},\ \widehat{g}\left(\left(\frac{d}{dt}\phi^t(x)\right)_{t=0}\right) = 1.$$

Notice that $(Y, \mathcal{F}, \widehat{g})$ is a Riemannian foliation with bundle-like metric $\widehat{g}$: this means that the foliation $(Y, \mathcal{F}, \widehat{g})$ is locally defined by Riemannian submersions or, equivalently, that if a geodesic is perpendicular to a leaf at one point then it is perpendicular to the leaves at all the other points.

LEMMA 4. (See [**13**, End of Section 3].) *Let Y be a path connected component of $(S(E_0), \mathcal{F})$. Then Y is either of the form $\mathbb{C} \times \mathbb{R}^{+*}$ or of the form $\mathbb{C} \times_{G_{\widehat{v}}} \mathbb{R}^{+*}$ where $\widehat{v}$ is a fixed element of $V_\xi\Gamma$ such that $\exists \nu \in \mathbb{Z}^*$, $(\xi^{-\nu} - 1)\widehat{v} \in V$ and $G_{\widehat{v}} = \{q^\nu \in q^{\mathbb{Z}} /\ (\xi^{-\nu} - 1)\widehat{v} \in V\}$ is a subgroup of $q^{\mathbb{Z}}$. In this latter case, the action of $q^\nu \in G_{\widehat{v}}$ on $\mathbb{C} \times \mathbb{R}^{+*}$ is defined by*

$$(z, t).q^\nu = (\xi^{-\nu}z + (\xi^{-\nu} - 1)\widehat{v}, tq^\nu).$$

PROOF. Consider an element $[z, \widehat{v}, x] \in Y$. If $\forall \nu \in \mathbb{Z}\setminus\{0\}$, $(\xi^{-\nu} - 1)\widehat{v} \notin V$, then Y is of the form $\mathbb{C} \times \mathbb{R}^{+*}$. If there exists $\nu \in \mathbb{Z}\setminus\{0\}$ such that $(\xi^{-\nu} - 1)\widehat{v} \in V$, then the reader can check easily that Y is of the form announced in the Lemma. □

REMARKS.

(1) If the reader wants to understand the arithmetic origin of the construction of Deninger's space $S(E_0)$, I suggest him to have a look at the end of the introduction of [**16**].

(2) As observed by Deninger (see [**8**, Section 4.7]), the correspondence $E_0 \to (S(E_0), \mathcal{F}, g, \phi^t)$ cannot be functorial because otherwise one should have a fiber functor to the category of finite dimensional $\mathbb{R}$-vector spaces. This latter fact should imply, for a supersingular elliptic curve E_0, the existence of a ring map

$$\operatorname{End} E_0 \otimes_{\mathbb{Q}} \mathbb{R} \to \operatorname{End} \mathrm{H}^1(E_0; \mathbb{R}) \simeq M_2(\mathbb{R}).$$

But such a map cannot exist since $\operatorname{End} E_0 \otimes_{\mathbb{Q}} \mathbb{R}$ is the field of quaternions ([**26**, p. 137]). The point here is that it is not possible to lift the four generators of $\operatorname{End} E_0$ all together to the same scheme in characteristic zero. In fact, if E_0 is supersingular then there is no canonical choice for $E_0 \to (S(E_0), \mathcal{F}, g, \phi^t)$, whereas there is one if E_0 is ordinary.

2.3. Reduced L^2-leafwise cohomology and its properties. Denote by $\mathcal{A}^j_{\mathcal{F}}(S(E_0))$ the set of real sections of the bundle $\wedge^j \mathrm{T}^*\mathcal{F} \to S(E_0)$ which are smooth along the leaves and continuous transversely. The metric g induces a metric $h_j(\cdot, \cdot)$ on the bundle $\wedge^j \mathrm{T}^*\mathcal{F}$ $(0 \leq j \leq 2)$; we then denote by $L^2(S(E_0); \wedge^j \mathrm{T}^*\mathcal{F})$ the completion of $\mathcal{A}^j_{\mathcal{F}}(S(E_0))$ with respect to the real scalar product

$$\forall \omega, \omega' \in \mathcal{A}^j_{\mathcal{F}}(S(E_0)),\ < \omega; \omega' >= \int_{S(E_0)} h_j(\omega, \omega')(\theta) d\mu(\theta). \tag{2.4}$$

Let $d^\dagger_{\mathcal{F}}$ denote the formal adjoint of the leafwise exterior derivative $d_{\mathcal{F}}$. Then one defines the reduced L^2-leafwise real cohomology groups $H^j_{L^2,\tau}(S(E_0), \mathbb{R})$, $(0 \leq j \leq 2)$ by

$$H^j_{L^2,\tau}(S(E_0), \mathbb{R}) = \ker(\, (d^\dagger_{\mathcal{F}})^* \cap (d^\dagger_{\mathcal{F}})^{**}),\ 0 \leq j \leq 2$$

where the unbounded operators $(d^\dagger_{\mathcal{F}})^*, (d^\dagger_{\mathcal{F}})^{**}$ both act on $L^2(S(E_0); \wedge^j \mathrm{T}^*\mathcal{F})$. I think that Deninger has known for a long time how to provide a Hodge theoretic interpretation of $H^j_{L^2,\tau}(S(E_0), \mathbb{R})$ as harmonic forms and identify it with a reduced leafwise L^2-cohomology group $\frac{\ker d_{\mathcal{F}}}{\overline{\operatorname{Im} d_{\mathcal{F}}}}$. The reader may reprove this fact by adapting the (Riemannian foliation) techniques of Alvarez-Lopez and Kordyukov ([**1**, Theorem 1.2]).

We shall set

$$H^j_{L^2,\tau}(S(E_0), \mathbb{C}) = H^j_{L^2,\tau}(S(E_0), \mathbb{R}) \otimes_{\mathbb{R}} \mathbb{C}.$$

Using (2.3) and the ϕ^t-invariance of μ one checks that $e^{-jt}(\phi^t)^*$ preserves the L^2-scalar product of $L^2(S(E_0); \wedge^j \mathrm{T}^*\mathcal{F})$. Thus, $(\phi^t)^*$ acts on $H^j_{L^2,\tau}(S(E_0), \mathbb{C})$, for instance, one has

$$(\phi^t)^*(x^{\log_q \overline{\xi} + \frac{2i\pi\nu'}{\log q}} d\overline{z}) = e^{t \log_q \overline{\xi} + t\frac{2i\pi\nu'}{\log q}} x^{\log_q \overline{\xi} + \frac{2i\pi\nu'}{\log q}} d\overline{z}$$

where $\log_q$ denotes the branch of the complex logarithm for the basis q determined by $-\pi < \operatorname{Arg} \alpha \leq \pi$.

One then obtains:

THEOREM 1. ([**13**, Theorem 3.1]) *Let* $\alpha \in C_c^\infty(\mathbb{R})$ *and set* $\Phi(s) = \int_{\mathbb{R}} e^{st}\,\alpha(t)dt$ *for* $s \in \mathbb{R}$. *Denote by* $(\phi^t)_j^*$ *the action of* ϕ^t *on* $H^j_{L^2,\tau}(S(E_0),\mathbb{C})$ $(0 \le j \le 2)$.

(0) *The family* $\{x^{\frac{2i\pi\nu}{\log q}},\ \nu \in \mathbb{Z}\}$ *defines an orthonormal basis of* $H^0_{L^2,\tau}(S(E_0),\mathbb{C})$. *Moreover,*

$$\mathrm{TR}\int_{\mathbb{R}} \alpha(t)(\phi^t)_0^* dt = \sum_{\nu\in\mathbb{Z}} \Phi\left(\frac{2\pi\nu i}{\log q}\right),$$

where TR *denotes the usual trace.*

(1) *The family* $\{x^{\log_q \xi + \frac{2i\pi\nu}{\log q}} dz,\ \ x^{\log_q \overline{\xi} + \frac{2i\pi\nu'}{\log q}} d\overline{z},\ \ \nu,\nu' \in \mathbb{Z}\}$ *defines an orthonormal basis of* $H^1_{L^2,\tau}(S(E_0),\mathbb{C})$. *Moreover,*

$$\mathrm{TR}\int_{\mathbb{R}} \alpha(t)(\phi^t)_1^* dt = \sum_{\rho\in\zeta^{-1}(0)} \Phi(\rho).$$

(2) *The family* $\{x^{1+\frac{2i\pi\nu'}{\log q}} dz \wedge d\overline{z},\ \nu' \in \mathbb{Z}\}$ *defines an orthonormal basis of* $H^2_{L^2,\tau}(S(E_0),\mathbb{C})$ *(we have used* $1 = \frac{2\log|\xi|}{\log q}$*). Moreover,*

$$\mathrm{TR}\int_{\mathbb{R}} \alpha(t)(\phi^t)_2^* dt = \sum_{\nu\in\mathbb{Z}} \Phi\left(1 + \frac{2\pi\nu i}{\log q}\right).$$

PROOF. (1) A 1-form $\omega = \alpha dz + \beta d\overline{z} \in L^2(S(E_0); \wedge^1 \mathrm{T}^*\mathcal{F})$ is given by two classes of V-invariant L^2_{loc}- functions on $\mathbb{C} \times V_\xi\Gamma \times \mathbb{R}^{+*}$ such that for any $q^\nu \in q^{\mathbb{Z}}$ one has

$$(q^\nu)^*\alpha = \xi^\nu \alpha,\ (q^\nu)^*\beta = \overline{\xi}^\nu \beta.$$

One checks that ω belongs to $H^1_{L^2,\tau}(S(E_0),\mathbb{C})$ if and only if the distributional partial derivatives of suitable representatives α and β with respect to x_1 and x_2 both vanish ($z = x_1 + ix_2$). Thus, α and β can be chosen to be independent of the $\mathbb{C}$-coordinates. Using the fact that the action by translation of V on $V_\xi\Gamma$ is ergodic with respect to μ_ξ, one checks that α and β are independent of the $V_\xi\Gamma$-coordinates as well. Finally, one observes that $\alpha \cdot x^{-\log_q \xi}$ and $\beta \cdot x^{-\log_q \overline{\xi}}$ define elements of $L^2(\mathbb{R}^{+*}/q^{\mathbb{Z}})$. Since the family $\{x^{\frac{2i\pi\nu}{\log q}},\ \nu \in \mathbb{Z}\}$ defines an orthonormal basis of $L^2(\mathbb{R}^{+*}/q^{\mathbb{Z}})$, one then gets easily the result. (0) and (2) are proved similarly. □

Now notice that $\lambda = x dx_1 \wedge dx_2$, ($z = x_1 + ix_2$) defines an oriented volume form along the leaves of $(S(E_0),\mathcal{F})$. Thus, $\frac{\mu}{\lambda} = \frac{\mu_\xi dx}{x^2}$ defines a transverse measure with associated Ruelle-Sullivan current $C(\frac{\mu}{\lambda})$, and we can pair sections of $\mathcal{A}^2_{\mathcal{F}}(S(E_0)) \otimes \mathbb{C}$ with $C(\frac{\mu}{\lambda})$. For instance, for any continuous function f on $S(E_0)$, one has $(f\lambda; C(\frac{\mu}{\lambda})) = \int_{S(E_0)} f\, d\mu$. Then one deduces from [**13**, Section 3] the following Poincaré duality result.

THEOREM 2. *One has the following two non degenerate Poincaré pairings* Tr, *defined using the basis of Theorem 1.*

(0) $H^0_{L^2,\tau}(S(E_0),\mathbb{C}) \times H^2_{L^2,\tau}(S(E_0),\mathbb{C}) \to \mathbb{C}$

$$\left(x^{\frac{2i\pi\nu}{\log q}},\, x^{1+\frac{2i\pi\nu'}{\log q}} dz \wedge d\overline{z}\right) \to \left(x^{1+\frac{2i\pi(\nu+\nu')}{\log q}} dz \wedge d\overline{z},\, C\left(\frac{\mu}{\lambda}\right)\right)$$
$$= \mathrm{Tr}\left(x^{\frac{2i\pi\nu}{\log q}} \cup x^{1+\frac{2i\pi\nu'}{\log q}} dz \wedge d\overline{z}\right)$$

and $\mathrm{Tr}(x^{\frac{2i\pi\nu}{\log q}} \cup x^{1+\frac{2i\pi\nu'}{\log q}} dz \wedge d\bar{z}) = 0$ *if and only if* $\nu + \nu' \neq 0$.

(1) $H^1_{L^2,\tau}(S(E_0), \mathbb{C}) \times H^1_{L^2,\tau}(S(E_0), \mathbb{C}) \to \mathbb{C}$

$$\left(x^{\log_q \xi + \frac{2i\pi\nu}{\log q}} dz,\, x^{\log_q \bar{\xi} + \frac{2i\pi\nu'}{\log q}} d\bar{z}\right) \to \left(x^{1+\frac{2i\pi(\nu+\nu')}{\log q}} dz \wedge d\bar{z}\,,\, C\left(\frac{\mu}{\lambda}\right)\right)$$

$$= \mathrm{Tr}\left(x^{\log_q \xi + \frac{2i\pi\nu}{\log q}} dz \cup x^{\log_q \bar{\xi} + \frac{2i\pi\nu'}{\log q}} d\bar{z}\right)$$

and $\mathrm{Tr}(x^{\log_q \xi + \frac{2i\pi\nu}{\log q}} dz \cup x^{\log_q \bar{\xi} + \frac{2i\pi\nu'}{\log q}} d\bar{z}) = 0$ *if and only if* $\nu + \nu' \neq 0$.

(2) *Moreover, for any* $t \in \mathbb{R}$ *and* $\omega, \omega' \in H^1_{L^2,\tau}(S(E_0), \mathbb{C})$ *(respectively,* $\omega \in H^0_{L^2,\tau}(S(E_0), \mathbb{C})$, $\omega' \in H^2_{L^2,\tau}(S(E_0), \mathbb{C})$*)*, $\mathrm{Tr}((\phi^t)^*\omega \cup (\phi^t)^*\omega') = e^t \mathrm{Tr}(\omega \cup \omega')$.

(3) *Set* $(\phi^t)^* = e^{t\Theta}$ *where* Θ *denotes the infinitesimal generator of the flow. For each* $j \in \{0, 1, 2\}$, Tr *induced an isomorphism between* $H^{2-j}_{L^2,\tau}(S(E_0), \mathbb{C})$ *and the smooth dual of* $H^j_{L^2,\tau}(S(E_0), \mathbb{C})$ *which by definition is the direct sum of the duals of the (finite dimensional) irreducible* $(\phi^t)^*$*-subspaces in a* $(\phi^t)^*$*-invariant decomposition of* $H^j_{L^2,\tau}(S(E_0), \mathbb{C})$. *The adjoint of* Θ *acting on* $H^j_{L^2,\tau}(S(E_0), \mathbb{C})$ *is* $\mathrm{Id} - \Theta$ *acting on* $H^{2-j}_{L^2,\tau}(S(E_0), \mathbb{C})$.

Next, the following cohomological formula is satisfied.

PROPOSITION 1. *For any* $s \in \mathbb{C}$,

$$\zeta_{E_0}(s) = \prod_{j=0}^{j=2} \left(\det{}_\infty \left(\frac{s\mathrm{Id} - \Theta}{2\pi} : H^j_{L^2,\tau}(S(E_0), \mathbb{R}) \right) \right)^{(-1)^{1+j}},$$

where $\det_\infty$ *denotes a regularized determinant whose definition is recalled in Section 6.*

PROOF. We recall formula (2.1). We shall simply prove that

$$\det{}_\infty \left(\frac{s\mathrm{Id} - \Theta}{2\pi} : H^0_{L^2,\tau}(S(E_0), \mathbb{R}) \right) = 1 - q^{-s},$$

leaving the rest to the reader. For $\nu \in \mathbb{Z}$, one has $\frac{s - \frac{2i\pi\nu}{\log q}}{2\pi} = \frac{-i}{\log q}(\frac{si\log q}{2\pi} + \nu)$, then using Theorem 1 we see that we just have to prove that $\prod_{\nu \in \mathbb{Z}}(\frac{-i}{\log q}(\frac{si\log q}{2\pi} + \nu)) = 1 - q^{-s}$. But this equality is a consequence of [**6**, Lemma 2.7]. □

Recall that one has a metric g along the (oriented) leaves of $(S(E_0), \mathcal{F})$ which satisfies (2.3), denote by $*$ the induced leafwise Hodge star operator acting on $\mathcal{A}^j_{\mathcal{F}}(S(E_0), \mathcal{F})$. One extends it to complex valued differential forms by the formula $*(\lambda + i\mu) = *\lambda - i * \mu$ for $\lambda, \mu \in \mathcal{A}^j_{\mathcal{F}}(S(E_0), \mathcal{F})$. For instance, one has $*(x^{\log_g \xi + \frac{2i\pi\nu}{\log q}} dz) = ix^{\log_g \bar{\xi} - \frac{2i\pi\nu}{\log q}} d\bar{z}$. Using (2.3) one sees that $(\phi^t)^* * = *(\phi^t)^*$ on $H^1_{L^2,\tau}(S(E_0), \mathbb{C})$. Moreover, it is clear that the hermitian scalar product (2.4) of $H^1_{L^2,\tau}(S(E_0), \mathbb{C})$ is given by $<\omega; \omega'>= \mathrm{Tr}(\omega \cup *\omega')$, then using Theorem 2.(1) one sees that

$$\forall \omega, \omega' \in H^1_{L^2,\tau}(S(E_0), \mathbb{C}),\ e^t < \omega; \omega' >=< (\phi^t)^*\omega; (\phi^t)^*\omega' > . \tag{2.5}$$

REMARKS.

(0) Recall that in order to construct the leafwise Riemann metric g on $(S(E_0), \mathcal{F})$ we have used Hasse's Theorem $\xi\overline{\xi} = q$, i.e., the zeroes of ζ_{E_0} have real part equal to $\frac{1}{2}$. Notice nevertheless that this fact is a formal consequence of Theorem 1 and (2.5). Indeed, Theorem 1.(1) shows that the eigenvalues of Θ acting on $H^1_{L^2,\tau}(S(E_0), \mathbb{C})$ coincide with the zeroes of ζ_{E_0}. Next, writing $(\phi^t)^* = e^{t\Theta}$ and taking the derivative at $t = 0$ in (2.5), one sees that $\Theta - \frac{1}{2}\mathrm{Id}$ is antisymmetric so that the zeroes of ζ_{E_0} have real part equal to $\frac{1}{2}$.

(1) It is an easy and instructive exercise to deduce the functional equation of ζ_{E_0} from Proposition 1, Theorem 2.(3) and Lemma 9 of Section 6. But notice that Hasse's Theorem implies the functional equation.

The following important result comes from [**13**, Proposition 3.3].

THEOREM 3.

(1) *There is a natural bijection between the set of valuations w of the function field $K(E_0)$ of E_0 and the set of primitive compact $\mathbb{R}$-orbits of ϕ^t on $S(E_0)$. It has the following property. If w corresponds to $\gamma = \gamma_w$, then*
$$l(\gamma_w) = \log N(w).$$

(2) *For each primitive compact orbit γ one can find $\widehat{v} \in V_\xi\Gamma$ such that*
$$G_{\widehat{v}} = \{q^\nu / \ \nu \in \mathbb{Z} \ \text{and} \ (\xi^{-\nu} - 1)\widehat{v} \in V\} \subset q^{\mathbb{Z}}$$
is a subgroup, $l(\gamma) = \log q \operatorname{card} \frac{q^{\mathbb{Z}}}{G_{\widehat{v}}}$. The path connected component of γ in $S(E_0)$ is identified with $\mathbb{C} \times_{G_{\widehat{v}}} \mathbb{R}^{+*} \subset S(E_0)$. Moreover, two distinct primitive compact $\mathbb{R}$-orbits of ϕ^t cannot belong to the same path connected component of $S(E_0)$.

PROOF. (1) The action of $\xi^{\mathbb{Z}}$ on $\mathbb{C} \times_V V_\xi\Gamma$ is defined by
$$\xi^n \cdot (z, \widehat{v}) = (\xi^n z, \xi^n \widehat{v}).$$
By definition of ϕ^t, it is clear that there is a bijection between closed ϕ^t-orbits of length $n \log q$ and ξ-orbits of length n on $\mathbb{C} \times_V V_\xi\Gamma$. Lemma 2 states that we have a natural ξ-equivariant isomorphism $\mathbb{C} \times_\Gamma T\Gamma \simeq \mathbb{C} \times_V V_\xi\Gamma$. Then we get a bijection between ξ-orbits of length n on $\mathbb{C} \times_V V_\xi\Gamma$ and ξ-orbits of length n on $\mathbb{C}/\Gamma$ (for such an orbit $[z]$, $z \in \mathbb{C}$, $\xi^n z - z \in \Gamma$). With the notations of Section 2.2, let $\overline{L}$ be the algebraic closure of the field L in $\mathbb{C}$, $\overline{R}$ the integral closure of R in $\overline{L}$. Fix a maximal ideal $\overline{\mathcal{M}}$ over $\mathcal{M}$ in $\overline{R}$, then $\frac{\overline{R}}{\overline{\mathcal{M}}} = \overline{\mathbb{F}_q}$. Recall that $E = \mathcal{E} \otimes_R L$. Since the curve $\mathcal{E}$ is proper over spec R, the natural map
$$\mathcal{E}(\overline{R}) \xrightarrow{\sim} E(\overline{L}) \tag{2.6}$$
is an isomorphism.

Denote by $E(\mathbb{C})_{\text{tors}}$ the set of torsion points of $E(\mathbb{C})$ (an elliptic curve is an abelian group). Since $E(\mathbb{C})_{\text{tors}}$ is algebraic over L, the inclusion $E(\overline{L}) \hookrightarrow E(\mathbb{C})$ induces an isomorphism
$$E(\overline{L})_{\text{tors}} \simeq E(\mathbb{C})_{\text{tors}}. \tag{2.7}$$
One has a notion of reduction map red : $\mathcal{E}(\overline{R}) \to \mathcal{E}(\overline{R}/\overline{\mathcal{M}})$ (see [**26**, Chapter VII]).

LEMMA 5. *One gets by composition a reduction map*

$$\text{red} : E(\mathbb{C})_{\text{tors}} \simeq \mathcal{E}(\overline{R})_{\text{tors}} \to \mathcal{E}(\overline{R}/\overline{\mathcal{M}})_{\text{tors}} = E_0(\overline{\mathbb{F}_q}).$$

It is equivariant with respect to the actions by $\Theta(\xi)$ on the left and by the q-Frobenius ϕ_0 on the right.

PROOF. First one composes the map red : $\mathcal{E}(\overline{R}) \to \mathcal{E}(\overline{R}/\overline{\mathcal{M}})$ on the left with the isomorphisms (2.6) and (2.7). Second one uses Lemma 1 and the equality $\overline{R}/\overline{\mathcal{M}} = \overline{\mathbb{F}}_q$ □

The group

$$E(\mathbb{C})^{\Theta(\xi^n)=1} = (\mathbb{C}/\Gamma)^{\xi^n=1} = (\xi^n - 1)^{-1}\Gamma/\Gamma.$$

has order $|\xi^n - 1|^2$ and thus is a torsion group. Therefore

$$E(\mathbb{C})_{\text{tors}}^{\Theta(\xi^n)=1} = E(\mathbb{C})^{\Theta(\xi^n)=1}.$$

Now, recall that $E_0(\mathbb{F}_{q^n}) = E_0(\overline{\mathbb{F}}_{q^n})^{\phi_0^n=1}$ and that ([**26**, p. 136])

$$\text{card}\, E_0(\mathbb{F}_{q^n}) = q^n + 1 - (\xi^n + \overline{\xi}^n) = |\xi^n - 1|^2.$$

Observe that Hasse's theorem $|\xi|^2 = q$ is used in this equality. Moreover, for any integer $N \geq 1$ prime to p (= char $\mathbb{F}_q$) the restriction of red to the following N-torsion groups induces an isomorphism

$$\text{red} : E(\mathbb{C})_N \xrightarrow{\sim} E_0(\overline{\mathbb{F}}_{q^n})_N \tag{2.8}$$

(see [**26**, IV Prop. 2.3 and VII Prop. 3.1]).

Now assume that E_0 is supersingular so that $E_0(\overline{\mathbb{F}_q})$ has no p-torsion and that $N = \#E_0(\mathbb{F}_{q^n})$ is prime to p. Then combining (2.8) and the previous Lemma one gets

$$\text{red} : E(\mathbb{C})^{\Theta(\xi^n)=1} = E(\mathbb{C})_N^{\Theta(\xi^n)=1} \simeq E_0(\overline{\mathbb{F}_q})_N^{\phi_0^n=1} = E_0(\mathbb{F}_{q^n}).$$

Since $E_0(\mathbb{F}_{q^n})$ is in bijection with the set of valuations of $K(E_0)$ of degree n, the result is proved when E_0 is supersingular.

Now we assume that E_0 is ordinary. In this case we take $\mathcal{E}$ (in Lemma 1) to be the canonical lifting of E_0 to $R = W(\mathbb{F}_q)$ the ring of Witt vectors.

LEMMA 6. *For any $N \in \mathbb{N}^*$, one has a split exact sequence*

$$0 \to \mu_{N_p}(\overline{R}) \to E(\overline{L})_N \overset{\text{red}}{\to} E_0(\overline{\mathbb{F}}_q)_N \to 0$$

where N_p is the largest power of p dividing N.

PROOF. Taking into account the isomorphism (2.8), we can assume that N is of the form p^n. The canonical (exact) sequence of p-divisible groups

$$0 \to \mathbb{G}_m(p) \to \mathcal{E}(p) \to \mathbb{Q}_p/\mathbb{Z}_p \to 0$$

over R is split. Hence we get a commutative diagram with compatible splittings:

$$\begin{array}{ccccccccc} 0 & \longrightarrow & \mu_{p^n}(\overline{R}) & \longrightarrow & \mathcal{E}_{p^n}(\overline{R}) & \overset{\longleftarrow}{\longrightarrow} & \mathbb{Z}/(p^n) & \longrightarrow & 0 \\ \downarrow & & \downarrow & & \downarrow \text{red} & & \downarrow = & & \downarrow \\ 0 & \longrightarrow & \mu_{p^n}(\overline{\mathbb{F}}_q) & \longrightarrow & (E_0)_{p^n}(\overline{\mathbb{F}}_q) & \overset{\longleftarrow}{\longrightarrow} & \mathbb{Z}/(p^n) & \longrightarrow & 0. \end{array}$$

Using this diagram one easily checks that the map red: $\mathcal{E}_{p^n}(\overline{R}) \to (E_0)_{p^n}(\overline{\mathbb{F}}_q)$ is surjective. We recall the isomorphism (2.6). Then we get the following exact sequence which proves the Lemma:

$$0 \to \mu_{p^n}(\overline{R}) \to E(\overline{L})_{p^n} \overset{\text{red}}{\to} E_0(\overline{\mathbb{F}}_q)_{p^n} \to 0.$$

□

We apply the previous Lemma with $N = |\xi^n - 1|^2 = \operatorname{card} E_0(\mathbb{F}_{q^n})$, since the exact sequence is split, we the get the following exact sequence:

$$0 \to \mu_{p^n}(\overline{R})^{\Theta(\xi)^n=1} \to E(\overline{L})_N^{\Theta(\xi)^n=1} \overset{\text{red}}{\to} E_0(\mathbb{F}_{q^n}) \to 0.$$

Using (2.7) one sees that $E(\overline{L})_N^{\Theta(\xi)^n=1}$ and $E_0(\mathbb{F}_{q^n})$ have the same order. Therefore,

$$\text{red} : E(\mathbb{C})^{\Theta(\xi)^n=1} = E(\overline{L})_N^{\Theta(\xi)^n=1} \to E_0(\mathbb{F}_{q^n})$$

is an isomorphism. Since $E_0(\mathbb{F}_{q^n})$ is in bijection with the set of valuations of degree n of $K(E_0)$, we get part (1) of the Theorem. Part (2) is left to the reader who may use Lemma 4. □

By combining Theorems 1 and 3 and the explicit formula (2.2), one gets the following

COROLLARY 1. *Denote by $(\phi^t)^*_j$ the operator $(\phi^t)^*$ acting on $H^j_{L^2,\tau}(S(E_0),\mathbb{R})$ for $0 \leq j \leq 2$. The following equality holds as distributions in $\mathbb{R}$:*

$$\sum_{j=0}^{j=2}(-1)^j \mathrm{TR}(\phi^t)^*_j = \sum_\gamma l(\gamma)\sum_{k\geq 1}\delta_{kl(\gamma)} + \sum_\gamma l(\gamma)\sum_{k\leq -1} e^{kl(\gamma)}\delta_{kl(\gamma)}.$$

REMARKS.

(0) We leave it as an instructive exercise to the reader to prove directly this corollary (admitting Theorem 3), starting from the left-hand side, by using Poisson's summation formula and the equalities

$$\forall n \in \mathbb{N},\ \operatorname{card} E_0(\mathbb{F}_{q^n}) = |1 - \xi^n|^2 = \sum_{\deg w|n} \deg w.$$

As explained by Deninger in [**13**], the dissymmetry between the coefficients of $\delta_{kl(\gamma)}$ for $k \geq 1$ and $k \leq -1$ is due to formula (2.3). Moreover, the reader can check by himself that one uses Hasse's Theorem ($|\xi|^2 = q$) in order to get the coefficient $e^{kl(\gamma)}$ for $k \leq -1$.

(1) Let $(X, \mathcal{F}, \phi^t)$ be a three dimensional compact foliated manifold endowed with a codimension one foliation and a flow ϕ^t permuting the leaves and transversal to them. Then $(X, \mathcal{F}, \phi^t)$ admits a bundle-like metric g_X and thus is a Riemannian foliation. Assume that the closed orbits γ of ϕ^t are non degenerate and denote by H^j ($0 \leq j \leq 2$) the L^2-leafwise cohomology groups. Then the following distributional Lefschetz trace formula holds in $\mathcal{D}'(\mathbb{R})$ and is proved by Alvarez-Lopez and Kordyukov ([**2**]).

$$\sum_{j=0}^{2}(-1)^j \mathrm{TR}((\phi^t)^* \mid H^j) = \chi_{\mathrm{Co}}(\mathcal{F},\Lambda)\delta_0 + \sum_\gamma l(\gamma)\sum_{k\in\mathbb{Z}^*}\epsilon_\gamma(k)\delta_{k\,l(\gamma)}, \tag{2.9}$$

where $\chi_{\text{Co}}(\mathcal{F}, \Lambda)$ denotes Connes' Euler characteristic of the foliation with respect to the transverse volume element for g_X and, for a closed orbit γ

$$\epsilon_\gamma(k) = \operatorname{sgn} \det \left(\mathrm{Id} - T_y \phi^{kl(\gamma)} \mid T_y \mathcal{F}\right) \in \{1, -1\}$$

does not depend on $y \in \gamma$ (the sign sgn is well-defined since γ is non degenerate). It should be possible (and hard) to prove Corollary 1 along the lines of [**2**].

(2) In formula (2.9) there is basically no dissymetry between the coefficients of $\delta_{kl(\gamma)}$ for $k \geq 1$ and $k \leq -1$. Indeed, it is known that a flow acting on a smooth compact Riemannian foliated manifold cannot satisfy the condition (2.3), it is thus necessary to consider **laminated** foliated spaces. To get a flavour of the problem, suppose that ϕ is a diffeomorphism acting on a smooth connected compact Riemannian manifold (N, g_1) such that $\phi^*(g_1) = C\, g_1$ for a constant $C > 0$. If C were $\neq 1$, then one would like to introduce the suspension $N \times_{\mathbb{Z}} \mathbb{R}^{+*}$ where (z, x) is identified with $(\phi^{-1}(z), Cx)$ and then to consider the metric $x g_1$ along the leaves of the foliation of $N \times_{\mathbb{Z}} \mathbb{R}^{+*}$ defined by the images of the sets $N \times \{x\}$. But, C is necessary equal to 1 since it is the absolute value of the degree of the diffeomorphism ϕ.

2.4. *L*-functions associated with unramified extensions of the function field of E_0. Let $\mathbb{F}_q \subset \mathbb{F}_{q^n}$ be a constant field extension with cyclic Galois G group of order n generated by the Frobenius automorphism $\phi_0 : z \to z^q$. It induces a Galois extension $K(E_0) \subset K_n(E_0)$ with Galois group G. Let χ be the character of a one dimensional representation of G and let $L(s, \chi)$ be the associated L-function

$$L(s, \chi) = \prod_{w \in |E_0|} \frac{1}{1 - \chi(\phi_0^{\deg w})(Nw)^{-s}}$$

(see [**23**, Chapter 9] for details). Let us consider the flat complex line bundle $\mathcal{L}_\chi$ over $S(E_0)$ defined by

$$(\mathbb{C} \times_V V_\xi \Gamma) \times_{q^{\mathbb{Z}}} (\mathbb{R}^{+*} \times \mathbb{C})$$

where $([z, \widehat{v}], x, \eta)$ is identified with $([\xi^{-\nu}, z, \xi^{-\nu}\widehat{v}], q^\nu x, \chi(\phi_0^\nu)\eta)$. One then defines, for each real t, a bundle map ψ^t by

$$(\phi^t)^* \left(\wedge^* T^* \mathcal{F} \otimes \mathcal{L}_\chi\right) \to \wedge^* T^* \mathcal{F} \otimes \mathcal{L}_\chi$$

$$\omega \otimes u \in \wedge^* T^* \mathcal{F}_{\phi^t(y)} \otimes (\mathcal{L}_\chi)_{\phi^t(y)} \to (T_y \phi^t)^*(\omega) \otimes u' = \psi^t(\omega \otimes u)$$

where $y \in S(E_0)$ is the class of $([z, \widehat{v}], x)$, $u \in (\mathcal{L}_\chi)_{\phi^t(y)}$ is the class of $([z, \widehat{v}], xe^t, \eta)$ and u' is the class of $([z, \widehat{v}], x, \eta)$. Let $\mathcal{A}^j_{\mathcal{F}}(S(E_0); \mathcal{L}_\chi))$ denote the set of the sections of the bundle $\wedge^* T^* \mathcal{F} \otimes \mathcal{L}_\chi$ which are continuous transversely and smooth along the leaves. Then one defines an endomorphism $(\psi^t)^*$ of $\mathcal{A}^j_{\mathcal{F}}(S(E_0); \mathcal{L}_\chi))$ by setting $(\psi^t)^*(w)(y) = \psi^t\left(w(\phi^t(y))\right)$.

Then we leave as an exercise to the reader the following two tasks:

(I) Define the L^2-leafwise cohomology groups with coefficients in $\mathcal{L}_\chi$: $H^j_{L^2,\tau}(S(E_0), \mathcal{L}_\chi)$ for $0 \leq j \leq 2$ and an action of $(\psi^t)^*$ on each of these groups.

(II) Prove a result analogous to Corollary 1 for $L(s, \chi)$, namely show that the explicit formula for $L(s, \chi)$ is identical to the Lefschetz trace formula for $(\psi^t)^*$ acting on $H^j_{L^2,\tau}(S(E_0), \mathcal{L}_\chi)$ for $0 \leq j \leq 2$.

It would be interesting to extend the results (I) and (II) to the case of a L-function associated to a *geometric* finite Galois extension F of $K(E_0)$ (i.e., $\mathbb{F}_q$ is algebraically closed in F) and to a character of the Galois group $\mathrm{Gal}\,(F, K(E_0))$.

3. Connection with Connes' results

3.1. Von Neumann algebras.

The results of this Section are new.

We refer to [**3**, Chapters I.4 and V] for the notions relevant to von Neumann algebras. First we recall, following [**3**, p. 51], the definition of the von Neumann algebra $W(S(E_0), \mathcal{F})$ of the foliation $(S(E_0), \mathcal{F})$. We endow $S(E_0)$ with the tribe $\mathcal{T}$ generated by the Borel tribe and the subsets A of $S(E_0)$ which are contained in a Borel subset of μ-measure 0. Let λ denote the leafwise volume Riemannian form along the leaves of $(S(E_0), \mathcal{F}, g)$ and let $\pi : S(E_0) \to \frac{S(E_0)}{\mathcal{F}}$ denote the projection onto the space of leaves.

DEFINITION 1.

(1) A measurable section η of the bundle of half-densities Hilbert spaces $(L^2(\pi(y), \sqrt{\lambda}))_{y\in S(E_0)}$ over $S(E_0)$ is given by a family

$$\eta = \left(\eta_y = f_y\sqrt{\lambda}(y) \in L^2(\pi(y), \sqrt{\lambda})\right)_{y\in S(E_0)}$$

such that the complex valued function $(y, y') \to f_y(y')$ induces a measurable map on $\{(y, y') \in S(E_0)^2 /\ y \in \pi(y')\}$.

(2) A random operator $q = (q_l)_{l\in \frac{S(E_0)}{\mathcal{F}}}$ is the data for each leaf $l \in \frac{S(E_0)}{\mathcal{F}}$ of a bounded operator q_l on $L^2(l, \sqrt{\lambda})$ such that for each measurable sections δ, η of the bundle $(L^2(\pi(y), \sqrt{\lambda}))_{y\in S(E_0)}$, the map

$$S(E_0) \to \mathbb{C} \tag{3.1}$$

$$y \to < q_{\pi(y)}(\delta_y); \eta_y > \tag{3.2}$$

is measurable with respect to the completed tribe $\mathcal{T}$. We then denote by $||q||$ the essential supremum of the $||q_l||$, $l \in \frac{S(E_0)}{\mathcal{F}}$.

(3) The classes of bounded random operators $q = (q_l)_{l\in \frac{S(E_0)}{\mathcal{F}}}$ define a von Neumann algebra $W(S(E_0), \mathcal{F})$. See [**3**, p. 52] for more details.

Next we define, following [**3**, Prop. 6], a trace on $W(S(E_0), \mathcal{F})$. Recall that we have a natural transverse measure $\frac{\mu}{\lambda}$ on $(S(E_0), \mathcal{F})$. Let $(U_\alpha)_{\alpha\in I}$ be a finite covering of $S(E_0)$ by domains of foliations charts $U_\alpha \simeq V_\alpha \times T_\alpha$ where V_α is an open disk of $\mathbb{C}$ and T_α is a local transversal. Let $(\psi_\alpha)_{\alpha\in I}$ be a continuous leafwise smooth partition of unity associated to this covering, $\mathrm{supp}\,\psi_\alpha \subset\subset U_\alpha$. Let $q = (q_l)_{l\in \frac{S(E_0)}{\mathcal{F}}}$ be a positive element of $W(S(E_0), \mathcal{F})$. For each plaque $\widetilde{p}$ of U_α, let $\mathrm{TR}\,(\psi_\alpha q_l)_{|\widetilde{p}}$ be the trace of the operator $\psi_\alpha^{\frac{1}{2}} q_l \psi_\alpha^{\frac{1}{2}}$ in $L^2(\widetilde{p}, \sqrt{\lambda}) \subset L^2(l, \sqrt{\lambda})$, where l is the leaf of $\widetilde{p}$. Recall that the transverse measure $\frac{\mu}{\lambda}$ is holonomy invariant and induces a measure on each T_α. Then the following number $\tau(q) \in [0, +\infty]$ is independent of the choices of U_α, ψ_α :

$$\tau(q) = \sum_{\alpha\in I} \int_{T_\alpha} \mathrm{TR}\,(\psi_\alpha^{\frac{1}{2}} q_l \psi_\alpha^{\frac{1}{2}})_{|\widetilde{p}}\ \frac{\mu}{\lambda}.$$

One checks easily that τ defines a semi-finite normal faithful trace on the von Neumann algebra $W(S(E_0), \mathcal{F})$. Moreover, observe that the flow ϕ^t induces an action, denoted $(\phi^t)^*$, on $W(S(E_0), \mathcal{F})$.

THEOREM 4.
(1) *The center of* $W(S(E_0), \mathcal{F})$ *is isomorphic to* $L^\infty(\frac{\mathbb{R}}{(\log q)\mathbb{Z}})$.
(2) $W(S(E_0), \mathcal{F})$ *is type* II_∞*-von Neumann algebra.*
(3) *The cross-product von Neumann algebra* $W(S(E_0), \mathcal{F}) \rtimes_{(\phi^t)^*} \mathbb{R}$ *is a type* $\mathrm{III}_{\frac{1}{q}}$*-factor.*

PROOF.
(1) This is a consequence of the fact that the additive action of V on $V_\xi\Gamma$ is ergodic with respect to μ_ξ.
(2) Using the fact that $(V_\xi\Gamma, \mu_\xi)$ has no atoms, one checks that when p runs over the set of self-adjoint projections of $W(S(E_0), \mathcal{F})$ then $\tau(p)$ takes all the values of $[0, +\infty]$. This implies immediately the result.
(3) We use the notations and results of [**3**, pp. 494, 495]. Using (2.3) and the ϕ^t-invariance of μ, one checks easily that $\tau \circ (\phi^t)^* = e^t\tau$, which implies that $W(S(E_0), \mathcal{F}) \rtimes_{(\phi^t)^*} \mathbb{R}$ is of type III. Moreover, the following second equality is clear:

$$S\left(W(S(E_0), \mathcal{F}) \rtimes_{(\phi^t)^*} \mathbb{R}\right) \cap \mathbb{R}^{+*} = \{\lambda > 0 / \ (\phi^{\log \lambda})^* = \mathrm{Id} = q^{\mathbb{Z}}.$$

Then [**3**, p. 473] implies the result. $\square$

Morally, $(S(E_0), \mathcal{F})$ is the noncommutative space of geometric points and the flow (ϕ^t) represents the continuous version of the geometric Frobenius morphism of E_0. Then
$W(S(E_0), \mathcal{F}) \rtimes \mathbb{R}$ represents the algebra of bounded measurable functions on the noncommutative space of closed points defined by the orbit space $S(E_0)/\{\phi^t\}$.

Now we explain briefly the connection with Connes' approach [**4**]. Let $C_{K(E_0)}$ be the Idele class group of the function field $K(E_0)$ defined as the field of rational functions on E_0. Assume E_0 absolutely irreducible. The Artin reciprocity map $C_{K(E_0)} \to \mathcal{G}$ is injective, where $\mathcal{G}$ denotes the abelian modified Galois group of the maximal abelian extension of $K(E_0)$ (see [**25**, p. 135]). Since E_0 has a rational point over $\mathbb{F}_q$, eg the origin, one has $\{\|u\| / \ u \in C_{K(E_0)}\} = q^{\mathbb{Z}}$ where $\|\|$ denotes the norm map. Then one has a non canonical group isomorphism $C_{K(E_0)} \simeq K_1 \times q^{\mathbb{Z}}$ where $K_1 = \{h \in C_{K(E_0)} / \ \|h\| = 1\}$ is compact. By classfield theory, the group $\frac{C_{K(E_0)}}{K_1} \simeq q^{\mathbb{Z}}$ acts on the constant field extensions (generated by the $\mathbb{F}_{q^{n!}} \supset \mathbb{F}_q$, $n \in \mathbb{N}^*$) of $K(E_0)$, this action corresponds basically to the one of the Frobenius endomorphism of E_0. In his approach to the zeta function $\zeta_{E_0}(s)$, Connes considers the noncommutative space $\frac{\mathbb{A}_{E_0}}{I_1}$ where $\mathbb{A}_{E_0}$ (resp. I_1) denotes the Adele ring of the function field $K(E_0)$ (resp. the group of ideles of norm 1), see [**4**, before Theorem 1]. Recall that, by the product formula, $K(E_0)^* \subset I_1$. The group I_1 acts by multiplication on $\mathbb{A}_{E_0}$ and preserves its (normalized) Haar measure $d\mu_{\mathbb{A}_{E_0}}$. Let $d\mu_{I_1}$ be a Haar measure of I_1. Denoting by $\int_{I_1} a_g \, d\mu_{I_1}(g)$ a "typical element" of the von Neumann algebra $W(E_0) = L^\infty(\mathbb{A}_{E_0}) \rtimes I_1$, one defines a trace on $W(E_0)$ by setting

$$\mathrm{TR}\left(\int_{I_1} a_g \, d\mu_{I_1}(g)\right) = \int_{\mathbb{A}_{E_0}} a_1(y) \, d\mu_{\mathbb{A}_{E_0}}(y).$$

One can check that TR defines a semi-finite normal faithful trace on $W(E_0)$. Now we fix a non canonical isomorphism $\mathcal{U} : \mathcal{I}_{E_0} \to I_1 \times q^{\mathbb{Z}}$ where $\mathcal{I}_{E_0}$ denotes the Idele group of $K(E_0)$. So we may write $\mathcal{U}^{-1}(\{1\} \times q^{\mathbb{Z}}) = \widehat{q}^{\mathbb{Z}}$ for a certain $\widehat{q} \in \mathcal{I}_{E_0}$. Then, we let $q \in q^{\mathbb{Z}}$ act on the von Neumann algebra $W(E_0) = L^\infty(\mathbb{A}_{E_0}) \rtimes I_1$ as the following automorphism Θ_q:

$$q \cdot \int_{I_1} a_g(\cdot)\, d\mu_{I_1}(g) = \Theta_q\Big(\int_{I_1} a_g(\cdot)\, d\mu_{I_1}(g)\Big) = \int_{I_1} a_g(\widehat{q}\cdot)\, d\mu_{I_1}(g)$$

Morally, $W(E_0) \rtimes q^{\mathbb{Z}}$ represents the algebra of bounded measurable functions on the noncommutative space of closed points $\left(\frac{\mathbb{A}_{E_0}}{I_1}\right)/q^{\mathbb{Z}}$. The next theorem shows that the approaches of Connes and Deninger provide, by two different ways, the same cross-product von Neumann algebra.

THEOREM 5.

(1) *The von Neumann algebra $W(E_0)$ is a type II_∞-factor.*
(2) *The von Neumann algebra $W(E_0) \rtimes q^{\mathbb{Z}}$ is a type $\mathrm{III}_{1/q}$-factor.*

PROOF.

(1) Since the action of I_1 on $\mathbb{A}_{E_0}$ is ergodic with respect to $\mu_{\mathbb{A}_{E_0}}$, $W(E_0)$ is a factor. Since $(\mathbb{A}_{E_0}, \mu_{\mathbb{A}_{E_0}})$ has no atoms, one can check that when p runs over the set of self-adjoint projections of $W(E_0)$, $\mathrm{TR}\,(p)$ takes all values in $[0, +\infty]$. This implies the result.

(2) One checks that $\mathrm{TR} \circ \Theta_q = \frac{1}{q}\mathrm{TR}$. Then the result follows from [**3**, Theorem 4.b, p. 474]. □

REMARK. Of course, the previous Theorem extends readily to any function field over $\mathbb{F}_q$ such that the associated curve has a rational point over $\mathbb{F}_q$.

3.2. Riemann Roch theorem. We begin with recalling the statement of the classical algebraic Riemann Roch theorem. Assume in this subsection that $\mathbb{F}_q$ is algebraically closed in $K(E_0)$. Let D be any divisor of E_0. Then, $\{f \in K(E_0)^* /\ \mathrm{div}\,(f) \geq -D\} \cup \{0\}$ is a finite dimensional $\mathbb{F}_q$-vector space whose dimension is denoted by $l(D)$. Let K_{E_0} be the canonical divisor of E_0. Then, since the genus of E_0 is equal to one, the Riemann Roch Theorem states that

$$l(D) - l(K_{E_0} - D) = \deg D.$$

We are going to explain its connection with Connes' measured foliated Riemann Roch theorem for $(S(E_0), \mathcal{F})$. Write $D = \sum_{j=1}^N n_j w_j$ where $n_j \in \mathbb{Z}$, $w_j \in E_0$ for any $j \in \{1, \ldots, N\}$. Following Theorem 3, we associate to each w_j $(0 \leq j \leq N)$ the compact primitive orbit γ_{w_j} of ϕ^t and its path connected component $\mathbb{C} \times_{G_{\widehat{v}_j}} \mathbb{R}^{+*}$ in $S(E_0)$ (see Theorem 3). By restriction the foliation $(S(E_0), \mathcal{F})$ induces a foliation $(\mathbb{C} \times_{G_{\widehat{v}_j}} \mathbb{R}^{+*}, \mathcal{F})$ on the non compact Borel subset $\mathbb{C} \times_{G_{\widehat{v}_j}} \mathbb{R}^{+*}$ of $S(E_0)$. The subgroup $G_{\widehat{v}_j}$ of $q^{\mathbb{Z}}$ is of the form $b_j^{\mathbb{Z}}$ for a suitable $b_j \in \mathbb{N}^* \cap G_{\widehat{v}_j}$. Denote by $\pi_{\widehat{v}_j}$ the projection map

$$\pi_{\widehat{v}_j} : \mathbb{C} \times \mathbb{R}^{+*} \to \mathbb{C} \times_{G_{\widehat{v}_j}} \mathbb{R}^{+*}$$

and denote by $D((m,n),2)$ the open disk of $\mathbb{C}$ of center $(m,n) \in \mathbb{Z}^2$ and radius 2. As (m,n,k) runs over $\mathbb{Z}^3$ the restrictions of $\pi_{\widehat{v}_j}$ to the following sets:

$$D((m,n),2) \times \left\{ b_j^{\frac{k}{2}}\theta \in \mathbb{R}^{+*} /\ \theta \in \left[b_j^{\frac{-1}{3}}, b_j^{\frac{1}{3}}\right] \right\}$$

define an atlas of foliated charts of $(\mathbb{C}\times_{G_{\widehat{v}_j}}\mathbb{R}^{+*},\mathcal{F})$. Since the Haar measure of $\mathbb{R}^{+*}$ is $G_{\widehat{v}_j}$-invariant, its restrictions to the various local transversals $\{b_j^{\frac{k}{2}}\theta/\ \theta\in[b_j^{\frac{-1}{3}},b_j^{\frac{1}{3}}]\}$ induce a holonomy-invariant transverse measure Λ_j (see Definition 3 of Section 8) on the foliation $(\mathbb{C}\times_{G_{\widehat{v}_j}}\mathbb{R}^{+*},\mathcal{F})$. We normalize Λ_j so that $\Lambda_j(\gamma_{w_j})=1$. Then one defines a transverse measure, still denoted Λ_j, on $(S(E_0),\mathcal{F})$ by setting for any Borel transversal T of $(S(E_0),\mathcal{F})$:

$$\Lambda_j(T)=\Lambda_j\left(T\cap(\mathbb{C}\times_{G_{\widehat{v}_j}}\mathbb{R}^{+*})\right).$$

Then we set $\Lambda=\sum_{j=1}^N\Lambda_j$. Next, one obtains a complex line bundle $\mathcal{E}\to S(E_0)$ whose restriction to each leaf L of $(S(E_0),\mathcal{F})$ is the holomorphic line bundle associated with the divisor $\sum_{j=1}^N n_j(L\cap\gamma_j)$ (recall that the curves γ_j are transverse to the foliation). Using the definition of the leafwise flat metric g of $(S(E_0),\mathcal{F})$ one checks that the Euler class of the bundle $T\mathcal{F}\to S(E_0)$ is zero. Then Connes' measured index theorem [**3**, p. 71] states that

$$\dim_\Lambda\left(\ker\overline{\partial}_{\mathcal{E}}\right)-\dim_\Lambda\left(\ker(\overline{\partial}_{\mathcal{E}})^*\right)=\sum_{j=1}^N n_j\Lambda(\gamma_j)=\deg D.$$

REMARKS.

(1) Notice that Λ defines a trace on the C^*-algebra of $(S(E_0),\mathcal{F})$ but not on $W(S(E_0),\mathcal{F})$ because the restriction of Λ to a transversal T to $(S(E_0),\mathcal{F})$ does not define a measure on $L^\infty(T,\frac{\mu}{\lambda})$.

(2) The leaves of $(S(E_0),\mathcal{F})$ have polynomial growth so that we could have considered the transverse measure C defined [**3**, p. 72] by an average process. But one can check that $C(\gamma_j)=0$ for any $j\in\{1,\dots,N\}$. Moreover, for a curve of genus ≥ 2, we expect that the leaves of the corresponding foliated laminated space should have exponential growth; think about the Poincaré half plane endowed with the standard metric.

4. The case of a general smooth projective variety Y over $\mathbb{F}_q$

4.1. Deninger's construction and partial results. We recall briefly how Deninger has associated to a smooth projective absolutely irreducible variety Y over $\mathbb{F}_q$ an infinite dimensional cohomology satisfying Poincaré Duality and allowing to recover the functional equation of the associated zeta function. We follow [**6**, Sections 2 and 3].

Set $\Theta=z\frac{d}{dz}$. By D.E.R.S. we denote the category of left $\mathbb{C}[z,z^{-1}][\Theta]$-modules D regular singular at $0,\infty$ which are free of finite rank over $\mathbb{C}[z,z^{-1}]$. There is an equivalence $\mathbb{H}$ between D.E.R.S. and the category of finite dimensional complex representations of $\mathbb{Z}$ given as follows. Let $e:\mathbb{C}\to\mathbb{C}^*$, $e(\tau)=\exp(2\pi i\tau)$ be the universal covering of $\mathbb{C}^*$. We shall view the composition of

$$\mathbb{C}[z,z^{-1}]\subset\mathcal{O}(\mathbb{C}^*)\text{ and }e^*:\mathcal{O}(\mathbb{C}^*)\hookrightarrow\mathcal{O}(\mathbb{C})$$

as an inclusion of $\mathbb{C}$-algebras. We endow $\mathcal{O}(\mathbb{C})$ with a structure of left $\mathbb{C}[z,z^{-1}][\Theta]$-module by letting Θ act on $\mathcal{O}(\mathbb{C})$ by the derivation $\frac{1}{2\pi i}\frac{d}{d\tau}$. The group $\mathbb{Z}$ acts $\mathbb{C}[z,z^{-1}][\Theta]$-linearly on $\mathcal{O}(\mathbb{C})$ by translations $(\nu^*\phi)(\tau)=\phi(\tau+\nu)$ for $\nu\in\mathbb{Z}$. Then we have

$$\mathbb{H}(D)=(D\otimes_{\mathbb{C}[z,z^{-1}]}\mathcal{O}(\mathbb{C}))^{\Theta'=0},$$

this is the kernel of $\Theta' = \Theta \otimes \mathrm{Id} + \mathrm{Id} \otimes \Theta$ on $D \otimes_{\mathbb{C}[z,z^{-1}]} \mathcal{O}(\mathbb{C})$ with the induced $\mathbb{Z}$-operation. We shall denote by F for the (inverse of the monodromy) automorphism of $\mathbb{H}(D)$ given by the action of $-1 \in \mathbb{Z}$.

PROPOSITION 2. *(see [**6**, Cor. 2.8]) Fix real numbers $\delta > 0$ and $q > 1$. Let D be an object in D.E.R.S. Then for any $s \in \mathbb{C}$ one has*

$$\det{}_\infty \left(\delta \left(s\mathrm{Id} \pm \frac{2\pi i}{\log q} \Theta \right) \mid D \right) = \det\left(\mathrm{Id} - q^{-s} F^{\mp 1} | \mathbb{H}(D)\right).$$

A standard decomposition (see Section 6) $D = D^+ \oplus D^-$ of D with respect to the operator $-\frac{2\pi i}{\log q}\Theta$ is also standard for any operator $\delta(s\mathrm{Id} - \frac{2\pi i}{\log q}\Theta)$ with $s \in \mathbb{C}$ and we have

$$\det{}_\infty \left(-\delta \left(s\mathrm{Id} - \frac{2\pi i}{\log q} \Theta \right) |D \right) = \epsilon(s) \det{}_\infty \left(\delta \left(s\mathrm{Id} - \frac{2\pi i}{\log q} \Theta \right) \mid D \right)$$

where

$$\epsilon(s) = \exp i\pi \left(\dim_\infty \left(\delta \left(s\mathrm{Id} - \frac{2\pi i}{\log q} \Theta \right) \mid D^+ \right) - \dim_\infty \left(\delta \left(s\mathrm{Id} - \frac{2\pi i}{\log q} \Theta \right) \mid D^- \right) \right)$$
$$= (-q^s)^{\dim \mathbb{H}(D)} \det\left(F | \mathbb{H}(D)\right)^{-1}.$$

Now we describe a particular quasi-inverse functor $\mathbb{D}$ to the equivalence of categories $\mathbb{H}$. A typical element of the group algebra $\mathbb{C}[\mathbb{C}]$ of $\mathbb{C}$ is written in the form $\sum r_\alpha e^\alpha$ with $\alpha, r_\alpha \in \mathbb{C}$ and a symbol e^α obeying the rule $e^{\alpha+\alpha'} = e^\alpha e^{\alpha'}$. We view $\mathbb{C}[\mathbb{C}]$ as a subalgebra of $\mathcal{O}(\mathbb{C})$ by mapping e^α to the function $\tau \to \exp(\alpha\tau)$. Then $\mathbb{C}[\mathbb{C}]$ inherits both a $\mathbb{Z}$ and $\mathbb{C}[z, z^{-1}][\Theta]$-action from $\mathcal{O}(\mathbb{C})$ as follows:

$$\forall \nu \in \mathbb{Z},\ \nu^*(\sum r_\alpha e^\alpha) = \sum r_\alpha \exp(\alpha\nu) e^\alpha;$$

$$\mathbb{C}[z, z^{-1}] = (\mathbb{C}[\mathbb{C}])^{\mathbb{Z}},\ \Theta_{\mathrm{alg}}(\sum r_\alpha e^\alpha) = \sum (2\pi i)^{-1} \alpha r_\alpha e^\alpha.$$

For a finite dimensional complex representation H of $\mathbb{Z}$ let F be the automorphism corresponding to the action -1. Decompose F into its semisimple and unipotent parts $F = F_s F_u$. Let H_s be the representation of $\mathbb{Z}$ on H where -1 acts by F_s. We set

$$D(\mathbb{H}) = (H_s \otimes_\mathbb{C} \mathbb{C}[\mathbb{C}])^{\mathbb{Z}}.$$

This is naturally a $\mathbb{C}[z, z^{-1}]$-module and it becomes a $\mathbb{C}[z, z^{-1}][\Theta]$-module by letting Θ act by

$$\frac{1}{2\pi i} \log F_u \otimes \mathrm{Id}_{\mathbb{C}[\mathbb{C}]} + \mathrm{Id}_{H_s} \otimes \Theta_{\mathrm{alg}}.$$

In this Section *we consider and fix* a d-dimensional absolutely irreducible smooth projective variety Y over $\mathbb{F}_q$ and a prime number l which is prime to q. Recall (see [**19**, p. 292]) that the geometric Frobenius Fr and $F = \mathrm{Fr}_q^* = (\mathrm{Id} \times ()^q)^{*-1}$ induce the same action on the l-adic cohomology groups $H^\bullet(\overline{Y}, \mathbb{Q}_l)$. We fix an embedding $\mathcal{I} : \mathbb{Q}_l \mapsto \mathbb{C}$ and we set

$$H^\bullet(Y \,|\, \mathbb{C}[z, z^{-1}]) = \mathbb{D}(H^\bullet(\overline{Y}, \mathbb{Q}_l) \otimes_{\mathbb{Q}_l, \mathcal{I}} \mathbb{C}),\ \overline{Y} = Y \times_{\mathbb{F}_q} \overline{\mathbb{F}_q}$$

where the l-adic cohomology is equipped with the action of $F = \mathrm{Fr}_q^* = (\mathrm{Id} \times ()^q)^{*-1}$. Since $\mathbb{D}$ is an exact tensor functor, the theory $H^\bullet(- \mid \mathbb{C}[z, z^{-1}])$ inherits all the properties of a cohomology theory. Let for any $j \in \mathbb{N}$

$$H^j(Y \,|\, \mathbb{C}[z, z^{-1}]) = H^j(Y \mid \mathbb{C}[z, z^{-1}])^+ \oplus H^j(Y \,|\, \mathbb{C}[z, z^{-1}])^-$$

be a standard decomposition of $H^j(Y \mid \mathbb{C}[z,z^{-1}])$ with respect to the operator $-\frac{2\pi i}{\log q}\Theta$. Proposition 2 implies:

PROPOSITION 3. *For any $j \in \mathbb{N}$ and $s \in \mathbb{C}$, we have*

$$\mathcal{I}\left(\det_{\mathbb{Q}_l}(\mathrm{Id} - q^{-s}\mathrm{Fr}_q^* \mid H^j(\overline{Y}, \mathbb{Q}_l))\right) = \det_\infty\left(s\mathrm{Id} - \frac{2\pi i}{\log q}\Theta \mid H^j(Y|\mathbb{C}[z,z^{-1}])\right)$$

and

$$\det_\infty\left(-s\mathrm{Id} + \frac{2\pi i}{\log q}\Theta \mid H^j(Y|\mathbb{C}[z,z^{-1}])\right) = \epsilon_j(s)\det_\infty\left(s\mathrm{Id} - \frac{2\pi i}{\log q}\Theta \mid H^j(Y|\mathbb{C}[z,z^{-1}])\right)$$

where $\epsilon_j(s)$ is equal to

$$\exp i\pi\left(\dim_\infty\left(s\mathrm{Id} - \frac{2\pi i}{\log q}\Theta \,|\, H^j(Y|\mathbb{C}[z,z^{-1}])^+\right) - \dim_\infty\left(s\mathrm{Id} - \frac{2\pi i}{\log q}\Theta \,|\, H^j(Y|\mathbb{C}[z,z^{-1}])^-\right)\right)$$

and also equal to

$$(-q^s)^{b_j}\mathcal{I}\left(\det_{\mathbb{Q}_l}(\mathrm{Fr}_q^* \,|\, H^j(\overline{Y}, \mathbb{Q}_l))\right)^{-1}, \text{ where } b_j = \dim_{\mathbb{Q}_l} H^j(\overline{Y}, \mathbb{Q}_l).$$

Now, recall that Poincaré duality gives a perfect pairing of objects:

$$\mathrm{Tr}_{\mathbb{C}[z,z^{-1}]} : H^j(Y \mid \mathbb{C}[z,z^{-1}]) \times H^{2d-j}(Y \,|\, \mathbb{C}[z,z^{-1}]) \to H^{2d}(Y \mid \mathbb{C}[z,z^{-1}]) \simeq \mathbb{C}[z,z^{-1}]$$

where $d = \dim Y$, such that for any real t

$$\mathrm{Tr}_{\mathbb{C}[z,z^{-1}]}(e^{\frac{2\pi i t}{\log q}\Theta}(\omega_1) \cup e^{\frac{2\pi i t}{\log q}\Theta}(\omega_2)) = e^{d\,t}\mathrm{Tr}_{\mathbb{C}[z,z^{-1}]}(\omega_1 \cup \omega_2).$$

Taking the derivative with respect to t at $t = 0$, one sees that $\frac{2\pi i}{\log q}\Theta$ acting on $H^j(Y \mid \mathbb{C}[z,z^{-1}])$ has the same eigenvalues with the same (algebraic) multiplicities as $d\mathrm{Id} - \frac{2\pi i}{\log q}\Theta$ acting on $H^{2d-j}(Y \mid \mathbb{C}[z,z^{-1}])$. Hence one obtains

$$\begin{aligned}\det_\infty\left(s\mathrm{Id} - \frac{2\pi i}{\log q}\Theta \mid H^j(Y \mid \mathbb{C}[z,z^{-1}])\right) &= \epsilon_j^{-1}(s)\det_\infty\left(-s\mathrm{Id} + \frac{2\pi i}{\log q}\Theta \mid H^j(Y \mid \mathbb{C}[z,z^{-1}])\right)\\ &= \epsilon_j^{-1}(s)\det_\infty\left((d-s)\mathrm{Id} - \frac{2\pi i}{\log q}\Theta \,|\, H^{2d-j}(Y \,|\, \mathbb{C}[z,z^{-1}])\right).\end{aligned}$$

This implies the functional equation for the zeta function $\zeta_Y(s)$ of Y in the form

$$\zeta_Y(s) = e^{i\pi(\chi^+(s) - \chi^-(s))}\zeta_Y(d-s)$$

where

$$\chi^\pm(s) = \sum_{j=0}^{2d}(-1)^j \dim_\infty\left(s\mathrm{Id} - \frac{2\pi i}{\log q}\Theta \mid H^j(Y \,|\, \mathbb{C}[z,z^{-1}])^\pm\right).$$

Note that by Deligne's Theorem the eigenvalues of $\frac{2\pi i}{\log q}\Theta$ acting on $H^j(Y \,|\, \mathbb{C}[z,z^{-1}])$ have real part $= \frac{j}{2}$. In the next subsection we shall in fact refer to $\mathrm{Tr} = L \circ \mathrm{Tr}_{\mathbb{C}[z,z^{-1}]}$

where $L : \mathbb{C}[z, z^{-1}] \to \mathbb{C}$ is the zero-coefficent linear form given by $L(\sum_{\nu\in\mathbb{Z}} a_\nu z^\nu) = a_0$.

4.2. A conjectural Riemannian laminated foliated space $(S_Y, \mathcal{F}, g, \phi^t)$. Motivated by the results of Section 2, we introduce the following well-known definition:

DEFINITION 2. A compact complex d-dimensional laminated space $\mathcal{L}$ is a compact topological space satisfying the following property. There exists a totally disconnected topological space Ω such that $\mathcal{L}$ admits a finite open cover $\cup_{i=1}^N A_i$ by local charts $f_i : A_i \to U_i \times W_i$ where U_i (resp. W_i) is an open subset of $\mathbb{C}^d$ (resp. Ω) such that each transition map $f_i \circ f_j^{-1}$ is of the form

$$f_j(A_i \cap A_j) \to U_i \times W_i \tag{4.1}$$

$$(z, \theta) \to (H(z, \theta), G(\theta)) \tag{4.2}$$

where $z \to H(z, \theta)$ is holomorphic and the maps $(z, \theta) \to \partial_z^\alpha H(z, \theta)$ $(\alpha \in \mathbb{N}^d)$, $\theta \to G(\theta)$ are continuous. Then the path connected components $\mathcal{C}$ of $\mathcal{L}$ carry a structure of d-dimensional complex manifold and constitute the leaves of a foliation.

The space $\frac{\mathbb{C}\times V_\xi \Gamma}{V}$ of Section 2.2 is an example of such a compact 1-dimensional complex laminated space. Recall that Y denotes a smooth projective absolutely irreducible variety over $\mathbb{F}_q$ admitting a rational point over $\mathbb{F}_q$. The results of Section 2 and of this Section lead us to state the following hypothesis. Of course, when the absolutely irreducible smooth projective variety Y is an elliptic curve over $\mathbb{F}_q$, the results of Section 2 show that this hypothesis is fully satisfied.

HYPOTHESIS 1. *There exists a compact complex d-dimensional laminated space $\mathcal{L}$ endowed with a leafwise holomorphic action of $q^{\mathbb{Z}}$ satisfying the following properties. Denote by Π the projection map $\Pi : \mathcal{L} \times \mathbb{R}^{+*} \to \frac{\mathcal{L}\times\mathbb{R}^{+*}}{q^{\mathbb{Z}}}$ where $q^{\mathbb{Z}}$ acts diagonally. Consider the foliated laminated space $(S_Y = \frac{\mathcal{L}\times\mathbb{R}^{+*}}{q^{\mathbb{Z}}}, \mathcal{F})$ whose (complex d-dimensional) leaves are the sets $\Pi(\mathcal{C} \times \{x\})$ where $\mathcal{C}$ is any leaf of $\mathcal{L}$. Then one has:*

(0) *There is a flow ϕ^t acting on $(S_Y, \mathcal{F})$, permuting (holomorphically) the leaves, whose trajectories are transverse to the leaves and compact orbits are non degenerate. Moreover, there is a natural bijection between the set of closed points w of Y and the set of compact primitive $\mathbb{R}$-orbits on S_Y. It has the following property. If w corresponds to $\gamma = \gamma_w$, then*

$$l(\gamma_w) = \log Nw$$

where Nw denotes the number of elements of the residue field of w.

(1) *There exists a Riemannian metric g along the (oriented) leaves of $(S_Y, \mathcal{F}, \phi^t)$ such that*

$$(\phi^t)^*(g) = e^t g. \tag{4.3}$$

We denote by $*$ the associated leafwise hodge star operator and by λ the associated leafwise Riemannian volume form along the leaves.

(2) *The foliated space $(S_Y, \mathcal{F})$ is endowed with a transverse positive measure Λ such that one defines a ϕ^t-invariant measure μ on S_Y by the formula $\mu(f) = (f\,\lambda; C(\Lambda))$ where $C(\Lambda)$ denotes the Ruelle-Sullivan current associated to Λ. For each $0 \le j \le 2d$, one defines a scalar product on the vector*

space of continuous sections of the bundle $\wedge^j T^*\mathcal{F} \to S_Y$ *by the formula* $< \omega; \omega' >= (\omega \cup \star\omega'; C(\Lambda))$*; we consider the associated* L^2*-completion. Define* H^j *to be the* L^2*-leafwise cohomology group*

$$H^j = \frac{\ker d_{\mathcal{F}}}{\operatorname{Im} d_{\mathcal{F}}}$$

where $d_{\mathcal{F}}$ *denotes the leafwise exterior derivative. Since* μ *is* ϕ^t*-invariant,* $(\phi^t)^*$ *acts on* H^j*: write* $e^{t\Theta_j} = (\phi^t)^*$*. Then* H^j *appear as a Hilbert completion of* $H^j(Y \mid \mathbb{C}[z, z^{-1}])$*; moreover,* $\frac{2\pi i}{\log q}\Theta$ *acting on* $H^j(Y \mid \mathbb{C}[z, z^{-1}])$ *and* Θ_j *have the same eigenvalues with the same (algebraic) multiplicities.*

(3) *The map*

$$\operatorname{Tr} : H^j \times \times H^{2d-j} \to \mathbb{C} \tag{4.4}$$

$$(\omega_1 \cup \omega_2) \to (\omega_1 \cup \omega_2; C(\Lambda)) \tag{4.5}$$

defines a non degenerate Poincaré duality pairing. When restricted from H^j *to* $H^j(Y \mid \mathbb{C}[z, z^{-1}])$*, this Poincaré pairing should induce the one above for* $H^j(Y \mid \mathbb{C}[z, z^{-1}])$*.*

REMARKS.

(0) The reason why (see [**7**, Section 5] and [**8**, Theorem 3.12]) one is looking for a space of the form $\frac{\mathcal{L}\times\mathbb{R}^{+*}}{q^{\mathbb{Z}}}$ is that the expected set of periods of ϕ^t is $\log q\,\mathbb{Z}$ and that a space of this form works for an elliptic curve E_0. But, in general, we expect that the flow will be more complicated than the one induced by $(l, x) \to (l, xe^t)$ because one cannot in general lift the Frobenius morphism of Y in characteristic zero and we expect that a path connected component of $\mathcal{L}$ should have much more holomorphic automorphisms than $\mathbb{C}$ (in the case of E_0).

(1) We recall part 2) of the Remarks following Lemma 2. Deninger has explained that the correspondence $Y \to (S_Y, \mathcal{F}, g, \phi^t)$, if it exists in such a generality, cannot be functorial because otherwise one should get a fiber functor to the category of finite dimensional $\mathbb{R}$-dimensional vectors spaces. Notice (see [**8**, Section 4.7]) that such a functor exists on the category of varieties Y whose associated motives are ordinary, which in some sense is the generic case. Recall that a simple motive M (over $\mathbb{F}_q$) is called ordinary if $\operatorname{End} M = \mathbb{Q}(\pi_M)$ where π_M is the Frobenius endomorphism of M. An arbitrary motive is called ordinary if it is the sum of simple ordinary motives. It is reasonable to expect that to each ordinary variety Y one could associate in a functorial way a canonical $(S_Y, \mathcal{F}, g, \phi^t)$.

(2) The results in Section 2 for $(S(E_0), \mathcal{F})$ lead to suggest that the transverse measure Λ of Hypothesis 1 might be constructed from a transverse measure on $\mathcal{L}$ and the Haar measure of $(\mathbb{R}^{+*}, \times)$.

(3) If, as we expect, the center of the von Neumann algebra of the foliated laminated space $\mathcal{L}$ is trivial, then Theorem 4 would extend immediately to $(S_Y, \mathcal{F})$.

When Y is a curve, we can content ourselves with the following *simplified* version of Hypothesis 1. Its main point is that the explicit formula for $\zeta_Y(s)$ should be read as a Lefschetz trace formula for ϕ^t.

HYPOTHESIS 2. *We assume that Y is a curve ($d = 1$).*

(A) *We keep the notations and assumptions 0), 1) and 2) of Hypothesis 1 but, we suppress assumption 3) and any reference to $H^j(Y \mid \mathbb{C}[z, z^{-1}])$ and to $\frac{2\pi i}{\log q}\Theta$.*

(B) *For any $\alpha \in C_c^\infty(\mathbb{R}, \mathbb{C})$ and $j \in \{0, 1, 2\}$, $\int_{\mathbb{R}} \alpha(t) e^{t\,\Theta_j} dt$ is a trace class operator acting on H^j and one has*

$$\sum_{j=0}^{2} (-1)^j \mathrm{TR} \int_{\mathbb{R}} \alpha(t) e^{t\,\Theta_j} dt = (2 - 2g) \log q\, \alpha(0) + \sum_{w} \log Nw \sum_{k \geq 1} \alpha(k \log Nw)$$
$$+ \sum_{w} \log Nw \sum_{k \leq -1} (Nw)^k \alpha(k \log Nw)$$

where w runs over the closed points of Y and Nw is the number of elements of the finite residue field of w.

(C) *One has*

$$\mathrm{TR} \int_{\mathbb{R}} \alpha(t) e^{t\,\Theta_0} dt = \sum_{\nu \in \mathbb{Z}} \Phi\left(\frac{2i\pi\nu}{\log q}\right)$$

where $\Phi(s) = \int_{\mathbb{R}} \alpha(t) e^{t\,s} dt$.

4.3. Consequences of Hypothesis 1 and 2.

In this subsection we assume Hypothesis 1 and 2.

LEMMA 7. *With the notations of Conjecture 1, one has*

$$\forall (t, \omega_1, \omega_2) \in \mathbb{R} \times H^j \times H^{2d-j},\ \mathrm{Tr}((\phi^t)^*(\omega_1) \cup (\phi^t)^*(\omega_2)) = e^{d\,t} \mathrm{Tr}(\omega_1 \cup \omega_2).$$

PROOF. Property (4.3) implies that $(\phi^t)^*(\lambda) = e^{d\,t}\lambda$. There exists a function f on S_Y such that $\omega_1 \cup \omega_2 = f\lambda$. Next we write

$$(\phi^t)^*(\omega_1) \cup (\phi^t)^*(\omega_2) = (\phi^t)^*(\omega_1 \cup \omega_2) = (\phi^t)^*(f\lambda).$$

Then the result follows from the ϕ^t-invariance of the measure μ introduced in Hypothesis 1.(2). □

Then Hypothesis 1 should provide a new proof of the Riemann Hypothesis for $\zeta_Y(s)$.

THEOREM 6. *(Deligne's Theorem). For any $j \in \{0, \ldots, 2d\}$, the zeroes of*

$$\det\big(\mathrm{Id} - q^{-s}\mathrm{Fr} : H^j(\overline{Y}; \mathbb{Q}_l)\big)$$

have real part equal to $\frac{j}{2}$.

PROOF. We follow [**7**, Section 3] and [**10**, Section 2]. Recall that $*$ denotes the leafwise Hodge star operator associated to the metric g. Property (4.3) shows that on the space of leafwise differential forms of degree j of $(S_Y, \mathcal{F})$, one has $(\phi^t)^* \circ * = e^{t(d-j)} * \circ (\phi^t)^*$ and hence $\Theta_{2d-j} \circ * = * \circ (d - j + \Theta_j)$. Then, using Lemma 7 and the fact that the Hilbert scalar product of H^j is given by

$$\forall \omega, \omega' \in H^j,\ < \omega, \omega' >= \mathrm{Tr}(\omega \cup *\omega'),$$

one checks easily that $\Theta_j - \frac{j\,\mathrm{Id}}{2}$ is skew symmetric. Thus, the eigenvalues of $\Theta_j - \frac{j\,\mathrm{Id}}{2}$ belong to $i\,\mathbb{R}$. Then the Theorem follows from Proposition 3 and Hypothesis 1.(2). □

When Y is a curve, Hypothesis 2 *alone* should provide also a new proof of the Riemann Hypothesis for $\zeta_Y(s)$.

THEOREM 7 (Weil's Theorem). *Assume that Y is a curve. Then the zeroes of $\zeta_Y(s)$ have real part $= \frac{1}{2}$.*

PROOF. Property (4.3) implies that $e^{t\,\Theta_2} \circ * = e^t * \circ\, e^{t\,\Theta_0}$ on H^0. Then using the isomorphism $* : H^0 \to H^2$ and Hypothesis 2.C) one gets

$$\mathrm{TR} \int_{\mathbb{R}} \alpha(t) e^{t\,\Theta_2} dt = \sum_{\nu \in \mathbb{Z}} \Phi\left(1 + \frac{2i\pi\nu}{\log q}\right).$$

Then using parts B) and C) of Hypothesis 2 and the standard explicit formula for $\zeta_Y(s)$, one checks that the spectrum of Θ_1 acting on H^1 coincides with $\zeta_Y^{-1}(0)$. But, proceeding as in the proof of the previous Theorem, one checks easily that the eigenvalues of Θ_1 have real part $= \frac{1}{2}$. This proves the result. □

4.4. Open questions and remarks. It could be possible (and interesting) to obtain results on zeta functions for examples of Riemannian laminated foliated spaces of the form

$$\left(S = \frac{\mathcal{L} \times \mathbb{R}^{*+}}{q^{\mathbb{Z}}}, \mathcal{F}, g, \phi^t\right)$$

where $\mathcal{L}$ is a 1-dimensional complex compact laminated space, g is a leafwise metric satisfying (4.3), the trajectories of ϕ^t are transverse to the leaves and its compact orbits are non degenerate. With this respect, we recall part 1) of the Remarks following Corollary 1 to the reader. In [**10**, Theorem 5.9], Deninger proved such a Lefschetz trace formula for an interesting specific example. For more general examples, one should use the Riemannian foliation techniques of [**2**] and take into account Deninger's remarks in [**10**, Section 5]. More precisely, Deninger suggests there that the coefficient of $\alpha(k \log Nw)$, $k \in \mathbb{Z}^*$, for the Lefschetz trace formula (such as the one expected in Hypothesis 2.B)) should be given by

$$l(\gamma_w) \frac{\det(\mathrm{Id} - T_y \phi^{kl(\gamma_w)} \mid T_y\mathcal{F})}{|\det(\mathrm{Id} - T_y \phi^{|k|\, l(\gamma_w)} \mid T_y\mathcal{F})|} \tag{4.6}$$

where y is any point of the closed orbit γ_w of ϕ^t; the fact that γ_w is non degenerate means simply that the denominator of (4.6) does not vanish. Property (4.3) implies that $T_y \phi^{kl(\gamma_w)} = e^{\frac{k}{2} l(\gamma_w)} O_k$ where O_k is an orthogonal endomorphism of the euclidean plane $T_y\mathcal{F}$. Deninger has observed that for $k = 1$, the coefficient (4.6) is equal to $l(\gamma_w)$ if and only if O_1 is a rotation. If O_1 is a rotation, then the reader can check that for each $k \geq 1$ the coefficient (4.6) is equal to $l(\gamma_w)$ whereas for each $k \leq -1$, it is equal to $l(\gamma_w) e^{kl(\gamma_w)}$ as expected.

5. About the Riemann zeta function $\widehat{\zeta}$

5.1. The explicit formula for $\widehat{\zeta}$. The Riemann zeta function is given by

$$\widehat{\zeta}(s) = 2^{-1/2} \pi^{-s/2} \Gamma(s/2) \prod_{p \in \mathcal{P}} \frac{1}{1 - p^{-s}}$$

where $\mathcal{P} = \{2, 3, 5, \ldots\}$ denotes the usual set of prime numbers. The following well-known explicit formulas express a connection between $\mathcal{P} \cup \{\infty\}$ and the zeroes of $\widehat{\zeta}$. Let $\alpha \in C_c^\infty(\mathbb{R},\mathbb{R})$ and for real s, set $\Phi(s) = \int_{\mathbb{R}} e^{st}\alpha(t)dt$. Then one has the following formula

$$\begin{aligned}(5.1)\quad &\Phi(0) - \sum_{\rho\in\widehat{\zeta}^{-1}\{0\}} \Phi(\rho) + \Phi(1) \\ &= \sum_{p\in\mathcal{P}} \log p \left(\sum_{k\geq 1}\alpha(k\log p) + \sum_{k\leq -1} p^k\alpha(k\log p)\right) + W_\infty(\alpha),\end{aligned}$$

where

$$W_\infty(\alpha) = \alpha(0)\log\pi + \int_0^{+\infty}\left(\frac{\alpha(t)+e^{-t}\alpha(-t)}{1-e^{-2t}} - \alpha(0)\frac{e^{-2t}}{t}\right)dt.$$

The left hand side $\Phi(0) - \sum_{\rho\in\widehat{\zeta}^{-1}\{0\}}\Phi(\rho) + \Phi(1)$ is reminiscent of a Lefschetz trace formula of the form

$$\mathrm{TR}\int_{\mathbb{R}}\alpha(t)e^{t\Theta_0}\,dt - \mathrm{TR}\int_{\mathbb{R}}\alpha(t)e^{t\Theta_1}\,dt + \mathrm{TR}\int_{\mathbb{R}}\alpha(t)e^{t\Theta_2}\,dt,$$

where $\Theta_0 = 0$ acts on $H^0 = \mathbb{R}$, $\Theta_2 = \mathrm{Id}$ acts on $H^2 = \mathbb{R}$ and Θ_1 acts on an infinite dimensional real vector space H^1. The idea of the proof of (5.1) is the following. Apply the residue theorem to

$$\left(\int_0^{+\infty}\sqrt{t}\,\alpha(\log t)\,t^s\frac{dt}{t}\right)\frac{\widehat{\zeta}'}{\widehat{\zeta}}(s)$$

on the interior of the rectangle of $\mathbb{C}$ defined by the four points

$$1+\epsilon+iT,\ -\epsilon+iT,\ -\epsilon-iT,\ 1+\epsilon-iT.$$

Then use the functional equation $\widehat{\zeta}(s) = \widehat{\zeta}(1-s)$ and the formula

$$\frac{\Gamma'}{\Gamma}\left(\frac{s}{2}+\frac{1}{4}\right) = \int_0^{+\infty}\left(\frac{e^{-u}}{u} - \frac{e^{-u(\frac{s}{2}+\frac{1}{4})}}{1-e^{-u}}\right)du;$$

finally, let T go to $+\infty$. In a series of interesting papers [**7**, Section 4], [**10**], [**9**, Section 5], [**11**, Section 3]), Deninger has compared the formula (5.1) with known (Lefschetz) trace formulas for a flow acting on a manifold; see also [**14**] for some basic material. His results and the proved results of Section 2 lead us to state the following geometric conjecture.

5.2. A conjectural Riemannian laminated foliated space $(S_{\mathbb{Q}}, \mathcal{F}, g, \phi^t)$.

CONJECTURE 1. *There exists a σ-compact (metric) complex 1-dimensional laminated space $\mathcal{L}$ endowed with a leafwise holomorphic action of $\mathbb{Q}^{+*}$ satisfying the following properties. Let Π denote the projection $\Pi : \mathcal{L}\times\mathbb{R}^{+*} \to \frac{\mathcal{L}\times\mathbb{R}^{+*}}{\mathbb{Q}^{+*}}$ where $\mathbb{Q}^{+*}$ acts diagonally. Consider the foliated laminated space $(S_{\mathbb{Q}} = \frac{\mathcal{L}\times\mathbb{R}^{+*}}{\mathbb{Q}^{+*}}, \mathcal{F})$ whose (complex 1-dimensional) leaves are the sets $\Pi(C\times\{x\})$ where C is any leaf of $\mathcal{L}$.*

Then one has:

(0) *There is a flow ϕ^t acting on $(S_{\mathbb{Q}}, \mathcal{F})$, permuting holomorphically the leaves, whose trajectories are transverse to the leaves and whose compact orbits are non degenerate. Its action on the "$\mathcal{L}$-leaf space" $\frac{\mathbb{R}^{+*}}{\mathbb{Q}^{+*}}$ is given by $\phi^t([x]) = [e^{-t}x]$. Moreover, there is a natural bijection between the set $\mathcal{P}$*

of prime numbers of $\mathbb{N}$ *and the set of primitive compact* $\mathbb{R}$*-orbits on* $S_{\mathbb{Q}}$*. It has the following property. If* $p \in \mathcal{P}$ *corresponds to* $\gamma = \gamma_p$*, then*

$$l(\gamma_p) = \log p.$$

(1) *There exists a compact subset* K *of* $\mathcal{L}$ *such that* $\mathbb{Q}^{+*} \cdot K = \mathcal{L}$*. The space* $\overline{S_{\mathbb{Q}}} = S_{\mathbb{Q}} \cup \mathcal{L}/\mathbb{Q}^{+*}$ *defines a compactification of* $S_{\mathbb{Q}}$ *to which the action of* ϕ^t *extends, and for each real* t ϕ^t *sends* $\mathcal{L}/\mathbb{Q}^{+*}$ *into itself. Moreover, there exists a point* $l_\infty \in \mathcal{L}/\mathbb{Q}^{+*}$*, corresponding to the archimedean absolute value of* $\mathbb{Q}$*, such that* $\forall t \in \mathbb{R}$*,* $\phi^t(l_\infty) = l_\infty$*.*

(2) *There exists a Riemannian metric* g *along the (oriented) leaves of* $(S_{\mathbb{Q}}, \mathcal{F}, \phi^t)$ *such that*

$$(\phi^t)^*(g) = e^t g. \tag{5.2}$$

Denote by $*$ *the associated leafwise Hodge star operator and by* λ *the associated leafwise Riemannian volume form along the leaves.*

(3) *The quotient space* $\mathcal{L}/\mathbb{Q}^{+*}$ *is foliated by its path connected components. The foliated space* $(\overline{S_{\mathbb{Q}}}, \mathcal{F})$ *is endowed with a holonomy invariant transverse positive measure* Λ *such that one defines a* ϕ^t*-invariant measure* μ *on* $S_{\mathbb{Q}}$ *by the formula* $\mu(f) = (f\lambda; C(\Lambda))$ *where* $C(\Lambda)$ *denotes the Ruelle-Sullivan current associated to* Λ*. For each* $0 \leq j \leq 2$*, one defines a scalar product on the vector space of continuous sections of the bundle* $\wedge^j T^*\mathcal{F} \to S_{\mathbb{Q}}$ *by the formula* $< \omega; \omega' >= (\omega \cup *\omega'; C(\Lambda))$*; we consider the associated* L^2*-completion. Define* $H^j(\overline{\operatorname{Spec}\mathbb{Z}}; \mathbb{R})$ *to be the* L^2*-leafwise cohomology group*

$$H^j(\overline{\operatorname{Spec}\mathbb{Z}}; \mathbb{R}) = \frac{\ker d_{\mathcal{F}}}{\overline{\operatorname{Im} d_{\mathcal{F}}}}$$

where $d_{\mathcal{F}}$ *denotes the leafwise exterior derivative. Since* μ *is* ϕ^t*-invariant,* $(\phi^t)^*$ *acts on* $H^j(\overline{\operatorname{Spec}\mathbb{Z}}; \mathbb{R})$*: write* $e^{t\Theta_j} = (\phi^t)^*$*. Then,* $e^{t\Theta_0} = \mathrm{Id}$ *on* $H^0(\overline{\operatorname{Spec}\mathbb{Z}}; \mathbb{R}) \simeq \mathbb{R}$ *and* $e^{t\Theta_2} = e^t\mathrm{Id}$ *on* $H^2(\overline{\operatorname{Spec}\mathbb{Z}}; \mathbb{R}) \simeq \mathbb{R}$*.*

(4) *One has the following Lefschetz trace formula: for any* $\alpha \in C^\infty(\mathbb{R}, \mathbb{R})$*,* $\sum_{j=0}^2 (-1)^j \mathrm{TR} \int_{\mathbb{R}} \alpha(t) e^{t\Theta_j}\, dt$ *is well-defined and is equal to the right-hand side of the explicit formula (5.1).*

REMARKS.

(0) Let $\mathbb{R}^{+*} \cup \{\iota\}$ be the standard compactification of $\mathbb{R}^{+*}$. Let $l \in \mathcal{L}$, $q \in \mathbb{Q}^{+*}$ and, $(x_n)_{n \in \mathbb{N}}$ be a sequence of positive reals such $\lim_{n \to +\infty} x_n = 0$. Then for each $n \in \mathbb{N}$ one has $(l, x_n) \simeq (q \cdot l, q x_n)$ and $(l, \frac{1}{x_n}) \simeq (q \cdot l, q \frac{1}{x_n})$ in $S_{\mathbb{Q}}$. Then one has

$$\lim_{n \to +\infty} (l, x_n) = \lim_{n \to +\infty} (l, \frac{1}{x_n}) = (l, \iota) \simeq (q \cdot l, \iota) \in \mathcal{L}/\mathbb{Q}^{+*}.$$

(1) Presumably, the transverse measure Λ might be constructed from the Haar measure of $(\mathbb{R}^{+*}, \times)$ and a transverse measure on the laminated space $\mathcal{L}$.

(2) The reason why Deninger is looking for a space of the form $\frac{\mathcal{L} \times \mathbb{R}^{+*}}{\mathbb{Q}^{+*}}$ is that the expected set of periods of ϕ^t is $\log \mathbb{Q}^{+*}$ (see [**8**, Theorem 3.12 and end of Section 4.2] and[**11**, Section 3]). Recall that in Section 2.2 the flow was defined by $([z, \widehat{v}], x) \to ([z, \widehat{v}], x e^t)$. Here, the requirement that ϕ^t should act on the "$\mathcal{L}$-leaf space" by $[x] \to [x e^{-t}]$ is simply a convention (see Section 5.5 below).

(3) Part of the assumptions of Conjecture 1.3) might be redundant, let us reproduce the argument of [**10**, Proof of Theorem 2.1]. It is reasonable to expect that $1 \in H^0(\overline{\operatorname{Spec}\mathbb{Z}};\mathbb{R}) \simeq \mathbb{R}$. If this is the case, then $(\phi^t)^*$ acts automatically as Id on $H^0(\overline{\operatorname{Spec}\mathbb{Z}};\mathbb{R}) \simeq \mathbb{R}$ and, using the isomorphism

$$* : H^0(\overline{\operatorname{Spec}\mathbb{Z}};\mathbb{R}) \simeq \mathbb{R} \to H^2(\overline{\operatorname{Spec}\mathbb{Z}};\mathbb{R}) \simeq \mathbb{R}$$

and the fact that $(\phi^t)^* \circ * = e^t * \circ (\phi^t)^*$ on $H^0(\overline{\operatorname{Spec}\mathbb{Z}};\mathbb{R})$, one gets that $(\phi^t)^*$ acts as $e^t\,\mathrm{Id}$ on $H^2(\overline{\operatorname{Spec}\mathbb{Z}};\mathbb{R}) \simeq \mathbb{R}$.

5.3. Comparison with Connes' approach [4]. In Connes' approach [**4**], which is rather different from Deninger's one, the noncommutative space associated to the Riemann zeta function is the quotient $\mathbb{A}_{\mathbb{Q}}/I_1$ where $\mathbb{A}_{\mathbb{Q}}$ denotes the space of $\mathbb{Q}$-adeles, and the group I_1 of ideles of norm 1 acts by multiplication on $\mathbb{A}_{\mathbb{Q}}$. Observe that we may write

$$\mathbb{A}_{\mathbb{Q}}/I_1 \simeq \frac{\mathcal{Z} \times \mathbb{R}^{+*}}{\mathbb{Q}^{+*}} \cup (\mathcal{Z}/\mathbb{Q}^{+*})$$

where $\mathcal{Z}$ is the quotient space $\mathbb{A}_f/I_{f,1}$ of the space of finite adeles by the group of finite ideles of norm 1. Note the similarity with Conjecture 1.(1). Since the group I_1 preserves the additive Haar measure $d\mu_{\mathbb{A}_{\mathbb{Q}}}$ of $\mathbb{A}_{\mathbb{Q}}$, one defines a trace τ of $L^\infty(\mathbb{A}_{\mathbb{Q}}) \rtimes I_1$ by setting for a typical element $\int_{I_1} a_g\, d\mu_{I_1}(g)$ of $L^\infty(\mathbb{A}_{\mathbb{Q}}) \rtimes I_1$:

$$\tau\left(\int_{I_1} a_g\, d\mu_{I_1}\right) = \int_{\mathbb{A}_{\mathbb{Q}}} a_1(y)\, d\mu_{\mathbb{A}_{\mathbb{Q}}}(y).$$

Any element of $\mathbb{A}_{\mathbb{Q}}$ is of the form (y_∞, y_f) with $y_\infty \in \mathbb{R}$ and y_f a finite adele. The group $(\mathbb{R},+)$ acts on $L^\infty(\mathbb{A}_{\mathbb{Q}}) \rtimes I_1$ by the formula

$$\forall t \in \mathbb{R},\ \psi_t\left(\int_{I_1} a_g(y_\infty, y_f)\, d\mu_{I_1}(g)\right) = \int_{I_1} a_g(e^{-t}y_\infty, y_f)\, d\mu_{I_1}(g).$$

The following result is an exercise in the context of [**4**]:

THEOREM 8.

(1) *The von Neumann algebra* $L^\infty(\mathbb{A}_{\mathbb{Q}}) \rtimes I_1$ *is a type* II_∞*-factor.*
(2) *The von Neumann algebra* $(L^\infty(\mathbb{A}_{\mathbb{Q}}) \rtimes I_1) \rtimes_{\psi_t} \mathbb{R}$ *is type* III_1*-factor.*

PROOF.

(1) Since the action of I_1 on $\mathbb{A}_Q$ is ergodic with respect to $\mu_{\mathbb{A}_{\mathbb{Q}}}$, $L^\infty(\mathbb{A}_{\mathbb{Q}}) \rtimes I_1$ is a factor. Using the fact that $(\mathbb{A}_Q, \mu_{\mathbb{A}_q})$ has no atoms, one checks that $\tau(p)$ takes all values in $[0,+\infty]$ when p runs over the set of self adjoint projections of $L^\infty(\mathbb{A}_{\mathbb{Q}}) \rtimes I_1$. Then by definition $L^\infty(\mathbb{A}_{\mathbb{Q}}) \rtimes I_1$ is of type II_∞.
(2) One has $\tau \circ \psi_t = e^t \tau$. With the notations and results of [**3**, pp. 473, 495], one first gets that $(L^\infty(\mathbb{A}_{\mathbb{Q}}) \rtimes I_1) \rtimes_{\theta_t} \mathbb{R}$ is of type III. Then since for any real $\lambda > 0$ the automorphism $\theta_\lambda = \psi_{\log\lambda}$ acts trivially on the (trivial!) center of $L^\infty(\mathbb{A}_{\mathbb{Q}}) \rtimes I_1$, one gets

$$S((L^\infty(\mathbb{A}_{\mathbb{Q}}) \rtimes I_1) \rtimes_{\psi_t} \mathbb{R}) \cap \mathbb{R}^{+*} = \mathbb{R}^{+*}.$$

This implies the result.

∎

Morally, the foliated space $(\overline{S_{\mathbb{Q}}}, \mathcal{F})$ of Conjecture 1 should be viewed as a noncommutative space which plays the role of the set of geometric points of $\overline{Spec\,\mathbb{Z}}$. The flow $(\phi^t)_{t\in\mathbb{R}}$ is the continuous version of the geometric Frobenius and $\overline{S_{\mathbb{Q}}}/(\phi^t)$ is the noncommutative space of closed points. The flow ϕ^t induces an action denoted $(\phi^t)^*$ on the von Neumann algebra $W(\overline{S_{\mathbb{Q}}}, \mathcal{F})$ of the foliation of $(\overline{S_{\mathbb{Q}}}, \mathcal{F})$, and $W(\overline{S_{\mathbb{Q}}}, \mathcal{F}) \rtimes_{(\phi^t)^*} \mathbb{R}$ represents the algebra of bounded measurable functions on the noncommutative orbit space $\overline{S_{\mathbb{Q}}}/(\phi^t)$.

Presumably, one could try to use (as in Section 3) the transverse measure Λ and the assumptions of Conjecture 1 to construct a trace τ on $W(\overline{S_{\mathbb{Q}}}, \mathcal{F})$ and to proceed along the lines of the proof of Theorem 8 in order to show that $W(\overline{S_{\mathbb{Q}}}, \mathcal{F}) \rtimes_{(\phi^t)^*} \mathbb{R}$ is a type III_1- factor. There is a difficulty due to the non compactness of $S_{\mathbb{Q}}$, we hope to come back on this elsewhere.

5.4. Arithmetic consequences of Conjecture 1 for $\widehat{\zeta}(s)$.

PROPOSITION 4. *Conjecture 1 implies the (expected) cohomological formula*

$$\widehat{\zeta}(s) = \prod_{j=0}^{2} \det{}_{\infty}\Big(\frac{s\mathrm{Id} - \Theta_j}{2\pi} : H^j(\overline{\mathrm{Spec}\,\mathbb{Z}}; \mathbb{R})\Big)^{(-1)^{1+j}}.$$

PROOF. From the Lefschetz trace formula of Conjecture 1.3) and the explicit formula (5.1), one gets that

$$\mathrm{Spec}\,\Theta_1 : H^1(\overline{\mathrm{Spec}\,\mathbb{Z}}; \mathbb{R}) \;=\; \widehat{\zeta}^{-1}(\{0\}). \tag{5.3}$$

Then using the formul

$$\widehat{\zeta}(s) = \frac{\prod_\rho \frac{1}{2\pi}(s-\rho)}{\frac{s}{2\pi}\frac{s-1}{2\pi}},$$

proved by Deninger [**5**, Theorem 3.3], one gets the result. □

THEOREM 9. *Conjecture 1 implies the Riemann hypothesis.*

PROOF. Proceeding as in the proof of Lemma 7 of Section 4.3 (i.e., using formula (5.2) and the ϕ^t-invariance of μ), one checks for any real t,

$$\forall(\omega_1, \omega_2) \in H^1(\overline{\mathrm{Spec}\,\mathbb{Z}}; \mathbb{R}),\ ((\phi^t)^*(\omega_1) \cup (\phi^t)^*(\omega_2);\ C(\Lambda)) = e^t(\omega_1 \cup \omega_2; C(\Lambda)). \tag{5.4}$$

Formula (5.2) shows that $e^{t\,\Theta_1} \circ * = * \circ e^{t\,\Theta_1}$ on $H^1(\overline{\mathrm{Spec}\,\mathbb{Z}}; \mathbb{R})$. Then, using formula (5.4), one obtains that the L^2-scalar product of $H^1(\overline{\mathrm{Spec}\,\mathbb{Z}}; \mathbb{R})$ defined in Conjecture 1. 3) satisfies

$$\forall t \in \mathbb{R},\ \forall \omega, \omega' \in\ H^1(\overline{\mathrm{Spec}\,\mathbb{Z}}; \mathbb{R}),\ e^t < \omega; \omega' > = < e^{t\,\Theta_1}(\omega); e^{t\,\Theta_1}\omega' > .$$

Then by taking the t-derivative at $t = 0$ one obtains that $\Theta_1 - \frac{1}{2}\mathrm{Id}$ is skew adjoint which, given (5.3), should prove the Riemann hypothesis. □

Moreover, since $*^2 = -\mathrm{Id}$ on $H^1(\overline{\mathrm{Spec}\,\mathbb{Z}}; \mathbb{R})$, the leafwise Hodge star $*$ defines a complex structure on $H^1(\overline{\mathrm{Spec}\,\mathbb{Z}}; \mathbb{R})$ which we may view as a complex vector space and one defines a hermitian scalar product by setting:

$$< \omega; \omega' >_h = (\omega \cup *\omega'; C(\Lambda)) + \sqrt{-1}(\omega \cup \omega'; C(\Lambda)).$$

Then $* \circ (\Theta_1 - \frac{1}{2}\mathrm{Id})$ becomes a hermitian operator whose eigenvalues coincide with the imaginary parts of the zeroes of $\widehat{\zeta}$. Recall that numerical investigations on the fluctuations of the spacings between consecutive zeroes of $\widehat{\zeta}$ have shown that their statistics resembles that of the fluctuations in the spacings of consecutive eigenvalues of random *hermitian* matrices, as opposed to the different statistics for random *real* symmetric matrices. Observe moreover that $e^{\Theta_1 - 1/2}$ is a complex unitary automorphism preserving the symplectic form (already considered by Deninger for other purposes): $(\omega, \omega') \to (\omega \cup \omega'; C(\Lambda))$. Recall that Katz-Sarnak's results ([**17**]) predict that the (infinite dimensional) complex unitary symplectic group should be a symmetry group for the family of L-functions $L(\chi, s)$, χ running over the family of quadratic characters ($\chi^2 = 1$), so that in particular $e^{\Theta_1 - 1/2}$ should be a complex unitary automorphism preserving a symplectic form.

5.5. Open problems and further explanations. It is an interesting challenge to prove a Lefschetz trace formula for a Riemannian foliated laminated spaces $(S, \mathcal{F}, g, \phi^t)$ satisfying the assumptions 1), 2) and 3) of Conjecture 1. We recall the Remarks following Corollary 1. Deninger's observation about (4.6) in Section 4.3 should apply also to the coefficient of $\alpha(k \log p)$ in formula (5.1) viewed as a Lefschetz trace formula. The term $W_\infty(\alpha)$ of (5.1) should correspond to the fixed point l_∞ of the flow ϕ^t (see [**10**, Section 5]). One obviously has

$$(W_\infty)_{|\mathbb{R}^{+*}} = \frac{1}{1 - e^{-2t}}, \quad (W_\infty)_{|\mathbb{R}^{-*}} = \frac{e^{-|t|}}{1 - e^{-2|t|}}.$$

Then Guillemin-Sternberg's formula (see [**7**, Proposition 4.1]) suggests strongly that the identity $(W_\infty)_{|\mathbb{R}^{+*}} = \frac{1}{1-e^{-2t}}$ should be a consequence of the fact that ϕ^t should act by e^{-2t} on $\frac{T_{l_\infty} \overline{S_\mathbb{Q}}}{T_{l_\infty}\mathcal{F}} \simeq \mathbb{R}$ for $t \geq 0$. Recall that the fixed point l_∞ belongs to a certain path connected component of $\mathcal{L}/\mathbb{Q}^{+*}$. In [**10**, Section 5], Deninger suggests that the restriction to $\mathbb{R}^*$ of the contribution (such as W_∞) of the fixed point l_∞ in the Lefschetz trace formula such as the one expected in Conjecture 1.3) should be given by

$$\frac{\det(\mathrm{Id} - T_{l_\infty}\phi^t \mid T_{l_\infty}\mathcal{F})}{|\det(\mathrm{Id} - T_{l_\infty}\phi^{|t|} \mid T_{l_\infty}\overline{S_\mathbb{Q}})|}. \tag{5.5}$$

Using (5.2) (on $\mathcal{L}/\mathbb{Q}^{+*}$), one checks easily that for any real t, $e^{\frac{-t}{2}} T_{l_\infty}\phi^t$ belongs to $\mathrm{SO}(T_{l_\infty}\mathcal{F})$. Then an easy computation shows that (5.5) coincides with $(W_\infty)_{|\mathbb{R}^*}$.

Now we recall briefly, following [**11**, Section 3] the arguments allowing to expect that the flow ϕ^t should act on the space $\mathbb{R}^{+*}/\mathbb{Q}^{+*}$ of $\mathcal{L}$-leaves by multiplication by e^{-t}. First of all, one has ([**6**, Proposition 5.2])

$$2^{-\frac{1}{2}}\pi^{-\frac{s}{2}}\Gamma(\frac{s}{2}) = \frac{1}{\det_\infty\left(\frac{1}{2\pi}(s\mathrm{Id} - \Theta_\mathbb{R}) \,:\, \mathbb{R}[z]\right)}$$

where $\Theta_\mathbb{R}$ is defined by

$$\forall P \in \mathbb{R}[z],\ \Theta_\mathbb{R}(P)(z) = \left(\frac{d}{dt}P(ze^{-2t})\right)_{t=0}.$$

Consider the action of $\Sigma_2 = \{1, -1\}$ by multiplication on $\mathbb{R}$. We view $\mathbb{R}^{\geq 0}$ as a C^ω-orbifold via the isomorphism

$$\mathbb{R}/\Sigma_2 \to \mathbb{R}^{\geq 0}$$

$$[r] \to r^2.$$

One defines a flow $\phi^t_\mathbb{R}$ acting on $\mathbb{R}^{\geq 0}$ by $\phi^t_\mathbb{R}(r') = r'e^{-2t}$. His work [**12**] on the cohomological and dynamical interpretation of the Γ-factors (in particular of

$2^{-\frac{1}{2}}\pi^{-\frac{s}{2}}\Gamma(\frac{s}{2}))$ has lead Deninger to argue that there should exist a ϕ^t-equivariant embedding

$$i_\infty : \mathbb{R}^{\geq 0} \to \overline{S_{\mathbb{Q}}}$$

with $i_\infty(0) = l_\infty$, smooth for the orbifold structure (i.e., basically $r \to i_\infty(r^2)$ is smooth), transversal to the leaves of $(S_{\mathbb{Q}}, \mathcal{F})$ and such that for each $r > 0$, $i_\infty(r^2)$ belongs to the leaf induced by $\mathcal{L} \times \{r\}$ in $(S_{\mathbb{Q}}, \mathcal{F})$. Then the tangent map

$$T_0 i_\infty : T_0\mathbb{R}^{\geq 0} \to \frac{T_{l_\infty}\overline{S_{\mathbb{Q}}}}{T_{l_\infty}\mathcal{F}}$$

is an isomorphism transforming the action of $T_0\phi^t_{\mathbb{R}}$ on $T_0\mathbb{R}^{\geq 0}$ into the action of $T_{l_\infty}\phi^t$ on $\frac{T_{l_\infty}\overline{S_{\mathbb{Q}}}}{T_{l_\infty}\mathcal{F}}$. Therefore, $T_{l_\infty}\phi^t$ should act as $e^{-2t}\,\mathrm{Id}$ on $\frac{T_{l_\infty}\overline{S_{\mathbb{Q}}}}{T_{l_\infty}\mathcal{F}}$ as expected above. Now for $[r] \in \mathbb{R}/\Sigma_2$ we set $i_\infty[r] = i_\infty(r^2)$. Then using the ϕ^t-equivariance of i_∞ (defined on $\mathbb{R}^{\geq 0}$), one checks that for any $r, s \in [0, +\infty[$: $i_\infty[r\, s] = \phi^{-\log r}(i_\infty[s])$. Setting $t = -\log r$, we get $i_\infty[e^{-t}\, s] = \phi^t(i_\infty[s])$. Therefore, the flow ϕ^t should act on the space $\mathbb{R}^{+*}/\mathbb{Q}^{+*}$ of $\mathcal{L}$-leaves by multiplication by e^{-t}.

There are reasons to believe (see end of [**10**]) that there should exist an infinite dimensional dynamical dissipative system with a semigroup action of $([1, \infty[, \times)$ whose compact finite dimensional global attractor should be the desired $(\overline{S_{\mathbb{Q}}}, \mathcal{F}, g, \phi^t)$. See [**22**] for the concept of global attractor.

We hope to say a little more on this geometric picture in a forthcoming paper.

REMARK. An Arakelov divisor of $\overline{\mathrm{Spec}\,\mathbb{Z}}$ is a formal sum $D = \sum_{p\in F} n_p \,|\,|_p + e^{x_\infty}\,|\,|_\infty$ where F is a finite set of prime numbers, the $n_p \in \mathbb{Z}$, x_∞ is a real number and $|\,|_\infty$ denotes the archimedean absolute value of $\mathbb{Q}$. The degree of D is defined to be

$$\deg D = -\sum_{p\in F} n_p \log p - x_\infty.$$

It would be interesting to extend the result of Section 3.2 and to prove a Riemann Roch measured index type theorem on $(\overline{S_{\mathbb{Q}}}, \mathcal{F}, g)$ for such a divisor D.

It would be interesting to test Deninger's programme and prove intermediate results along the following lines. One should associate, in a **functorial** way[1], to any number field K a (canonical) Riemannian laminated foliated dynamical system $(\overline{S_K}, \mathcal{F}, g, \phi^t)$ such that the following properties should be satisfied. If K is a finite Galois extension of $\mathbb{Q}$ with Galois group G, let p_i, $1 \leq i \leq l$, be the prime numbers that are ramified in K and, for each $i \in \{1, \ldots, l\}$ let P_i^j, $1 \leq j \leq f_i$, be the prime ideals of $\mathcal{O}_K$ that lie over p_i. Then G should act on $(\overline{S_K}, \mathcal{F}, g, \phi^t)$ and commutes with ϕ^t. Moreover, to the injection $\mathbb{Q} \mapsto K$ one should associate a branched foliated covering map $\pi_K : \overline{S_K} \to \overline{S_{\mathbb{Q}}}$ of degree $\dim_{\mathbb{Q}} K$, with branched set upstairs (resp. downstairs) the closed orbits associated with the P_i^j (resp. the p_i). Let $\rho : G \to GL_{\mathbb{C}}(V)$ be a complex representation of finite degree. Consider the (leafwise) constructible sheaf $\mathcal{V}_\rho$ whose space of sections over an open subset $U \subset \overline{S_{\mathbb{Q}}}$ is the space of continuous leafwise locally constant G-invariant functions on $\pi^{-1}(U)$ with values in V. Notice that outside the branched set downstairs, $\mathcal{V}_\rho$ is defined by a leafwise flat complex vector bundle. Then the vector space $C^0(\overline{S_K}; \wedge^* T^*\mathcal{F} \otimes V)^G$ may be identified with the space of global (continuous) sections over $\overline{S_{\mathbb{Q}}}$ of the sheaf

[1] Notice the contrast with part 1) of the Remarks following Hypothesis 1.

$\wedge^* T^* \mathcal{F} \otimes \mathcal{V}_\rho$. I thank W. Lueck for helpful conversations on this point. Now there should exist an action

$$\psi^t : (\phi^t)^*(\mathcal{V}_\rho) \to \mathcal{V}_\rho,$$

$t \in \mathbb{R}$, opposite to the one of ϕ^t. Then the vector spaces $C^0(\overline{S_K}; \wedge^* T^* \mathcal{F} \otimes V)^G$ should allow to define L^2-leafwise cohomology groups which should be relevant for the study of the Artin L-function $L_\rho(s)$ (see [**20**, p. 535]) via a suitable Lefschetz trace formula. Let us recall that (see [**7**]) Deninger's infinite dimensional cohomological formalism applies more generally to motives (of which Artin representations are a particular case) and should provide an explanation of the Artin and Beilinson conjectures.

Andre Reznikov told me during my stay at Bar-Ilan University that there are many interesting automorphic L-functions which cannot be associated (via the Langlands correspondence) to motivic L-functions. They come from Maass forms, with eigenvalue not $1/4$; for a discussion see Section 1 of [**24**]. Thus, in order to state a Langlands correspondence which should take into account all the automorphic L-functions, one should find a suitable class of geometric L-functions which should be much larger than the class of motivic L-functions. If Deninger's program works perfectly well, then it might be possible that, to such an automorphic function $L(s)$, one could associate a leafwise constructible sheaf $\mathcal{E}$ over $\overline{S_\mathbb{Q}}$, which is defined outside a certain branched set by a leafwise flat complex vector bundle, endowed with an action

$$\psi^t : (\phi^t)^*(\mathcal{E}) \to \mathcal{E},$$

$t \in \mathbb{R}$, opposite to the one of ϕ^t. Roughly speaking, the complex of leafwise differential forms twisted by $\mathcal{E}$ should allow to define L^2-leafwise cohomology groups which should be relevant for the study of the L-function $L(s)$ via a suitable Lefschetz trace formula. Of course, at this stage, this is a pure speculation.

Appendix A. Regularized determinants

We follow [**6**] to which we refer for more details. Let Θ be an endomorphism of a complex vector space V of countable dimension. Assume the following two conditions:

(1) $V = \oplus_{\alpha \in S} V_\alpha$ where S is a countable subset of V, the V_α are finite dimensional Θ-invariant subspaces such that α is the only eigenvalue of $\Theta_{|V_\alpha}$.

(2) Let $\mathrm{Sp}(\Theta)$ be the set of eigenvalues of Θ with their algebraic multiplicities. We assume that the Dirichlet series

$$\sum_{\alpha \in \mathrm{Sp}\,\Theta} \frac{1}{\alpha^s}, \text{ with } \alpha^{-s} = |\alpha|^{-s} e^{-is\mathrm{Arg}\alpha}, \quad -\pi < \mathrm{Arg}\alpha \leq \pi$$

converges absolutely for $\mathrm{Re}\, s >> 0$ and has an analytic continuation denoted $\zeta_\Theta(s)$ to the half plane $\mathrm{Re}\, s > -\epsilon$ which is holomorphic in a neighborhood of $s = 0$.

Under these two conditions, we set $\dim_\infty(\Theta|V) = \dim V_0 + \zeta_{\dot\Theta}(0)$ where $\zeta_{\dot\Theta}$ is the induced endomorphism of V/V_0 and

$$\det{}_\infty(\Theta|V) = \exp(-\zeta'_{\dot\Theta}(0)) \text{ if } 0 \notin \mathrm{Sp}(\Theta), \ \det{}_\infty(\Theta|V) = 0 \text{ if } 0 \in \mathrm{Sp}(\Theta).$$

REMARK. If Θ appears as an unbounded operator acting on a Hilbert space H which admits a subvector space V such that conditions **(1)** and **(2)** are satisfied, then we shall simply write $\det_\infty(\Theta)$ for $\det_\infty(\Theta|V)$.

Note that for any positive real number $\delta > 0$, we have

$$\det_\infty(\delta\Theta|V) = \delta^{\dim_\infty(\Theta|V)}\det_\infty(\Theta|V).$$

In order to deal with the functional equation, we shall need the case $\delta = -1$ which will involve the regularized superdimension $\mathrm{sdim}_\infty(\Theta)$ defined as follows.

Assume that for a pair (V, Θ) satisfying condition **(1)** above we are given a decomposition $V = V^+ \oplus V^-$ into Θ-invariant subspaces. Then $(V^\pm, \Theta^\pm)$ with $\Theta^\pm = \Theta_{V^\pm}$ also satisfy condition **(1)**. We say that $\mathrm{sdim}_\infty(\Theta)$ exists with respect to this decomposition, if the Dirichlet series attached to $\Theta^\pm$

$$\sum_{\alpha \in \mathrm{Sp}\,\Theta^\pm} \frac{1}{\alpha^s}, \text{ with } \alpha^{-s} = |\alpha|^{-s} e^{-is\mathrm{Arg}\alpha}$$

converge absolutely for $\mathrm{Re}s >> 0$ and have analytic continuations $\zeta_{\Theta^\pm}(s)$ to $\mathrm{Re}\, s > -\epsilon$ for some $\epsilon > 0$ with at most first poles at $s = 0$. Writing

$$\zeta_{\Theta^\pm}(s) = \frac{\lambda^\pm}{s} + H^\pm(s),$$

where $\lambda^\pm \in \mathbb{C}$ and $H^\pm$ is holomorphic in a neighborhood of $s = 0$, we set

$$\mathrm{sdim}_\infty(\Theta) = (\dim V_0^+ + H^+(0)) - (\dim V_0^- + H^-(0)).$$

We call a decomposition $V = W^+ \oplus W^-$ into Θ-invariant subspaces commensurable with (V^+, V^-) if $V^+ \cap W^+$ (resp. $V^- \cap W^-$) is of finite codimension in V^+ and W^+ (resp. in V^- and W^-).

LEMMA 8. *The regularized superdimension of Θ exists with respect to $V = V^+ \oplus V^-$ if and only if it exists with respect to $V = W^+ \oplus W^-$. In case the two superdimensions exist, their difference is an even integer.*

For any pair (Θ, V) satisfying conditions **(1)** and **(2)**, we define a Θ-invariant decomposition by

$$V^+ = \oplus V_\alpha \text{ where } \alpha = 0 \text{ or } \alpha \neq 0 \text{ and } -\pi < \mathrm{Arg}\,\alpha \leq 0$$

$$V^- = \oplus V_\alpha \text{ where } \alpha \neq 0 \text{ and } 0 < \mathrm{Arg}\,\alpha \leq \pi.$$

A decomposition of V is called standard if it is commensurable to this one. The next Lemma is the main result of this section.

LEMMA 9. *Let (V, Θ) be a pair satisfying conditions* **(1)** *and* **(2)** *such that $\mathrm{sdim}_\infty(\Theta)$ exists with respect to a standard decomposition. Then $\det_\infty(-\Theta)$ is well-defined and we have*

$$\det_\infty(-\Theta) = e^{i\pi(\mathrm{sdim}_\infty(\Theta))}\det_\infty(\Theta),$$

where $\mathrm{sdim}_\infty(\Theta))$ is made up with respect to any standard decomposition of V.

Appendix B. Closed points and Frobenius morphism of a curve over a finite field

For more details we refer the reader to [**15**, Chapters I.6 and II.2] and [**18**, Chapters VI.9 and VII.3].

We begin with a concrete example. Let $P(X_0, \ldots, X_d) \in \mathbb{F}_q[X_0, \ldots, X_d]$ be an irreducible polynomial. Set $H = \{x = (x_0, \ldots, x_d) \in \overline{\mathbb{F}_q}^{d+1} / \ P(x) = 0\}$ and assume that for any $x \in H$, $(\partial_{x_0}P(x), \ldots, \partial_{x_d}P(x)) \neq (0, \ldots, 0)$. Then H defines a smooth affine hypersurface of $\overline{\mathbb{F}_q}^{d+1}$. Let $x = (x_0, \ldots, x_d) \in H$ and denote by d^0x

the smallest integer $k \in \mathbb{N}$ such that $x_j \in \mathbb{F}_{q^k}$ for $0 \leq j \leq d$. Consider the ring homomorphism $\Psi : \mathbb{F}_q[X_0, \dots, X_d] \to \overline{\mathbb{F}_q}$ defined by $\Psi(X_j) = x_j$ for $0 \leq j \leq d$. Then $\ker \Psi = y$ is a maximal ideal (containing (P)) defining a closed point y of H. Observe that if in the definition of Ψ one replaces x by $\mathrm{Fr}^k(x) = (x_0^{q^k}, \dots, x_d^{q^k})$ for $0 \leq k < d^0 x$ then one still gets *the same* maximal ideal y. Thus, one may identify the closed point y with the orbit under the action of the Frobenius morphism Fr of the geometric point x.

Now let Y be a d-dimensional smooth projective absolutely irreducible variety over $\mathbb{F}_q$. This means that the set of geometric points $\overline{Y} = Y \otimes_{\mathbb{F}_q} \overline{\mathbb{F}_q}$ is irreducible. The (geometric) $\overline{\mathbb{F}_q}$-linear Frobenius morphism $\mathrm{Fr} : \overline{Y} \to \overline{Y}$ is defined in an affine chart by

$$x = (x_0, \dots, x_d) \to \mathrm{Fr}^k(x) = (x_0^{q^k}, \dots, x_d^{q^k})$$

(see [**15**, p. 454]). A closed point of y is defined to be an orbit of Fr, denote by $d^0 y$ the number of elements of the orbit y and by $|Y|$ the set of closed points of Y. For each $r \in \mathbb{N}^*$ denote by N_r the number of points of $\overline{Y}$ whose coordinates belong to $\mathbb{F}_{q^r}$ (i.e., which are fixed by Fr^r). The Hasse-Weil zeta function of Y is then defined by

$$\zeta_Y(s) = \exp\left(\sum_{r=1}^{+\infty} \frac{N_r}{r} q^{-rs}\right).$$

On the other hand, Y is endowed with a structural sheaf $\mathcal{O}$ such that for each $y \in |Y|$ the stalk $\mathcal{O}_y$ is a local ring with maximal ideal m_y and the residue field $\frac{\mathcal{O}_y}{m_y}$ is a finite extension of $\mathbb{F}_q$ of degree $d^0 y$: $\#\frac{\mathcal{O}_y}{m_y} = q^{d^0 y}$. In the particular case where Y admits an open affine subset of the form H above, then

$$\mathcal{O}(H) = \frac{\mathbb{F}_q[X_0, \dots, X_d]}{(P)}.$$

Observing that $N_r = \sum_{k|r} k d_k$ where d_k is the number of closed points of degree k, one checks immediately that

$$\zeta_Y(s) = \prod_{y \in |Y|} \frac{1}{1 - q^{-s\, d^0 y}}.$$

Now assume that Y is a curve (i.e., $d = 1$) and denote by $K(Y)$ the function field of Y; $K(Y)$ is by definition the field of rational functions on Y and is the field of fractions of $\mathcal{O}_y$ for any $y \in |Y|$. A valuation of $K(Y)$ is by definition a non trivial discrete valuation ring of $K(Y)$. It is known that the map $y \to \mathcal{O}_y$ defines a bijection between $|Y|$ and the set of valuations of $K(Y)$.

Appendix C. Transverse measures and Ruelle-Sullivan currents on foliated spaces

For more details we refer to [**3**, Chapter I.5] and [**27**].

In this section, we fix a compact foliated smooth manifold $(X, \mathcal{F})$ such that the integrable sub-bundle of TX: $T\mathcal{F} \to X$ is oriented and of rank r. We denote by $C^\infty(X, (\wedge^r T\mathcal{F})^+)$ the set of smooth positive sections of the oriented line bundle $\wedge^r T\mathcal{F} \to X$. Recall that a (Borel) transversal is a Borel subset whose intersection with each leaf is at most countable. The interested reader can extend easily these definitions to the more general context of compact foliated laminated spaces working with spaces of sections which are continuous and leafwise smooth.

DEFINITION 3. Let $(U_i)_{1\leq i\leq N}$ be a finite open cover of $(X,\mathcal{F})$ by foliated charts $h_i : U_i \to V_i \times T_i$ with transition maps:

(C.1) $$h_{i,j} = h_j \circ h_i^{-1} : h_i(U_i \cap U_j) \to V_j \times T_j$$

(C.2) $$(y,\theta) \to (f_{i,j}(y,\theta), g_{i,j}(\theta))$$

where V_i is an open ball in $\mathbb{R}^r$, the Jacobian det $\frac{\partial f_{i,j}(y,\theta)}{\partial y}$ is positive and the local transversal T_i is an open ball of $\mathbb{R}^{\dim X - r}$. Then a holonomy invariant transverse measure Λ is given by the data for each $i \in \{1,\dots,N\}$ of a measure Λ_i on T_i which takes finite non negative values on the compact subsets and such that for each Borel subset B $(\subset T_i)$ of the domain of definition of $g_{i,j}$ one has $\Lambda_i(B) = \Lambda_j(g_{i,j}(B))$.

Such a measure Λ induces on each Borel transversal a measure still denoted Λ which is holonomy invariant. One can check that Λ satisfies the following stronger property which we may take as a definition of a holonomy invariant transverse measure.

PROPOSITION 5. *Let Λ be as in the previous definition. Then Λ induces a σ-additive map still denoted Λ: $B \to \Lambda(B)$ on the set of Borel transversals with values in $[0,+\infty]$ such that*

(1) *If B_1 and B_2 are two Borel transversals and $\psi : B_1 \to B_2$ is a Borelian bijection such that $\psi(y)$ belongs to the leaf of y for any $y \in B_1$, then $\Lambda(B_1) = \Lambda(B_2)$.*
(2) *For any compact subset K of a Borel transversal, $\Lambda(K) < +\infty$.*

DEFINITION 4.

(1) A foliated current C of dimension r on $(X,\mathcal{F})$ is a continuous linear map on $C^\infty(X, \wedge^r T^* X)$ such that for each $\omega \in C^\infty(X, \wedge^r T^*X)$ vanishing on $C^\infty(X, \wedge^r T\mathcal{F})$ one has $(\omega; C) = 0$.
(2) A foliated current C of dimension r is closed if for any $\eta \in C^\infty(X, \wedge^{r-1} T^*X)$ one has $(d\eta; C) = 0$.
(3) A foliated current C of dimension r is said to be positive if $(\omega; C) \geq 0$ for any $\omega \in C^\infty(X, \wedge^r T^*X)$ whose pairing with each section of $C^\infty(X, (\wedge^r T\mathcal{F})^+)$ gives a non negative function.

REMARK. We could have considered a slightly different definition with currents defined as continuous linear forms on $C^\infty(X, \wedge^r T^*\mathcal{F})$.

DEFINITION 5. Let Λ be as in Definition 3 and $(\psi_i)_{1\leq i\leq N}$ a partition of unity subordinated to the covering $(U_i)_{1\leq i\leq N}$: $\operatorname{supp}\psi_i \subset\subset U_i$ for $1 \leq i \leq N$. The Ruelle-Sullivan current $C(\Lambda)$ associated to Λ is the positive foliated current of dimension r defined as follows. For any $\omega \in C^\infty(X, \wedge^r T^*X)$,

$$(\omega; C(\Lambda)) = \sum_{i=1}^{N} \int_{T_i} \left(\int_{V_i \times \{\theta\}} (h_i^{-1})^*(\psi_i\, \omega) \right) d((h_i^{-1})^*\Lambda)(\theta)$$

where $(h_i^{-1})^*\Lambda$ denotes the measure induced on T_i by Λ via h_i^{-1}; this measure is independent of the point of V_i by the holonomy invariance of Λ. This definition of $C(\Lambda)$ depends neither on the choice of the covering charts $(U_i)_{1\leq i\leq N}$ nor on that of the partition of unity $(\psi_i)_{1\leq i\leq N}$.

PROPOSITION 6. *Let C be a positive foliated current of dimension r on $(X, \mathcal{F})$. Then there exists a holonomy invariant transverse measure Λ such that C is the Ruelle-Sullivan current associated to Λ.*

PROOF. One may assume that the open covering $(U_i)_{1 \leq i \leq N}$ of Definition 3 is such that each local transversal T_i contains an open ball T_i' such that $X = \cup_{1 \leq i \leq N} h_i^{-1}(V_i \times T_i')$. Then, for each $i \in \{1, \ldots, N\}$, choose $\omega_i \in C_c^\infty(V_i \times T_i), \wedge^r T^* V_i)$ such that for each $\theta \in T_i'$, $\int_{V_i} \omega_i(y, \theta) = 1$. Define a measure Λ_i on T_i' by

$$\forall f \in C_c^0(T_i', \mathbb{R}), \; \Lambda_i(f) = (f\omega_i; C).$$

If ω_i' is another such differential form, then one can find $\beta_i \in C_c^\infty(V_i \times T_i, \wedge^{r-1} T^* V_i)$ such that $d_{\mathcal{F}} \beta_i = \omega_i - \omega_i'$. Then, using the fact that the leafwise current C is closed, one easily checks that Λ_i does not depend on the choice of ω_i. Now the required holonomy invariant transverse measure is defined by the data of the Λ_i $(1 \leq i \leq N)$. □

EXAMPLE. Let Γ be a discrete finitely generated group and $\pi : \widetilde{M} \to M$ a Γ-normal covering of a smooth compact r-dimensional manifold M. Let B be a smooth compact manifold on which Γ acts (on the right) by diffeomorphisms. Consider the projection map p onto the quotient space

$$p : \widetilde{M} \times B \to \frac{\widetilde{M} \times B}{\Gamma} = Y,$$

where Γ acts on $\widetilde{M} \times B$ by $(z, b).\cdot\gamma = (z\cdot\gamma, b\cdot\gamma)$. The sets $p(\widetilde{M} \times \{b\})$, $(b \in B)$ define the leaves of a foliation $\mathcal{F}$ on Y. Fix $\phi \in C_c^\infty(\widetilde{M}, \mathbb{R})$ such that $\sum_{\gamma\in\Gamma} \phi(\cdot\gamma) \equiv 1$. Let μ be a Γ-invariant measure on B. Then μ defines a holonomy invariant transverse measure, still denoted μ, on $(Y, \mathcal{F})$ whose Ruelle-Sullivan current is defined by

$$\forall \omega \in C^\infty(Y, \wedge^r T^* Y), \; (\omega; C(\mu)) = \int_B \left(\int_{\widetilde{M}} p^*(\omega)\, \phi \right) d\mu(\theta).$$

REMARK. With the notations of Section 2.2, the reader can check that

$$\mathbb{C} \times_V V_\xi \Gamma \simeq \mathbb{C} \times_\Gamma \mathrm{T}\Gamma,$$

and that the Haar measure of $\mathrm{T}\Gamma$ induces a transverse measure on the 1-dimensional complex laminated space $\mathbb{C} \times_V V_\xi \Gamma$.

Acknowledgements

I am very happy to thank Christopher Deninger since I have learnt a lot of beautiful mathematics by talking with him and reading his papers. I am also grateful to the referees whose comments have allowed to improve the paper. This report is based on lectures that I gave in Bar Ilan University; I take this opportunity to thank Louis Rowen and Steve Shnider for their kind invitation and very warm hospitality. It is also a pleasure to thank my friends Misha Katz and Uri Onn for very helpful comments on this paper. A part of this paper was written while I was visiting the University of Savoie, I am very grateful to K. Kurdyka for his kind hospitality. Finally, I thank B. Jacob, W. Lueck, P. Gérardin and G. Skandalis for useful conversations.

References

1. J. A. Alvarez Lopez and Y. Kordyukov, *Long time behaviour of leafwise heat flow for Riemannian foliations*, Compositio Math. **125** (2001), 129-153.
2. ______, *Distributional Betti numbers of transitive foliations of codimension one*, Foliations: Geometry and Dynamics (Warsaw, 2000), World Sci. Publishing, River Edge, NJ, (2002), pp. 159-183.
3. A. Connes, *Noncommutative Geometry*, Academic Press, 1994.
4. ______ *Trace formula in noncommutative geometry and the zeroes of the Riemann zeta function*, Selecta Math. (N.S.) **5** (1999), no. 1, 29-106.
5. C. Deninger, *Local L-factors of motives and regularized determinants*, Inv. Math. **107** (1992), 135-150.
6. ______, *Motivic L-functions and regularized determinants*, Proc. Symp. Pure Math. **55**, (1994), no. 1, 707-743.
7. ______, *Some analogies between number theory and dynamical systems onf foliated spaces*, Doc. Math. J. DMV, Extra volume ICM I, (1998), 23-46.
8. ______, *On dynamical systems and their possible significance for Arithmetic Geometry*, Regulators in Analysis, Geometry and Number Theory (A. Reznikov, N. Schappacher, eds.), Progress in Mathematics **171**, (1999), Birkhäuser, pp. 29-87.
9. ______, *Some ideas on dynamical systems and the Riemann zeta function*, preprint, 2000.
10. ______, *Number theory and dynamical systems on foliated spaces*, Jahresberichte der DMV. **103**, (2001), no. 3, 79-100.
11. ______, *A note on arithmetic topology and dynamical systems*, Algebraic Number Theory and Algebraic Geometry, Contemp. Math., **300** (2002), 99-114.
12. ______, *On the Γ-factors of motives. II*, Doc. Math. (electronic) (2001), 69-97.
13. ______, *On the nature of explicit formulas in analytic number theory, a simple example*, Number Theoretic Methods (Iizuka, 2001), Dev. Math. **8**, Kluwer Acad. Publ., Dordrecht, (2002), 97-118.
14. C. Deninger and W. Singhof, *A note on dynamical trace formulas*, Dynamical, Spectral and Arithmetic Zeta Functions (San Antonio, TX, 1999), Contemp. Math. **290** (2001), pp. 41-55.
15. R. Hartshorne, *Algebraic Geometry*, Graduate Texts in Math., vol. 52, 1977.
16. Y. Ihara, *On $(\infty \times p)$-adic coverings of curves (the simplest example)*, Trudy Mat. Inst. Steklov **132** (1973), 133-148.
17. N. Katz and P. Sarnak, *Zeta functions and symmetry*, Bull. Amer. Math. Soc. (N.S.) **36** (1999), no. 1, 1–26.
18. D. Lorenzini, *An Invitation to Arithmetic Geometry*, Graduate Studies in Math., vol. 9, Amer. Math. Soc., 1995.
19. J. Milne, *Étale cohomology*, Princeton Mathematical Series, vol. 33, 1980.
20. J. Neukirch, *Algebraic Number Theory*, Springer, vol. 322, 1999.
21. F. Oort, *Lifting an endomorphism of an elliptic curve to characteristic zero*, Indag. Math. **35** (1973), 466-470.
22. J. Robinson, *Infinite-dimensional dynamical systems*, Cambridge Texts in Applied Mathematics, 2001.
23. M. Rosen, *Number Theory in Function Fields*, Graduate Texts in Math., vol. 210, 2000.
24. P. Sarnak, *Spectra of hyperbolic surfaces*, Notes for Baltimore Lectures 2003; `http://www.math.princeton.edu/ sarnak`.
25. J-P. Serre, *Groupes algébriques et corps de classe*, Editions Hermann, 1959.
26. J. Silverman, *The arithmetic of elliptic curves*, Graduate Texts in Math., vol. 106, 1992.
27. S. Vassout, *Feuilletages et résidu non commutatif longitudinal.* Thèse de doctorat, Université Paris VI, `http://www.math.jussieu.fr/~vassout/`, 2001.

INSTITUT DE JUSSIEU ET CNRS, ETAGE 7E, 175 RUE DU CHEVALERET, 75013, PARIS, FRANCE

Contemporary Mathematics
Volume **387**, 2005

Some More Non-arithmetic Rigid Groups

Alexander Lubotzky

Dedicated to the memory of Bob Brooks

ABSTRACT. We simplify and generalize the method of [**1**] and present many more counterexamples to Platonov's conjecture, i.e., linear groups which are super rigid (a' la Margulis) but not of arithmetic type.

0. Introduction

Let Γ be a finitely generated group. Γ is called **rigid** if for every $n \geq 1$, Γ admits only finitely many isomorphism classes of irreducible representations into $GL_n(\mathbf{C})$. Platonov (cf. [**9**, p. 437]) conjectured that if Γ is a linear rigid group then Γ is of "arithmetic type" (i.e., commensurable with a product of S-arithmetic groups).

In [**1**], H. Bass and the author gave counter-examples to Platonov's conjecture. The examples Γ there are very special; they are subgroups of $L \times L$ when L is a uniform (arithmetic) lattice in the rank one simple Lie group $F_4^{(-20)}$. The proof there relies on four main ingredients:

(1) The super-rigidity (á la Margulis) of L, which was proved by Corlette [**2**] and Gromov-Schoen [**4**].
(2) The Ol'shanskiĭ-Rips theorem ([**8**]) asserting that L being hyperbolic has a finitely presented infinite quotient H with no proper finite index subgroup.
(3) Grothendieck's theorem [**5**] which says that once the inclusion of Γ to $L \times L$ induces an isomorphism of the pro-finite completions $\hat{\Gamma}$ and $\hat{L} \times \hat{L}$, the representation theory of Γ is the same as that of $L \times L$.
(4) The vanishing of the second Betti number $\beta_2(L') = \dim H^2(L', \mathbf{R}) = 0$ for every finite index subgroup L' of L (proved by Kumareson and Vogan-Zuckerman (cf. [**11**])).

In this note we give a simplified version of the proof in [**1**]. We eliminate the use of ingredients (3) and (4). Not using (3) makes the proof more elementary, but avoiding (4) is even more significant: our proof now works also when L is any

2000 *Mathematics Subject Classification.* Primary 22E40.
The author is indebted to the Israel Science Foundation and the U.S.-Israel Binational Science Foundation for their support.

uniform lattice in $Sp(n,1)$ (for every $n \geq 2$), groups for which (4) does not hold (as follows from ([**6**, Cor. 6.5]). We thus have many new examples which are Zariski dense in $Sp(n,1) \times Sp(n,1)$.

Just as in [**1**], the counter examples to Platonov's conjecture constructed here are even super-rigid (see §1). In summary:

THEOREM. *Let G be either $Sp(n,1), n \geq 2$, or $F_4^{-(20)}$ and let L be a cocompact lattice in G. Then $L \times L$ contains a subgroup Γ of infinite index which is Zariski dense in $G \times G$, super-rigid and not of arithmetic type.*

We end the note with two suggestive remarks to be explained in §4:

We show that if L satisfies the congruence subgroup property, then a further simplification is possible, avoiding also the use of (1) and (2), i.e., the works of Corlette, Gromov-Schoen, Ol'shanskii and Rips.

Furthermore, if a single uniform lattice L in $Sp(n,1)$ or $F_4^{(-20)}$ satisfies the congruence subgroup property, then there exists a hyperbolic group á la Gromov which is not residually-finite. Moreover, there exists a hyperbolic group with no proper finite index subgroup. So, answering the congruence subgroup problem in the affirmative, for one such L would settle in the negative the long standing open problem on the residual-finiteness of hyperbolic groups.

1. FAb fibre products of rigid groups are rigid

Throughout the paper, Γ is a finitely generated group. Γ is **FAb** if for every finite index subgroup Λ of Γ, $\Lambda^{ab} := \Lambda/[\Lambda,\Lambda]$ is finite. Γ is **rigid** if for every $n \geq 1$, it has only finitely many non-equivalent irreducible n-dimensional complex representations. Finally, Γ is **super-rigid** if $A(\Gamma)^0$, the connected component of $A(\Gamma)$ is finite-dimensional algebraic group. Here, $A(\Gamma)$ is the pro-algebraic completion of Γ. There is a homomorphism $i : \Gamma \to A(\Gamma)$ such that for every representation $\rho : \Gamma \to GL_n(\mathbf{C})$, there exists a unique algebraic representation $\tilde{\rho} : A(\Gamma) \to GL_n(\mathbf{C})$ with $\tilde{\rho} \circ i = \rho$.

It is not difficult to prove:

PROPOSITION 1.1. *super-rigid $\Rightarrow$ rigid $\Rightarrow$ FAb.*

PROPOSITION 1.2. *If Λ is a finite index subgroup of Γ, then Λ is super-rigid (resp. rigid, FAb) iff Γ is.*

We also have:

PROPOSITION 1.3. *Assume Γ has FAb. Then given n, there exists a finite index subgroup $\Lambda = \Lambda(n)$ of Γ such that for every n-dimensional representation $\rho : \Gamma \to GL_n(\mathbf{C})$, $\overline{\rho(\Lambda)}$, the Zariski closure, is a connected group. If ρ is irreducible then $\overline{\rho(\Lambda)}$ is semisimple.*

PROOF. Denote $H = \overline{\rho(\Gamma)}$, then there exists a finite subgroup F of H such that $H = H^0 \cdot F$ ([**12**, 10.10]). By Jordan's Theorem ([**12**, 9.2]), F has an abelian normal subgroup of index at most $J(n)$. Thus the same applies to H/H^0. Let now $\tilde{\Lambda}$ be the intersection of all the normal subgroups of Γ of index at most $J(n)$ and $\Lambda = [\tilde{\Lambda}, \tilde{\Lambda}]$. It follows that Λ has finite index in Γ and for every n-dimensional representation of Γ, $\rho(\Lambda) \subseteq \overline{\rho(\Gamma)}^0$. This also implies that $\rho(\Lambda)$ is dense in $\overline{\rho(\Gamma)}^0$ since Λ has finite index in Γ and $\overline{\rho(\Gamma)}^0$ has no finite index subgroups. Finally, if ρ

is irreducible, then $\overline{\rho(\Gamma)}^0$ is reductive, but actually semisimple as it has no abelian quotient. □

PROPOSITION 1.4. *A finitely generated FAb group Γ is not rigid iff there exists a finite index subgroup Λ of Γ and a simple algebraic group H of adjoint type (i.e., $Z(H) = 1$) such that Λ has infinitely many non-conjugate homomorphisms $\rho : \Lambda \to H$ with $\rho(H)$ Zariski dense in H.*

PROOF. Assume there are infinitely many non-equivalent irreducible n-dimensional representations ρ of Γ and let $\Lambda = \Lambda(n)$ as in Proposition 1.3. Then $\overline{\rho(\Lambda)}$ is a connected semisimple group for every such ρ. There are only finitely many conjugacy classes of semisimple connected subgroups of $GL_n(\mathbf{C})$, so we may assume that $H = \overline{\rho(\Lambda)}$ is fixed. From Clifford theorem [W, Theorem 1.7], it follows that there are infinitely many non-equivalent Λ-representations with Zariski-dense image in H. Infinitely many of them are still non-equivalent when H is divided by its finite center. We can therefore assume H is of adjoint type and a direct product of its simple components. There are still infinitely many non-equivalent homomorphisms of Λ to one of the simple factors of H. Replacing H by this factor gives the result. The other direction is easy as $(\mathrm{Aut}(H) : \mathrm{Inn}(H))$ is finite. □

We now show how new rigid (resp. super-rigid) groups can be obtained as a fibre product of rigid (resp. super-rigid) groups.

For $i = 1, 2$, let L_i be a finitely generated group with epimorphisms ρ_i from L_i onto the same finitely presented group D, and $R_i = \mathrm{Ker}\ \rho_i$. Let

$$\Gamma = L_1 \times_D L_2 = \{(x, y) \in L_1 \times L_2 \mid \rho_1(x) = \rho_2(y)\}$$

be the **fibre product** of L_1 and L_2 over D.

The projections π_i of Γ to L_i are onto with kernels $(1, R_2)$ and $(R_1, 1)$. Also, one can easily see that Γ is finitely generated since D is finitely presented.

PROPOSITION 1.5. *If L_1 and L_2 are rigid and Γ has FAb, then Γ is also rigid.*

PROOF. If not, by Proposition 1.4, there exists a finite index subgroup Λ of Γ and a simple algebraic group H that Λ has infinitely many non-equivalent homomorphisms $\rho : \Lambda \to H$ with dense image. Let $R_1' = (R_1, 1) \cap \Lambda$ and $R_2' = (1, R_2) \cap \Lambda$. Then $[R_1', R_2'] = 1$ and both are normal in Λ. Thus $\overline{\rho(R_1')}$ and $\overline{\rho(R_2')}$ are commuting normal subgroups of H, so one of them must be trivial. Hence for infinitely many ρ's, either $\rho(R_1')$ is trivial or $\rho(R_2')$ is. In the first case we have that ρ factors through π_2, i.e., we have a representation of $\Lambda / R_1' = \Lambda / (R_1, 1) \cap \Lambda \simeq \Lambda(R_1, 1)/(R_1, 1)$ which is commensurable to L_2. Hence, this group is not rigid and by Proposition 1.2 also L_2 is not rigid, a contradiction. The case with $\rho(R_2') = 1$ is treated similarly and the Proposition is proved. □

It is somewhat more difficult to prove, but under an additional hypothesis it is also true, that when we start with L_i super-rigid and reductive, then Γ is super-rigid.

Recall that a group is called reductive if every representation of it is fully reducible.

PROPOSITION 1.6. *Assume L_1 and L_2 are super-rigid and reductive, D has no finite index subgroup and Γ is FAb then Γ is super-rigid and reductive, in fact, $A(\Gamma)^0 = A(L_1)^0 \times A(L_2)^0$*

PROOF. Let $\rho : \Gamma \to GL_n(\mathbf{C})$ be a representation. Replacing Γ by Λ as in Proposition 1.3, we can assume $H = \overline{\rho(\Lambda)}$ is connected and it is equal to its own commutator subgroup (since Γ is FAb).

We claim that H is semi-simple. If not, it has a non-trivial unipotent radical U. Dividing by $[U, U]$ and further if needed, we can assume that U is a simple H-module. So $H = U \cdot S$, a semi-direct product when S is semisimple and U a simple S-module. Clearly, this is not the trivial module, since H has no abelian quotient.

As in the proof of (1.5), $\overline{\rho(R_1')}$ and $\overline{\rho(R_2')}$ are commuting normal subgroups. Moreover, $\Lambda/(R_1' \times R_2') \simeq D$ and D has no finite index subgroups, hence no finite dimensional representations. It follows that $\overline{\rho(R_1' \times R_2')} = \overline{\rho(R_1')} \cdot \overline{\rho(R_2')} = H$.

If $\overline{\rho(R_1')}$ does not contain U, then $H/\overline{\rho(R_1')}$ has non-trivial unipotent radical and so Λ/R_1', which is commensurable to L_2 is not reductive, a contradiction. Hence $\overline{\rho(R_1')}$, and similarly $\overline{\rho(R_2')}$, contains U. But $[\overline{\rho(R_1)}, \overline{\rho(R_2')}] = 1$, since $[R_1', R_2'] = 1$. This implies that $[U, \overline{\rho(R_2')}] = [\overline{\rho(R_1')}, U] = 1$ and hence H acts trivially on U, a contradiction. Hence U is trivial and H is semisimple. Again $\overline{\rho(R_1')}$ and $\overline{\rho(R_2')}$ are normal and commuting. This implies that (after dividing by the center of H) $H = \overline{\rho(R_1')} \times \overline{\rho(R_2')}$ but then $H/\overline{\rho(R_i')}$ is an homomorphic image of $A(L_{3-i})^0$. This proves that $A(\Gamma)^0 = A(L_1)^0 \times A(L_2)^0$. □

2. When is a fibre product FAb?

In this section we show some sufficient conditions for a fibre product to be FAb.

Let L be a finitely generated group, $\rho : L \to D$ a homomorphism onto a finitely presented group D with kernel R, and $\Gamma = L \times_D L$ the fibre product of L over D,

$$\Gamma = \{(x, y) \in L \times L \mid \rho(x) = \rho(y)\}.$$

Note that $\Gamma = (R, 1)\Delta(L) = (1, R)\Delta(L)$ when $\Delta(L)$ is the diagonal embedding of L into Γ, $\Delta(L) = \{(x, x) \mid x \in L\}$.

LEMMA 2.1. *If L is FAb and if for every finite index normal subgroup L_0 of L, and every finite index subgroup R_0 of R which is normal in L, $[L_0, R_0]$ is of finite index in R_0, then Γ is FAb.*

PROOF. Let Γ_0 be a finite index normal subgroup of Γ, $L_0 = \Gamma_0 \cap \Delta(L)$ and $R_0 = (R, 1) \cap \Gamma_0$. By abuse of notation we will consider R_0 as a subgroup of L and at the same time a subgroup of $(L, 1)$. From our assumption, it follows that $[R_0, L_0]$ is of finite index in R_0 and hence in $(R, 1)$. As L is FAb, $[L_0, L_0]$ is of finite index in $\Delta(L)$. This implies that $[R_0L_0, R_0L_0]$ is of finite index in Γ. As $\Gamma_0 \supseteq R_0L_0$, we get that $[\Gamma_0, \Gamma_0]$ is of finite index in Γ. □

Let now L be a finitely generated group with FAb. Assume L has an infinite finitely presented quotient D with kernel R, such that $\hat{D} = \{1\}$ when $\hat{D}$ is the profinite completion of D, i.e., D has no proper finite index subgroup. Clearly $D^{ab} = \{1\}$, i.e., D is a perfect group. It has therefore a universal central extension

$$(*) \qquad 1 \to H_2(D) \to \bar{D} \to D \to 1$$

(cf. [**7**, §5]).

LEMMA 2.2. *Let*

$$(**) \qquad 1 \to C \to E \to D \to 1$$

be a central extension of D *such that* E^{ab} *is finite. Then* $rk(C) \leq rk(H_2(D))$. *(For an abelian group* A, *we denote* $rk(A) = \dim_{\mathbf{Q}}(A \otimes_{\mathbf{Z}} \mathbf{Q})$*).*

PROOF. There is a homomorphism ψ from the universal central extension $(*)$ to $(**)$. We claim that $\psi(\bar{D})$ is of finite index in E. Indeed, $C\psi(\bar{D}) = E$ and C is central. Hence $\psi(\bar{D}) \lhd E$ and the quotient is abelian. But E^{ab} is finite, hence $(E : \psi(\bar{D})) < \infty$. This implies that $\psi(H_2(D))$ is of finite index in C and hence $rk(C) \leq rk(H_2(D))$. □

LEMMA 2.3. *For a finite index normal subgroup* L_0 *of* L, *denote* $R_0 = R \cap L_0$. *Then* $rk(R_0/[L_0, R_0]) \leq rk(H_2(D))$ *for every* $L_0 \lhd L$ *of finite index.*

PROOF. The map $L_0 \to D$ is onto since $\hat{D} = 1$ and so:

$$1 \to R_0/[L_0, R_0] \to L_0/[L_0, R_0] \to D \to 1$$

is a central extension of D. As L is FAb, L_0^{ab} is finite and our Lemma follows from Lemma 2.2. □

Given L and $\rho : L \to D$ with $R = \operatorname{Ker} \rho$ as before, choose now L_0 of finite index with $R_0 = R \cap L_0$ and $rk(R_0/[L_0, R_0])$ maximal among all possible such L_0. As $\hat{D} = \{1\}$, it follows that $L_0/R_0 \simeq D$ and we have a central extension

$$(2.1) \qquad 1 \to R_0/[L_0, R_0] \to D_0 := L_0/[L_0, R_0] \to D = L_0/R_0 \to 1.$$

Now, $R_0/[L_0, R_0]$ is dense in the profinite topology of D_0 since $\hat{D} = \{1\}$, hence $\hat{D}_0$ is abelian. But, L_0 is FAb, so $\hat{D}_0$ is finite. Replace L_0 by a finite index subgroup L_0' so that $D_1 = L_0'/([L_0, R_0] \cap L_0')$ satisfies $\hat{D}_1 = \{1\}$ and in particular $D_1^{ab} = \{1\}$. Let us now rename and call $L := L_0'$, $\tilde{D} := D_1$, $\tilde{R} = \operatorname{Ker}(L \to \tilde{D})$ and $R = \operatorname{Ker}(L \to D)$. So $\tilde{D}$ is a central (possibly infinite!) extension of the original D and $\hat{\tilde{D}} = \{1\}$. We have two exact sequences

$$(2.2) \qquad 1 \to R \to L \to D \to 1$$

and

$$(2.3) \qquad 1 \to \tilde{R} \to L \to \tilde{D} \to 1,$$

where $\tilde{R} \subseteq R$, $[L, R] \subseteq \tilde{R}$, and $R/\tilde{R}$ is a finitely generated abelian group.

The crucial point is:

CLAIM. For every normal subgroup L_1 of finite index in L and every R_1 of finite index in $\tilde{R}$ which is normal in L, $[L_1, R_1]$ is of finite index in R_1.

PROOF. The group $\tilde{R}/[L_1, R_1]$ is a finitely generated virtually abelian group and $R/[L_1, R_1]$ is therefore a virtually polycyclic group. Hence $L/[L_1, R_1]$ is an extension of the form $1 \to C \to L/[L_1, R_1] \to D \to 1$ where C is virtually polycyclic and $\hat{D} = \{1\}$. Moreover, L is FAb. So, all this implies that $(\widehat{L/[L_1, R_1]})$ is finite. Hence, whenever $L/[L_1, R_1]$ is mapped into a residually-finite group, its image is finite. This applies, in particular, to the image of $L/[L_1, R_1]$ in its action (by conjugation) on $R/[L_1, R_1]$. (Recall that $C = R/[L_1, R_1]$ is virtually polycyclic and its automorphism group is, therefore, residually finite, as it is for every finitely generated residually finite group.) Thus for some finite index subgroup L_2 of L_1, $[L_2, R] \subseteq [L_1, R_1]$. By the maximality choice of L_0, it follows that $[L_2, R]$ is of finite index in $\tilde{R}$ and hence $[L_1, R_1]$ is also of finite index there. □

To summarize, by replacing D by $\tilde{D}$ (and the original L by a subgroup of finite index), we get an exact sequence

$$1 \to \tilde{R} \to L \to \tilde{D} \to 1 \tag{2.4}$$

which satisfies the claim. Hence by Lemma 2.1, we have:

COROLLARY 2.4. *The fibre product* $\Gamma = L \times_{\tilde{D}} L$ *is FAb.*

3. The main result

In this section, we pick the fruits of the preparations in the previous two sections and construct non-arithmetic super-rigid groups.

Propositions 1.5 and 1.6 show how to get new (super) rigid groups from old ones. The standard examples of rigid groups are irreducible lattices in semisimple groups of higher rank, but these do not have infinite quotients, so fibre products of them are commensurable to other arithmetic groups. This leads us to lattices in $Sp(n,1)$ and $F_4^{(-20)}$. They are super-rigid ([**2**],[**4**]) and at the same time have many infinite quotients.

So, from now on in this section, let L be a torsion-free uniform (=cocompact) lattice in one of the groups $Sp(n,1)$ or $F_4^{(-20)}$. It is a hyperbolic group and hence by a result proved independently by Ol'shanskii and Rips (see [**8**]), it has a finitely presented infinite quotient $\rho : L \to D$ where D has no proper finite index subgroups. Replace D now by $\tilde{D}$ (and L by a finite index subgroup, also called L) as in §2. Let $\tilde{\rho} : L \to \tilde{D}$ be the new map and $\Gamma = L \times_{\tilde{D}} L$ the fibre product. (The reader may note that at this point our method differs from the one in [**1**], where $L \times_D L$ was used; this change is exactly what enables us to avoid the use of ingredient (4) in the introduction and to have the result applied also for lattices in $Sp(n,1)$ and not merely for $F_4^{(-20)}$.)

THEOREM 3.1. Γ *is a super-rigid group. In fact,* $A(\Gamma)^0 = A(L)^0 \times A(L)^0$.

PROOF. By Corollary 2.4, Γ is FAb. Hence Proposition 1.6 can be applied to deduce that Γ is super-rigid (and hence rigid) and $A(\Gamma)^0 = A(L)^0 \times A(L)^0$. □

The proof actually shows that every representation of Γ can be extended, on a finite index subgroup, to a representation of $L \times L$. One can now repeat the standard argument given in ([**1**, pp. 1171–1172]) to show that Γ is not (virtually) isomorphic to any lattice in a product of linear algebraic groups (over archimedean or non-archimedean fields). So Γ is not "of arithmetic type," giving the desired counter-example to Platonov's conjecture.

4. Remarks on the congruence subgroup property

Let L be a uniform lattice in $Sp(n,1)$ or $F_4^{(-20)}$ as in Section 3. Such an L is an arithmetic lattice ([**4**]). The question whether L satisfies the congruence subgroup property (CSP, for short) is still open. Serre's conjecture, posed in [**10**], suggests that lattices in rank one groups do not have CSP, while lattices in higher rank simple groups do. So by this conjecture, L is not expected to have the CSP. But this conjecture was made in 1970. Since then, it has been shown that in spite of $Sp(n,1)$ and $F_4^{(-20)}$ being rank one groups, lattices in them behave in many ways (property T, super-rigidity, arithmeticity, etc.) like higher rank lattices. One may, therefore, suggest that they do have the CSP. As of now, the answer is not known for any single such L.

We make here two short remarks showing that an affirmative answer for the CSP for L would have two interesting corollaries:

REMARK 4.1. If L has the CSP, then the above construction (in §3) of D (and hence of Γ) can be done without appealing to the work of Ol'shanskii and Rips. In fact, if N is any infinite normal subgroup of L (which is finitely generated as normal subgroup) of infinite index, then $D = L/N$ has only finitely many finite index subgroups. Indeed, if not, it has infinitely many finite index normal subgroups. Pulling them back to L, we get infinitely many normal **congruence** subgroups of L all containing N. But it is not difficult to prove that for L being an arithmetic subgroup of a simple algebraic group, the intersection of any infinite collection of normal congruence subgroups must be finite and central. This shows that $D = L/N$ has only finitely many finite index subgroups. We can replace D by their intersection (and L by the preimage) to get the desired finitely presented quotient without any finite index subgroup.

It is of interest to recall that the CSP implies super-rigidity. Hence, if L has CSP, we can produce a counter-example without appealing to the work of [**2**] and [**4**]. So the proof would not use any of the four ingredients (1)-(4) listed in the introduction.

REMARK 4.2. If there is one such uniform lattice L satisfying CSP, then there exists a hyperbolic group without any finite index subgroup and in particular, a non-residually-finite hyperbolic group.

Indeed, L is hyperbolic (but residually finite). For "many" $g \in L$, the normal closure N of g in L is an infinite subgroup of infinite index and $D = L/N$ is also hyperbolic (see [**3**]). Repeating the argument from Remark 4.1, we deduce that D has a finite index subgroup (hence also hyperbolic) with no finite index subgroup. Thus, answering the congruence subgroup problem in the affirmative for one uniform lattice in $Sp(n,1)$ or $F_4^{(-20)}$ would solve the long standing problem of the existence of a non-residually finite hyperbolic group.

References

1. H. Bass and A. Lubotzky, *Nonarithmetic superrigid groups: counterexamples to Platonov's conjecture*, Ann. of Math. **151** (2000), 1151–1173.
2. K. Corlette, *Archimedean superrigidity and hyperbolic geometry*, Ann. of Math. **135** (1992), 165–182.
3. T. Delzant, *Sous-groupes distingueés et quotients des groupes hyperboliques*, Duke Math J. **83** (1996), 661–682.
4. M. Gromov and R. Schoen, *Harmonic maps into singular spaces and p-adic superrigidity for lattices in groups of rank one*, Inst. Hautes Études Sci. Publ. Math. **76** (1992), 165–246.
5. A. Grothendieck, *Représentations linéaires et compactification profinie des groupes discrets*, Manuscripta Math. **2** (1970), 375–396.
6. Jian-Shu Li, *Nonvanishing theorems for the cohomology of certain arithmetic quotients*, J. Reine Angew. Math. **428** (1992), 177–217.
7. J. Milnor, *Introduction to Algebraic-K-Theory*, Princeton University Press, Princeton, NJ, 1971.

8. A. Yu. Ol'shanskiĭ, *On the Bass-Lubotzky question about quotients of hyperbolic groups*, J. Algebra **226** (2000), 807–817.
9. V. Platonov and A. Rapinchuk, *Algebraic Groups and Number Theory*, Pure and Applied Mathematics, Academic Press Inc., Boston, MA, 1994.
10. J.-P. Serre, *Le problème des groupes de congruence pour* SL_2, Ann. of Math **92** (1970), 489–527.
11. D.A. Vogan Jr. and G.J. Zuckerman, *Unitary representations with nonzero cohomology*, Compositio Math. **53** (1984), 51–90.
12. B.A.F. Wehrfritz, *Infinite Linear Groups*, Springer, Berlin, 1973.

INSTITUTE OF MATHEMATICS, HEBREW UNIVERSITY, JERUSALEM 91904, ISRAEL
E-mail address: alexlub@math.huji.ac.il

Contemporary Mathematics
Volume **387**, 2005

On Domain Monotonicity for the Principal Eigenvalue of the Laplacian with a Mixed Dirichlet-Neumann Boundary Condition

Ross G. Pinsky

ABSTRACT. Let $\Omega \subset R^d$ be a bounded domain with smooth boundary and let $A \subset\subset \Omega$ be a smooth, compactly embedded subdomain. Consider the operator $-\frac{1}{2}\Delta$ in $\Omega - \bar{A}$ with the Dirichlet boundary condition at ∂A and the Neumann boundary condition at $\partial\Omega$, and let $\lambda_0(\Omega, A) > 0$ denote its principal eigenvalue. We discuss the question of monotonicity of $\lambda_0(\Omega, A)$ in its dependence on the domain Ω.

The main point of this note is to suggest an open problem that is in the spirit of Chavel's question concerning domain monotonicity for the Neumann heat kernel. Let $\Omega \subset R^d$ be a bounded domain with smooth boundary and let $A \subset\subset \Omega$ be a smooth, compactly embedded subdomain. Consider the operator $-\frac{1}{2}\Delta$ in $\Omega - \bar{A}$ with the Dirichlet boundary condition at ∂A and the Neumann boundary condition at $\partial\Omega$, and let $\lambda_0(\Omega, A) > 0$ denote its principal eigenvalue. If instead of the Neumann boundary condition, one imposes the Dirichlet boundary condition at $\partial\Omega$, then it's easy to see that $\lambda_0(\Omega, A)$ is monotone decreasing in Ω and increasing in A. Similarly, in the case at hand, it is clear that $\lambda_0(\Omega, A)$ is monotone increasing in A; however, the question of monotonicity in Ω is not easily resolved. The impetus for studying this question arose in part from a recent paper [**5**] in which one can find the asymptotic behavior of $\lambda_0(\Omega, A)$ when A is a ball that shrinks to a point, and in part from the work of Chavel [**2**], Kendall [**3**] and Bass and Burdzy [**1**], where a monotonicity property of the Neumann heat kernel was studied.

Let $B_\epsilon(x)$ denote the ball of radius $\epsilon > 0$ centered at $x \in \Omega$, let $|\Omega|$ denote the volume of Ω and let ω_d denote the volume of the unit ball in R^d. In [**5**], it was shown that

$$\lim_{\epsilon \to 0} \epsilon^{2-d} \lambda_0(\Omega, B_\epsilon(x)) = \frac{d(d-2)\omega_d}{2|\Omega|}, \text{ if } d \geq 3, \tag{1-a}$$

2000 *Mathematics Subject Classification.* Primary 60C05; Secondary 60F05.

Key words and phrases. principal eigenvalue, Neumann boundary condition, Dirichlet boundary condition, mixed boundary condition, domain monotonicity.

This research was supported by the Fund for the Promotion of Research at the Technion and by the V.P.R. Fund.

and

$$\lim_{\epsilon\to 0}(-\log\epsilon)\lambda_0(\Omega, B_\epsilon(x)) = \frac{\pi}{|\Omega|}, \text{ if } d=2. \tag{1-b}$$

In particular, then we obtain from (1) the following proposition.

PROPOSITION 1. *If $\Omega_1 \subsetneq \Omega_2$, then for each $x \in \Omega_1$ there exists an $\epsilon_0 = \epsilon_0(x) > 0$ such that*

$$\lambda_0(\Omega_1, B_\epsilon(x)) > \lambda_0(\Omega_2, B_\epsilon(x)), \text{ for } \epsilon \in (0, \epsilon_0).$$

REMARK. A careful look at the proof of (1) shows that $\epsilon_0(x)$ in Proposition 1 may be chosen uniformly for x away from $\partial\Omega_1$.

The question we pose here is this:

QUESTION 1. *Under what "generic" conditions on A, Ω_1 and Ω_2, satisfying $A \subset\subset \Omega_1 \subset \Omega_2$, is it true that $\lambda_0(\Omega_1, A) \geq \Omega_0(\Omega_2, A)$?*

The following well-known probabilistic representation of $\lambda_0(\Omega, A)$ gives some useful intuition for the problem

$$\lim_{t\to\infty} \frac{1}{t} \log P_x(\tau_A > t) = -\lambda_0(\Omega, A), \tag{2}$$

where $x \in \Omega - \bar{A}$, and $P_x(\tau_A > t)$ denotes the probability that a Brownian motion starting from $x \in \Omega$ and normally reflected at the barrier $\partial\Omega$ will not reach the set A by time t.

The following example shows that monotonicity in Ω does not hold in complete generality, and that a certain convexity requirement is reasonable. Let $\Omega_1 \subset R^2$ be the skewed, barbell-shaped region pictured below and defined as follows: let $\Omega_1 = B_1((-2,0)) \cup B_1((2,0)) \cup T_\delta$, where $\delta \in (0, \frac{1}{4})$ and T_δ is the cone-like region bounded by the line connecting $(-\frac{3}{2}, \sqrt{\frac{3}{4}})$ to $(1+\delta, \sqrt{1-(1-\delta)^2})$ and the line connecting $(-\frac{3}{2}, -\sqrt{\frac{3}{4}})$ to $(1+\delta, -\sqrt{1-(1-\delta)^2})$. Let $\Omega_2 = B_4((0,0))$ and let $A = B_\rho((0,0))$, where $\rho > 0$ is chosen sufficiently small so that $A \subset\subset \Omega_1$ for every $\delta \in (0, \frac{1}{4})$. Then $\lambda_0(\Omega_2, A) > 0$ and doesn't depend on δ, but as is easy to understand from (2) and as is not hard to show rigorously, $\lambda_0(\Omega_1, A)$ approaches 0 as $\delta \to 0$.

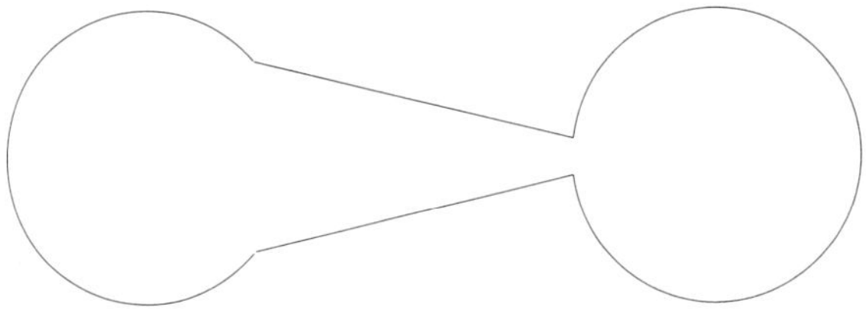

We prove the following result.

THEOREM 1. *Let $\Omega_1 \subset R^d$ be convex and let $B_{r_0}(x_0) \subset\subset \Omega_1$. Let Ω_2 satisfy the following condition: there exists an R such that $\Omega_1 \subset B_R(x_0) \subset \Omega_2$. Then*

$$\lambda_0(\Omega_1, B_{r_0}(x_0)) \geq \lambda_0(\Omega_2, B_{r_0}(x_0)).$$

REMARK. In words, the theorem indicates that monotonicity holds if the inner domain A is a ball, and if it is possible to impose a ball which is concentric to A between the two boundaries, $\partial\Omega_1$ and $\partial\Omega_2$.

The method of proof is similar to that used in [**2**] to prove a certain monotonicity property of the Neumann heat kernel. Before giving the proof, we describe the conjecture raised with regard to the Neumann heat kernel, the results from [**2**] and [**3**], and the example constructed in [**1**].

Consider the operator $-\frac{1}{2}\Delta$ in Ω with the Neumann boundary condition at $\partial\Omega$. Let $p_\Omega(t,x,y)$ denote the corresponding heat kernel. As a function of y, $p_\Omega(t,x,\cdot)$ is the density of the probability distribution corresponding to the position at time t of a Brownian motion starting from x and normally reflected at the barrier $\partial\Omega$; that is,

$$p_\Omega(t,x,y) = P_x(X(t) \in dy). \tag{3}$$

It is well-known that $\lim_{t\to\infty} p_\Omega(t,x,y) = \frac{1}{|\Omega|}$. Thus, if $\Omega_1 \subsetneq \Omega_2$, it follows that for each $x,y \in \Omega_1$, there exists a $t_0 = t_0(x,y)$ such that $p_{\Omega_1}(t,x,y) > p_{\Omega_2}(t,x,y)$, for $t > t_0$. We paraphrase the question posed in [**2**] as follows:

QUESTION 2. *Under what generic conditions on Ω_1 and Ω_2 satisfying $\Omega_1 \subset \Omega_2$ is it true that $p_{\Omega_1}(t,x,y) \geq p_{\Omega_2}(t,x,y)$ for all t and all $x,y \in \Omega_1$?*

It is easy to see that some convexity is needed. Indeed, consider the case that Ω_2 is a square and Ω_1 is a very thin L-shaped subset of Ω_2 running along the lower and left boundaries of Ω_2. Then the intuition gleaned from (3) suggests that for fixed t and thin enough Ω_1, the above inequality will be violated if x is chosen from the lower right hand corner of the square and y is chosen from the upper left hand corner.

Using a straightforward integration by parts, Chavel showed in [**2**] that $p_{\Omega_1}(t,x,y) \geq p_{\Omega_2}(t,x,y)$ when Ω_1 is a convex domain containing x and y, and Ω_2 is a ball centered at either x or y. Building on Chavel's result, Kendall [**3**] gave a nice argument using couplings of reflected Brownian motions to show that $p_{\Omega_1}(t,x,y) \geq p_{\Omega_2}(t,x,y)$ when Ω_1 is a convex domain containing x and y, and Ω_2 is such that one can fit a ball B, centered at either x or y, between the two domains; that it, $\Omega_1 \subset B \subset \Omega_2$. Note that in their results, the conditions on the domains depend on the points x,y.

What happens if one dispenses with Kendall's assumption concerning the fitting of a ball between Ω_1 and Ω_2? In [**1**], Bass and Burdzy gave an example showing that the above inequality does not always hold if one only assumes that Ω_1 is convex. They obtained the reverse inequality for a certain pair of domains Ω_1, Ω_2 and for a certain set of points t_0, x_0, y_0. Bass and Burdzy used probabilistic methods starting from (3).

Returning to the question in the present paper, we propose the following problem:

OPEN PROBLEM. *Give an example of a triple A, Ω_1, Ω_2 such that Ω_1 is convex, $A \subset\subset \Omega_1 \subset \Omega_2$ and $\lambda_0(\Omega_2, A) > \lambda_0(\Omega_1, A)$, or alternatively, show that the reverse inequality always holds.*

We suspect that the inequality $\lambda_0(\Omega_2, A) \leq \lambda_0(\Omega_1, A)$ does not always hold when Ω_1 is convex, but we also suspect that it is more difficult to find an example here than it was for the Neumann heat kernel problem. Bass and Burdzy gave an example for a specific (short) time t_0. In the present situation, of course there is no time parameter, and the eigenvalue depends on the entire infinite time interval $[0,\infty)$.

PROOF OF THEOREM 1. We will prove that

$$\lambda_0(\Omega_1, B_{r_0}(x_0)) \ge \lambda_0(B_R(x_0), B_{r_0}(x_0)) \tag{4}$$

and that

$$\lambda_0(B_R(x_0), B_{r_0}(x_0)) \ge \lambda_0(\Omega_2, B_{r_0}(x_0)). \tag{5}$$

Without loss of generality, we will assume that $x_0 = 0$.

We first consider (4). Let $\phi_{0,1}, \phi_{0,2} > 0$ denote the principal eigenfunctions corresponding respectively to $\lambda_0(\Omega_1, B_{r_0}(0))$ and $\lambda_0(B_R(0), B_{r_0}(0))$. Using the fact that $\frac{1}{2}\Delta\phi_{0,1} = -\lambda_0(\Omega_1, B_{r_0}(0))\phi_{0,1}$ and $\frac{1}{2}\Delta\phi_{0,2} = -\lambda_0(B_R(0), B_{r_0}(0))\phi_{0,2}$ for the first equality below, and using integration by parts, the fact that $\phi_{0,1}, \phi_{0,2}$ vanish on $\partial B_{r_0}(0)$, and the fact that $\nabla\phi_{0,1}\cdot n = 0$ on $\partial\Omega_1$ for the second one, we have

$$\begin{aligned}&(\lambda_0(\Omega_1, B_{r_0}(0)) - \lambda_0(B_R(0), B_{r_0}(0)) \int_{\Omega_1 - B_{r_0}(0)} \phi_{0,1}\phi_{0,2}dx \\ &= \frac{1}{2}\int_{\Omega_1 - B_{r_0}(0)} (\phi_{0,1}\Delta\phi_{0,2} - \phi_{0,2}\Delta\phi_{0,1})dx = \frac{1}{2}\int_{\partial\Omega_1} \phi_{0,1}\nabla\phi_{0,2}\cdot n,\end{aligned}$$

where n is the outward unit normal to Ω_1 at $\partial\Omega_1$. In light of this, to complete the proof of (4) it suffices to show that $\nabla\phi_{0,2}\cdot n \ge 0$ on $\partial\Omega_1$.

By symmetry, $\phi_{0,2}$ depends only on $|x|$, so we will write $\phi_{0,2}(r)$. By the convexity of Ω_1, in order to show that $\nabla\phi_{0,2}\cdot n \ge 0$ on $\partial\Omega_1$, it suffices to show that $\phi_{0,2}$ is nondecreasing in r. If the contrary were true, then there would exist an $r_1 \in (r_0, R)$ such that $\phi'_{0,2}(r_1) = 0$. But then $\phi_{0,2}$ would constitute a positive eigenfunction corresponding to a positive eigenvalue for the Neumann Laplacian in the annulus $B_R(0) - \bar{B}_{r_0}(0)$. However, this contradicts the fact that the only positive eigenfunctions for the Neumann Laplacian in the annulus are the constant functions, corresponding to the eigenvalue 0. Thus, we conclude that $\phi_{0,2}$ is nondecreasing in r.

We now turn to (5). We could use Kendall's construction, but there is no need; (5) follows immediately from the variational formula for $\lambda_0(\cdot,\cdot)$. Indeed, we have

$$\lambda_0(\Omega, A) = \inf \frac{\frac{1}{2}\int_{\Omega-A} |\nabla u|^2 dx}{\int_{\Omega-A} u^2 dx}, \tag{6}$$

where the infimum is over those $0 \ne u \in C^1(\bar{\Omega} - A)$ satisfying $u = 0$ on ∂A. The infimum is attained at $u = u_0 \ge 0$, where u_0 is the eigenfunction corresponding to $\lambda_0(\Omega, A)$. In particular, when $\Omega = B_R(0)$ and $A = B_{r_0}(0)$, this eigenfunction is radially symmetric and positive on the interior of its domain of definition, hence equal to a positive constant c_0 on $\partial B_R(0)$. Let

$$u_1(x) = \begin{cases} u_0(x), & \text{if } x \in \bar{B}_R(0) - B_{r_0}(0) \\ c_0, & \text{if } x \in \Omega_2 - B_R(0). \end{cases}$$

Now u_1 is not in $C^1(\overline{\Omega} - B_{r_0}(0))$ because of a discontinuity in its derivative at $\partial B_R(0)$. However, as is well-known, the infimum in the variational formula (6) may be taken over a wider class of functions which includes u_1. Then by the variational formula,

$$\lambda_0(\Omega_2, B_{r_0}(0)) = \inf \frac{\frac{1}{2}\int_{\Omega_2 - B_{r_0}(0))} |\nabla u|^2 dx}{\int_{\Omega_2 - B_{r_0}(0))} u^2 dx} \leq \frac{\frac{1}{2}\int_{\Omega_2 - B_{r_0}(0))} |\nabla u_1|^2 dx}{\int_{\Omega_2 - B_{r_0}(0))} u_1^2 dx}$$
$$\leq \frac{\frac{1}{2}\int_{B_R(0) - B_{r_0}(0))} |\nabla u_0|^2 dx}{\int_{B_R(0) - B_{r_0}(0))} u_0^2 dx} = \lambda_0(B_R(0), B_{r_0}(0)).$$

□

We conclude with an explicit sufficient condition for monotonicity in the case of smooth, bounded two-dimensional, star-shaped domains. A bounded C^1-domain $\Omega \subset R^2$ is star-shaped with respect to the origin if and only if it can be represented in polar coordinates in the form $\Omega = \{(r, \theta) : r < R(\theta)\}$, where $R > 0$ is a C^1 function on the circle.

THEOREM 2.

(i) *Let Ω_1, Ω_2 and A be two-dimensional, star-shaped C^1-domains with respect to the origin, satisfying $A \subset\subset \Omega_1 \subset \Omega_2$. Let $\Omega_i = \{(r, \theta) : r < R_i(\theta)\}$, $i = 1, 2$. Let*

$$f = \frac{R_2}{R_1}.$$

Assume that

$$\frac{(f')^2}{f} \leq (f-1)^2. \tag{7}$$

Then

$$\lambda_0(\Omega_1, A) \geq \lambda_0(\Omega_2, A).$$

Furthermore, if the strict inequality $R_2 > R_1$ holds, then $\lambda_0(\Omega_1, A) > \lambda_0(\Omega_2, A)$.

(ii) *Let $A \subset R^2$ be a star-shaped C^1-domain with respect to the origin and let $\{\Omega_t,\ t \in [0, 1]\}$ be a one-parameter family of star-shaped domains with respect to the origin, satisfying $A \subset\subset \Omega_s \subset \Omega_t$, for $s < t$, and defined by $\Omega_t = \{(r, \theta) : r < R(\theta, t)\}$, where $R(\theta, t)$ is a C^2-function. If*

$$\left| \left(\frac{R_t}{R} \right)_\theta \right| \leq \frac{R_t}{R},$$

then $\lambda_0(\Omega_t, A)$ is nonincreasing in t.

REMARK. Note that in contrast to Theorem 1, no condition is imposed on A in Theorem 2 beyond its being star-shaped and smooth.

EXAMPLE. Let $R, \bar{R} > 0$ be C^1 functions. Define $\Omega_0 = \{(r, \theta) : r < R(\theta)\}$ and $\Omega_t = \{(r, \theta) : r < R(\theta) + t\bar{R}(\theta)\}$, $t > 0$. Let $A \subset\subset \Omega_0$ be star-shaped with respect to the origin and C^1. From Theorem 2(ii) it follows that $\lambda_0(\Omega_t, A)$ is nonincreasing in t if $|\frac{\bar{R}'}{\bar{R}} - \frac{R'}{R}| \leq 1$. A particular case of this is when $\bar{R} = R$, so that the Ω_t's all have the same shape, differing only in magnification.

PROOF. (i) Assume for now that $R_2 > R_1$. Let $A = \{(r,\theta) : r < a(\theta)\}$. Let $\phi \geq 0$ denote the eigenfunction corresponding to $\lambda_0(\Omega_1, A)$. Of course, $\phi = 0$ on ∂A. Extend ϕ to all of Ω_1 by defining $\phi(x) = 0$ for $x \in A$. Let

$$u(r,\theta) = \begin{cases} \phi(\frac{R_1}{R_2}(\theta)r,\theta), & \text{if } (\frac{R_1}{R_2}(\theta)r,\theta) \in \bar{\Omega}_1 - A \\ 0, & \text{otherwise.} \end{cases}$$

Since $\phi(r,\theta) = 0$ for $r \leq a(\theta)$, it follows that $u(r,\theta) = 0$ for $r \leq \frac{R_2(\theta)}{R_1(\theta)} a(\theta)$. Thus, since $R_2 > R_1$, we have $u = 0$ on $\bar{A}$. Note that u cannot be the eigenfunction corresponding to $\lambda_0(\Omega_2, A)$ because it vanishes at all points in $\Omega_2 - A$ which are sufficiently close to A, whereas the eigenfunction is strictly positive in $\Omega_2 - \bar{A}$. Since the right hand side of (6) is attained only when u is equal to the eigenfunction, it then follows from the variational formula (6) that

$$\lambda_0(\Omega_2, A) < \frac{\frac{1}{2}\int_{\Omega_2 - A} |\nabla u|^2(r,\theta)drd\theta}{\int_{\Omega_2 - A} u^2(r,\theta)drd\theta}. \tag{8}$$

We have

$$|\nabla u|^2(r,\theta) = u_r^2(r,\theta) + \frac{1}{r^2}u_\theta^2(r,\theta) =$$

$$\phi_r^2\left(\frac{R_1}{R_2}(\theta)r,\theta\right)\left(\frac{R_1}{R_2}(\theta)\right)^2 + \frac{1}{r^2}\left(\phi_r\left(\frac{R_1}{R_2}(\theta)r,\theta\right) r \left(\frac{R_1}{R_2}(\theta)\right)' + \phi_\theta\left(\frac{R_1}{R_2}(\theta)r,\theta\right)\right)^2.$$

Thus, making the change of variables $s = \frac{R_1}{R_2}(\theta)r$, we obtain

$$\begin{aligned} \int_{\Omega_2 - A} |\nabla u|^2(r,\theta)drd\theta &= \int_0^{2\pi} d\theta \int_{\frac{R_2(\theta)}{R_1(\theta)}a(\theta)}^{R_2(\theta)} dr \left(u_r^2 + \frac{1}{r^2}u_\theta^2\right)(r,\theta) \\ &= \int_0^{2\pi} d\theta \int_{a(\theta)}^{R_1(\theta)} dr \left[\phi_r^2(s,\theta)\frac{R_1}{R_2}(\theta) + \phi_r^2(s,\theta)\left(\left(\frac{R_1}{R_2}(\theta)\right)'\right)^2 \frac{R_2}{R_1}(\theta)\right. \\ &\quad \left. + \frac{1}{s^2}\phi_\theta^2(s,\theta)\frac{R_1}{R_2}(\theta) + \frac{2}{s}\phi_r(s,\theta)\phi_\theta(s,\theta)\left(\frac{R_1}{R_2}(\theta)\right)'\right]. \end{aligned} \tag{9}$$

Let $\delta(\theta) > 0$ be an as yet unspecified function. Using the inequality $2ab \leq a^2 + b^2$, we have

$$\begin{aligned} &\frac{2}{s}\phi_r(s,\theta)\phi_\theta(s,\theta)\left(\frac{R_1}{R_2}(\theta)\right)' \\ &\quad \leq \frac{1}{\delta(\theta)}\phi_r^2(s,\theta)\left(\left(\frac{R_1}{R_2}(\theta)\right)'\right)^2 \frac{R_2}{R_1}(\theta) + \frac{1}{s^2}\delta(\theta)\phi_\theta^2(s,\theta)\frac{R_1}{R_2}(\theta). \end{aligned} \tag{10}$$

Substituting (10) in (9), we have

$$\int_{\Omega_2 - A} |\nabla u|^2(r,\theta)drd\theta \leq \int_0^{2\pi} d\theta \int_{a(\theta)}^{R_1(\theta)} dr \ [\phi_r^2(s,\theta)C(\theta) + \frac{1}{s^2}\phi_\theta^2(s,\theta)D(\theta)], \tag{11}$$

where $C(\theta) = \frac{R_1}{R_2}(\theta) + \left(\left(\frac{R_1}{R_2}(\theta)\right)'\right)^2 \frac{R_2}{R_1}(\theta)\left(1 + \frac{1}{\delta(\theta)}\right)$

and $D(\theta) = \frac{R_1}{R_2}(\theta)(1 + \delta(\theta))$.

From (11), we conclude that

$$\int_{\Omega_2-A}|\nabla u|^2(r,\theta)drd\theta \le \int_{\Omega_1-A}|\nabla\phi|^2(r,\theta)drd\theta, \text{ if } C(\theta),D(\theta)\le 1. \tag{12}$$

The same change of variables also shows that

$$\begin{aligned}\int_{\Omega_2-A}u^2(r,\theta)drd\theta &= \int_0^{2\pi}d\theta\int_{a(\theta)}^{R_1(\theta)}dr\ \phi^2(s,\theta)\frac{R_2}{R_1}(\theta)\\ &\ge \int_0^{2\pi}d\theta\int_{a(\theta)}^{R_1(\theta)}dr\ \phi^2(s,\theta) = \int_{\Omega_1-A}\phi^2 drd\theta.\end{aligned} \tag{13}$$

Since $\lambda_0(\Omega_1,A) = \frac{\frac{1}{2}\int_{\Omega_1-A}|\nabla\phi|^2(r,\theta)drd\theta}{\int_{\Omega_1-A}\phi^2(r,\theta)drd\theta}$, it follows from (8), (12) and (13) that $\lambda_0(\Omega_1,A) > \lambda_0(\Omega_2,A)$, if $C(\theta),D(\theta)\le 1$. To obtain $D(\theta)\le 1$, we need

$$\delta(\theta) \le \frac{R_2}{R_1}(\theta) - 1. \tag{14}$$

Doing a little algebra, we conclude that to obtain $C(\theta)\le 1$, we need

$$\delta(\theta) \ge \frac{((\frac{R_1}{R_2}(\theta))')^2}{\frac{R_1}{R_2}(\theta) - (\frac{R_1}{R_2}(\theta))^2 - ((\frac{R_1}{R_2}(\theta))')^2}; \tag{15-a}$$

$$\frac{R_1}{R_2}(\theta) - \left(\frac{R_1}{R_2}(\theta)\right)^2 - \left(\left(\frac{R_1}{R_2}(\theta)\right)'\right)^2 > 0. \tag{15-b}$$

Recall that by assumption, R_1 is strictly smaller than R_2. Thus, in order that there exist a function $\delta > 0$ satisfying (14) and (15-a), it is necessary and sufficient that

$$\frac{((\frac{R_1}{R_2}(\theta))')^2}{\frac{R_1}{R_2}(\theta) - (\frac{R_1}{R_2}(\theta))^2 - ((\frac{R_1}{R_2}(\theta))')^2} \le \frac{R_2}{R_1}(\theta) - 1. \tag{16}$$

After a little algebra, one finds that (15-b) and (16) together are equivalent to (15-b) and the inequality

$$\left(\left(\frac{R_1}{R_2}\right)'\right)^2 + 2\left(\frac{R_1}{R_2}\right)^2 \le \frac{R_1}{R_2} + \left(\frac{R_1}{R_2}\right)^3. \tag{17}$$

It is easy to see that if (15-b) does not hold, then (17) does not hold. Thus, (17) is necessary and sufficient to guarantee the existence of a $\delta(\theta) > 0$ such that $C(\theta),D(\theta)\le 1$. Note that (17) is equivalent to (7) with f replaced by $g \equiv \frac{1}{f}$. It is easy to see that (7) holds for f if and only if it holds for $g = \frac{1}{f}$. This completes the proof in the case that $R_1 < R_2$.

Now assume that $R_1 \le R_2$. Let $R_{2,\delta} = R_2 + \delta$, for $\delta > 0$, and let $\Omega_{2,\delta}$ denote the corresponding domain. Using the variational formula (6), it is easy to see that $\lambda_0(\Omega_{2,\delta},A)$ is lower semicontinuous in $\delta > 0$. Since $R_{2,\delta} > R_1$, we have $\lambda_0(\Omega_1,A) > \lambda_0(\Omega_{2,\delta},A)$, and thus $\lambda_0(\Omega_1,A) \ge \lambda_0(\Omega_2,A)$.

(ii) It suffices to show that for each $t \in [0,1)$, there exists an $\epsilon_0 = \epsilon_0(t) > 0$ such that

$$\lambda_0(\Omega_s,A) \le \lambda_0(\Omega_t,A), \text{ for } s \in (t,t+\epsilon_0].$$

For $\epsilon > 0$, we write

$$R(\theta, t+\epsilon) = R(\theta, t) + R_t(\theta, t)\epsilon + \frac{1}{2}R_{tt}(\theta, t)\epsilon^2 + o(\epsilon^2), \text{ as } \epsilon \to 0. \tag{18}$$

The assumption that R is C^2 guarantees that the term $o(\epsilon^2)$ in (18) is uniform in t and θ. We now apply the result in part (i) with R_1 replaced by $R(\cdot, t)$ and R_2 replaced by $R(\cdot, t+\epsilon)$, for $\epsilon > 0$. The function denoted by f in the statement of part (i) is given in the present context by $f_\epsilon(\theta) = \frac{R(\theta,t+\epsilon)}{R(\theta,t)}$. Thus, from (18), we have

$$f_\epsilon(\theta) = 1 + \frac{R_t(\theta, t)}{R(\theta, t)}\epsilon + \frac{1}{2}\frac{R_{tt}(\theta, t)}{R(\cdot, t)}\epsilon^2 + o(\epsilon^2), \text{ as } \epsilon \to 0. \tag{19}$$

Substituting f_ϵ for f in (7) and equating powers of ϵ, one finds that the leading order non-vanishing term is ϵ^2 and that (7) will hold for sufficiently small $\epsilon > 0$ if $((\frac{R_t}{R})_\theta)^2 < (\frac{R_t}{R})^2$, or equivalently, if $|(\frac{R_t}{R})_\theta)| < \frac{R_t}{R}$. The limiting case, $|(\frac{R_t}{R})_\theta)| \le \frac{R_t}{R}$, is treated as in part (i). $\square$

Acknowledgement.

I thank Yehuda Pinchover for suggesting the simple proof that $\phi_{0,2}$ is nondecreasing in r, in place of my more complicated one in the original draft.

References

1. R. F. Bass, and K. Burdzy, *On domain monotonicity of the Neumann heat kernel*, J. Funct. Anal. **116** (1993), 215–224.
2. I. Chavel, *Heat diffusion in insulated convex domains*, J. London Math. Soc. **34** (1986), 473–478.
3. W. S. Kendall, *Coupled Brownian motions and partial domain monotonicity for the Neumann heat kernel*, J. Funct. Anal. **86** (1989), 226–236.
4. R. G. Pinsky, *On the convergence of diffusion processes conditioned to remain in a bounded region for large time to limiting positive recurrent diffusion processes*, Ann. Probab. **13** (1985), 363–378.
5. ———, *Asymptotics of the principal eigenvalue and expected hitting time for positive recurrent elliptic operators in a domain with a small puncture*, J. Funct. Anal. **200** (2003), 177–197.

Department of Mathematics, Technion-Israel Institute of Technology, Haifa, 32000 Israel

E-mail address: `pinsky@math.technion.ac.il`

Contemporary Mathematics
Volume **387**, 2005

The Sharp Form of the Strong Szegő Theorem

Barry Simon

In memoriam, Robert Brooks (1952–2002)

ABSTRACT. Let f be a function on the unit circle and $D_n(f)$ be the determinant of the $(n+1)\times(n+1)$ matrix with elements $\{c_{j-i}\}_{0\le i,j\le n}$ where $c_m = \hat{f}_m \equiv \int e^{-im\theta} f(\theta)\frac{d\theta}{2\pi}$. The sharp form of the strong Szegő theorem says that for any real-valued L on the unit circle with L, e^L in $L^1(\frac{d\theta}{2\pi})$, we have

$$\lim_{n\to\infty} D_n(e^L)e^{-(n+1)\hat{L}_0} = \exp\biggl(\sum_{k=1}^\infty k|\hat{L}_k|^2\biggr)$$

where the right side may be finite or infinite. We focus on two issues here: a new proof when $e^{i\theta}\to L(\theta)$ is analytic and known simple arguments that go from the analytic case to the general case. We add background material to make this article self-contained.

1. Introduction

Let $\{c_m\}_{m=-\infty}^\infty$ be a two-sided sequence of complex numbers. A *Toeplitz matrix* is a finite matrix constant along diagonals:

$$T_{n+1} = \begin{pmatrix} c_0 & c_1 & c_2 & \dots & c_n \\ c_{-1} & c_0 & c_1 & \dots & c_{n-1} \\ \vdots & & & & \vdots \\ c_{-n} & c_{-n+1} & & \dots & c_0 \end{pmatrix} \tag{1.1}$$

It turns out that the natural way to label T is in terms of the Fourier transform of c, that is,

$$f(\theta) = \sum_{m=-\infty}^\infty c_m e^{im\theta} \tag{1.2}$$

on $\partial\mathbb{D}$ ($\mathbb{D} = \{z \mid |z|<1\}$, $\partial\mathbb{D} = \{z \mid |z|=1\}$). f is called the *symbol* of the Toeplitz matrix.

2000 *Mathematics Subject Classification.* 47B35, 05E35, 30E05.

Supported in part by NSF grant DMS-0140592.

One can define a symbol as a distribution so long as $|c_m|$ is polynomially bounded in m, but we will discuss the case where there is a signed measure, $d\mu$, so that

$$c_n = \int e^{-in\theta}\, d\mu(\theta) \equiv \hat{\mu}_n \tag{1.3}$$

As usual, for $f \in L^1(\partial\mathbb{D}, \frac{d\theta}{2\pi})$, we define $\hat{f}_n$ to be the Fourier coefficients of the measure $f\frac{d\theta}{2\pi}$. We will most often discuss the case where $d\mu$ is absolutely continuous, that is, $d\mu = w(\theta)\frac{d\theta}{2\pi}$ and where $w \geq 0$ or even that $w = e^L$.

$D_n(d\mu)$ is the determinant of T_{n+1}. The strong Szegő theorem says that if $L, e^L \in L^1$ with L real, then

$$\log D_n\biggl(e^L\,\frac{d\theta}{2\pi}\biggr) \sim (n+1)\hat{L}_0 + \sum_{k=1}^\infty |k|\,|\hat{L}_k|^2 \tag{1.4}$$

There are a number of remarkable aspects of (1.4). The first term was found in 1915 and the second in 1952. Despite the 37-year break, they were both found by Szegő — the twenty-year old in 1915 [**30**] and the 57-year old in 1952 [**34**]! You might wonder about whether (1.4) is the leading term in a systematic $1/n$ series. In fact, if L is real-valued and if $e^{i\theta} \mapsto L(\theta)$ is analytic in the neighborhood of $\partial\mathbb{D}$, then the error in (1.4) is $O(e^{-Bn})$ — there are no more terms in the series (this follows from (2.21) and (5.17) below). Lest you be shocked by this, we note that for many models in statistical mechanics, the free energy has a volume term, a surface term, and then, if the interaction is short-range, exponentially small errors.

A second remarkable aspect is the subtlety. Why should $\log w$ enter at all, and then in both linear and quadratic terms? There is a fascination with this subject among mathematicians who have extended the result both in the context of function algebras [**38, 15, 22**] and in the context of pseudodifferential operators on manifolds [**39, 14**] (see [**28**] for literally dozens of papers on each aspect).

A third aspect is that there are a remarkable number of applications of this result. Szegő returned to find the second term because of a question raised by Onsager who ran into Toeplitz determinants in his work on the Ising model (see [**23, 5**] for a discussion of this). They enter in the study of some Coulomb gases [**20, 21, 10**] and in electrical engineering applications [**18, 3**]. And they have had a surge of interest recently because of their role in random matrix theory [**24**].

When Szegő [**34**] proved (1.4), he assumed L was $C^{1+\varepsilon}$. There were many papers on this subject which improved this incrementally until Ibragimov [**16**], fifteen years later, proved the following sharp form:

THEOREM 1.1 ([**16, 12**]). *Let L be a real-valued function on $\partial\mathbb{D}$ so that $L, e^L \in L^1(\partial\mathbb{D}, \frac{d\theta}{2\pi})$. Then*

$$\lim_{n\to\infty} D_n\biggl(e^L\,\frac{d\theta}{2\pi}\biggr)e^{-(n+1)\hat{L}_0} = \exp\biggl(\sum_{k=1}^\infty |k|\,|\hat{L}_k|^2\biggr) \tag{1.5}$$

Thus, (1.4) holds whenever the right side makes sense, that is, L in $H^{1/2}$, the Sobolev space of order $\frac{1}{2}$. This should be supplemented with a result of Golinskii-Ibragimov [**12**]:

THEOREM 1.2. *If $d\mu = e^L\frac{d\theta}{2\pi} + d\mu_{\rm s}$ with $d\mu_{\rm s}$ singular is a positive measure on $\partial\mathbb{D}$ and $\lim_{n\to\infty} D_n(d\mu)e^{-(n+1)\hat{L}_0} < \infty$, then $d\mu_{\rm s} = 0$.*

The combination of these two theorems has a spectral theory consequence that links it up to the theme of the conference and to Bob Brooks' interests. As we will see in Section 2, probability measures on $\partial\mathbb{D}$ have associated parameters $\{\alpha_n(d\mu)\}_{n=0}^\infty$ called Verblunsky coefficients. It can be shown using Theorems 1.1, 1.2, and 2.4 that

THEOREM 1.3. *Let $d\mu$ be a probability measure on $\partial\mathbb{D}$ and $\{\alpha_n(d\mu)\}_{n=0}^\infty$ its Verblunsky coefficients. Then the following are equivalent:*

(i) $\sum_{n=0}^\infty n|\alpha_n|^2 < \infty$
(ii) $d\mu_{\rm s} = 0$ *and* $d\mu = e^L \frac{d\theta}{2\pi}$ *where* $\sum_{k=1}^\infty k|\hat{L}_k|^2 < \infty$.

This is one of those gems of spectral theory that give necessary and sufficient conditions relating properties of a measure and its inverse spectral parameters.

We have two main themes in this article. First, we wish to note that despite it taking fifteen years to go from $L \in C^{1+\varepsilon}$ to $L \in H^{1/2}$, there is an elegant and simple argument to do this jump. This combines arguments of Golinskii-Ibragimov [**12**] and Johansson [**17**] whose proofs have "easy" halves that handle opposite sides of the extension. It does not appear to be widely appreciated that their arguments can be combined in this way. In fact, these results reduce (1.4) to the case where the $\hat{L}_k$ decay exponentially, that is, $e^{i\theta} \mapsto L(\theta)$ is real analytic on $\partial\mathbb{D}$.

Second, we have a new proof of (1.4) in this real analytic case that is perhaps less mysterious than the elaborate calculation in Szegő [**34**]. From our point of view, the two terms in (1.4) come from two terms in the Christoffel-Darboux formula for $z = e^{i\theta}$.

While the arguments of [**12, 17**] are simple, they depend on considerable general machinery relating Toeplitz determinants to orthogonal polynomials on the unit circle (OPUC) on the one hand, and to the statistical mechanics of Coulomb gases on the other (both themes go back to Szegő's early work: [**30**] has the Coulomb gas representation and the main point of [**31, 32**] is to discuss the connection to OPUC), so this article, in attempting to be self-contained, provides this background.

Sections 2 and 3 discuss the basics of OPUC. In Section 4, we get the leading term in (1.4), not only for its own sake, but to define in Section 5 the Szegő function which will play a critical role in many aspects of the remainder. These preliminaries allow us to present the Golinskii-Ibragimov half of the extension in Section 6. After proving the Coloumb gas representation in Section 7, we can prove the Johansson half of the extension in Section 8. The final three sections provide the proof of (1.4) in the analytic case: Section 9 has a preliminary proving the Christoffel-Darboux formula, and the last two sections finish the proof.

I have written a comprehensive book on OPUC [**28, 29**] and everything in this paper appears there, but it seemed sensible, given the fact that the material is spread through a long book, to pull out exactly what is needed to prove Ibragimov's theorem.

While (1.4) is sharp in one sense, it is not the end of the story by any means. First, there is a simple argument of Johansson [**17**] that drops the requirement of reality from L: if $e^L, L \in L^1$ and $L \in H^{1/2}$, then an extension (1.4) holds in the sense that

$$e^{-(n+1)\hat{L}_0} D_n\biggl(e^L \frac{d\theta}{2\pi}\biggr) \to \exp\biggl(\sum_{k=1}^\infty |k|\hat{L}_k\hat{L}_{-k}\biggr)$$

There are also subtleties in extending (1.4) to allow matrix-valued symbols, to allow complex w's with nonzero winding number, and to determine the leading behavior when $L \notin L^1$. The reader can consult [**28**] for references on all these issues.

Over the course of studying asymptotics of Toeplitz determinants, I have learned a lot in discussions with Percy Deift, Rowan Killip, and Irina Nenciu, and I would like to thank them for their insights.

Bob Brooks was a substantial mathematician and wonderful person. We lost him too soon. I'm glad to dedicate this article to his memory.

2. Verblunsky Coefficients and Toeplitz Matrices

If c_n are the moments of a measure, μ, it is natural to form the monic orthogonal polynomials, $\Phi_n(z;d\mu)$, defined by

$$\Phi_n(z) = z^n + \text{ lower order} \tag{2.1}$$

$$\langle z^j, \Phi_n \rangle_\mu = 0 \qquad \text{if } j = 0, 1, \dots, n-1 \tag{2.2}$$

where

$$\langle f, g \rangle = \int \overline{f(e^{i\theta})}\, g(e^{i\theta})\, d\mu(\theta) \tag{2.3}$$

is the $L^2(\partial\mathbb{D}, d\mu)$ inner product. In order to form Φ_n for all n, we need the measures $d\mu$ to be nontrivial, that is, not supported on a finite set of points.

The matrix elements $c_{k-\ell} = \int e^{i(\ell-k)\theta}\, d\mu = \langle z^k, z^\ell \rangle_\mu$, so $D_n(d\mu)$ is a Gram determinant. Such determinants allow change of basis, that is, if $P_k(z) = z^k +$ lower order, $\det(\langle P_k, P_\ell \rangle)_{0 \leq k,\ell \leq n} = D_n(d\mu)$. We can take $P_k = \Phi_k$, in which case, $\langle P_k, P_\ell \rangle$ is a diagonal matrix (!), and so,

Theorem 2.1.

$$D_n(d\mu) = \prod_{j=0}^{n} \|\Phi_j\|_{d\mu}^2 \tag{2.4}$$

Φ_j is the orthogonal projection in L^2 of z^j onto the orthogonal complement of $[1, \dots, z^{j-1}]$, the span of $\{1, \dots, z^{j-1}\}$. Since multiplication by z is unitary, $z\Phi_j$ is the projection of z^{j+1} onto $[z, \dots, z^j]^\perp$ while Φ_{j+1} is the projection of z^{j+1} onto $[1, \dots, z^j]^\perp$, so

$$\|\Phi_{j+1}\| \leq \|z\Phi_j\| = \|\Phi_j\| \tag{2.5}$$

Thus, $\|\Phi_j\|$ is decreasing in j. It follows that

Theorem 2.2.

$$\text{(a)} \quad \lim D_n(d\mu)^{1/n+1} = \lim \frac{D_{n+1}(d\mu)}{D_n(d\mu)} = \lim_{n\to\infty} \|\Phi_n\|^2 \tag{2.6}$$

$$\text{(b)} \quad \frac{D_{n+1}}{D_n} \leq \frac{D_n}{D_{n-1}} \tag{2.7}$$

These ideas all go back to Szegő [**31, 32**]. The next idea, which is a set of recursion relations of the Φ_n, was first written down by Szegő [**33**], but the basic parameters occurred in a related context in Verblunsky [**36, 37**]. To state them, we need to define the reversed polynomials

$$\Phi_n^*(z) = z^n\, \overline{\Phi_n(1/\bar{z})} \tag{2.8}$$

THEOREM 2.3. *For any nontrivial measure, $d\mu$, there exists a sequence of numbers $\{\alpha_n\}_{n=0}^\infty$ so that*

$$\Phi_{n+1}(z) = z\Phi_n(z) - \bar{\alpha}_n \Phi_n^*(z) \tag{2.9}$$

Moreover, $\alpha_n \in \mathbb{D}$ and

$$\|\Phi_{n+1}\|^2 = (1 - |\alpha_n|^2)\|\Phi_n\|^2 \tag{2.10}$$

and

$$\|\Phi_n\|^2 = c_0(d\mu) \prod_{j=0}^{n-1} (1 - |\alpha_j|^2) \tag{2.11}$$

REMARKS.

(1) In the next section, we will see that $\mu \mapsto \{\alpha_n\}_{n=0}^\infty$ is one-one for μ's which are normalized.
(2) There is a converse going back to Verblunsky [**36**] (see [**28**] for many other proofs) that the map from probability measures to $\times_{n=0}^\infty \mathbb{D}$ by $\mu \mapsto \{\alpha_n\}_{n=0}^\infty$ is onto.
(3) The α_n are called the *Verblunsky coefficients* for $d\mu$.
(4) Applying $Q^*(z) = z^{n+1}\, \overline{Q(1/\bar{z})}$ to (2.9) yields

$$\Phi_{n+1}^*(z) = \Phi_n^*(z) - \alpha_n z \Phi_n(z) \tag{2.12}$$

(5) The proof below is a variant of one of Atkinson [**1**]. Szegő's proof first proves the Christoffel-Darboux formula (see Section 9) and uses that to prove (2.9).

PROOF. Let $V_n g = z^n \bar{g}$ on $L^2(d\mu)$. V_n is anti-unitary, maps $\mathcal{P}_n$, the polynomials of degree n, to themselves, and maps Φ_n to Φ_n^*. Since Φ_n is the unique element (up to constants) of $\mathcal{P}_n$ orthogonal to $\{1, z, \dots, z^{n-1}\}$ and V_n is anti-unitary, Φ_n^* is the unique element of $\mathcal{P}_n$ orthogonal to $\{V_n 1, \dots, V_n z^{n-1}\} = \{z^n, z^{n-1}, \dots, z\}$.

Now, for $j = 1, \dots, n$,

$$\langle z^j, z\Phi_n \rangle = \langle z^{j-1}, \Phi_n \rangle = 0$$

and clearly, $\langle z^j, \Phi_{n+1} \rangle = 0$. Thus,

$$\langle z^j, \Phi_{n+1} - z\Phi_n \rangle = 0$$

for $j = 1, \dots, n$. Since Φ_n and Φ_{n+1} are monic, $\Phi_{n+1} - z\Phi_n \in \mathcal{P}_n$. So, by the first part of the proof, (2.9) holds for a suitable constant α_n. Thus

$$\alpha_n = -\overline{\Phi_{n+1}(0)} \tag{2.13}$$

Since $\Phi_n^* \perp \Phi_{n+1}$, we have

$$\begin{aligned} \|\Phi_n\|^2 = \|z\Phi_n\|^2 &= \|\Phi_{n+1} + \bar{\alpha}_n \Phi_n^*\|^2 \\ &= \|\Phi_{n+1}\|^2 + |\alpha_n|^2 \|\Phi_n\|^2 \end{aligned}$$

which implies (2.10). (2.10) in turn implies $|\alpha_n| < 1$. (2.11) follows by induction and $\|\Phi_0\|^2 = \|1\|^2 = c_0(d\mu)$. □

This leads to

THEOREM 2.4. *Suppose $\int d\mu = 1$. We have*

$$F(d\mu) = \lim_{n\to\infty} \frac{D_{n+1}(d\mu)}{D_n(d\mu)} = \prod_{j=0}^{\infty} (1 - |\alpha_j|^2) \tag{2.14}$$

where the product always converges although the limit may be zero.

If $F(d\mu) > 0$, then

$$G_n = \frac{D_n}{F^{n+1}} \tag{2.15}$$

obeys

$$G_{n+1} \geq G_n \tag{2.16}$$

The limit always exists (but may be infinite) and is given by

$$G(d\mu) = \lim_{n\to\infty} G_n(d\mu) = \prod_{j=0}^{\infty} (1 - |\alpha_j|^2)^{-j-1} \tag{2.17}$$

REMARKS.

(1) In particular,

$$F > 0 \Leftrightarrow \sum_{n=0}^{\infty} |\alpha_n|^2 < \infty \tag{2.18}$$

$$G < \infty \Leftrightarrow \sum_{n=0}^{\infty} (n+1)|\alpha_n|^2 < \infty \tag{2.19}$$

(2) (2.14) in a sense goes back to Verblunsky [**37**]. (2.17) seems to have only been noted by Baxter [**4**] many years later.

(3) If $c_0 \neq 1$, $F(d\mu) = c_0 \prod_{j=0}^{\infty}(1 - |\alpha_j|^2)$ while $G(d\mu)$ is still given by (2.17). Indeed, $G_n(d\mu) = G_n(d\mu/\int d\mu)$.

(4) (2.7) says $\log D_n$ is concave in n. The monotonicity of G_n is a standard fact about concave functions with finite asymptotics.

PROOF. By (2.4) and then (2.11),

$$\frac{D_{n+1}}{D_n} = \|\Phi_{n+1}\| = \prod_{j=0}^{n} (1 - |\alpha_j|^2) \tag{2.20}$$

(if $\|\Phi_1\|^2 = c_0 = 1$). From this and $|\alpha_j| < 1$, (2.14) is immediate. If F is nonzero,

$$\frac{D_{n+1}}{D_n}\frac{1}{F} = \prod_{j=n+1}^{\infty} (1 - |\alpha_j|^2)^{-1}$$

so that

$$\begin{aligned} G_n &= \frac{c_0}{F} \prod_{k=1}^{n} \left[\frac{D_k}{D_{k-1}}\frac{1}{F}\right] \\ &= \prod_{j=0}^{\infty} (1 - |\alpha_j|^2)^{-\min(n,j)-1} \end{aligned} \tag{2.21}$$

from which $G_{n+1} \geq G_n$ and (2.17) are immediate. $\square$

3. Bernstein-Szegő Approximations

Given a nontrivial probability measure, $d\mu$, with Verblunsky coefficients $\{\alpha_n\}_{n=0}^{\infty}$, we will identify the measures, $d\mu^{(N)}$, with

$$\alpha_j(d\mu^{(N)}) = \begin{cases} \alpha_j(d\mu) & j = 0, 1, \dots, N-1 \\ 0 & j \geq N \end{cases} \tag{3.1}$$

and see $d\mu^{(N)} \to d\mu$ weakly. In many ways, the general proof of the strong Szegő theorem will play off this approximation and the distinct approximation obtained by truncating the Fourier series for L in $e^L \frac{d\theta}{2\pi}$. As a preliminary, we need

THEOREM 3.1. *$\Phi_n(z)$ has all its zeros in $\mathbb{D}$. $\Phi_n^*(z)$ has all its zeros in $\mathbb{C}\backslash\bar{\mathbb{D}}$.*

REMARK. This proof is due to Landau [**19**]. See [**28**] for many other proofs of this fact.

PROOF. Let $\Phi_n(z_0) = 0$. Then $\pi_{n-1} = \Phi_n(z)/(z - z_0)$ is a polynomial of degree $n-1$, and so $\langle \pi_{n-1}, \Phi_n \rangle = 0$. Since $(z - z_0)\pi_{n-1} = \Phi_n$,

$$\begin{aligned} \|\pi_{n-1}\|^2 &= \|z\pi_{n-1}\|^2 = \|z_0\pi_{n-1} + \Phi_n\|^2 \\ &= |z_0|^2 \|\pi_{n-1}\|^2 + \|\Phi_n\|^2 \end{aligned}$$

or

$$(1 - |z_0|^2)\|\pi_{n-1}\|^2 = \|\Phi_n\|^2 \tag{3.2}$$

from which we conclude $|z_0| < 1$, that is, the zeros of Φ_n lie in $\mathbb{D}$. Since $\Phi_n^*(z_0) = 0$ if and only if $\Phi_n(1/\bar{z}_0) = 0$, the zeros of Φ_n^* lie in $\mathbb{C}\backslash\bar{\mathbb{D}}$. □

Given μ, define a measure, $\tilde{\mu}^{(N)}$, by

$$d\tilde{\mu}^{(N)} = \frac{d\theta}{2\pi|\Phi_N(e^{i\theta}; d\mu)|^2} \tag{3.3}$$

which can be defined since $\Phi_N(e^{i\theta}) \neq 0$ for all θ by Theorem 3.1. We have

LEMMA 3.2. *For $j \geq 0$,*

$$\Phi_{N+j}(z; d\tilde{\mu}^{(N)}) = z^j \Phi_N(z; d\mu) \tag{3.4}$$

Moreover,

$$\alpha_\ell(d\tilde{\mu}^{(N)}) = 0 \tag{3.5}$$

for $\ell \geq N$.

PROOF. Since

$$|\Phi_N(e^{i\theta})|^2 = e^{-iN\theta}\Phi_N(e^{i\theta})\Phi_N^*(e^{i\theta}) \tag{3.6}$$

we have for $k \in \mathbb{Z}$, $k \leq N-1$ (includes $k < 0$),

$$\begin{aligned} \int_0^{2\pi} e^{-ik\theta} \frac{\Phi_N(e^{i\theta})}{|\Phi_N(e^{i\theta})|^2} \frac{d\theta}{2\pi} &= \frac{1}{2\pi i} \oint_{|z|=1} z^{N-k} \frac{dz}{z\Phi_N^*(z)} \\ &= 0 \end{aligned}$$

since $N-k-1 \geq 0$ and $\Phi_N^*(z)^{-1}$ is analytic in a neighborhood of $\bar{\mathbb{D}}$ by Theorem 3.1. This says

$$\langle z^\ell, z^j \Phi_N \rangle_{\tilde{\mu}^{(N)}} = 0 \tag{3.7}$$

for $\ell = 0, 1, \dots, N+j-1$. Since $z^j \Phi_N$ is monic, (3.4) holds.

By (2.13), if $\Phi_{k+1}(0) = 0$, then $\alpha_k = 0$, so (3.4) implies $\Phi_{N+j}(0; d\tilde\mu^{(N)}) = 0$, which in turn implies (3.5). □

THEOREM 3.3 (Geronimus [**11**]). *Let $d\mu, d\nu$ be two nontrivial measures on $\partial\mathbb{D}$. Suppose that for some fixed N,*

$$\Phi_N(z; d\mu) = \Phi_N(z; d\nu) \tag{3.8}$$

Then

$$\Phi_j(z; d\mu) = \Phi_j(z; d\nu) \qquad j = 0, 1, \dots, N-1 \tag{3.9}$$

$$\alpha_j(d\mu) = \alpha_j(d\nu) \qquad j = 0, 1, \dots, N-1 \tag{3.10}$$

$$\frac{c_j(d\mu)}{c_0(d\mu)} = \frac{c_j(d\nu)}{c_0(d\nu)} \qquad j = 0, 1, \dots, N \tag{3.11}$$

$$\|\Phi_j\|^2_{d\mu} = c_0(d\mu) c_0(d\nu)^{-1} \|\Phi_j\|^2_{d\nu} \qquad j = 0, 1, \dots, N \tag{3.12}$$

PROOF. (2.9) and (2.12) can be written in matrix form

$$\begin{pmatrix} \Phi_{j+1}(z) \\ \Phi^*_{j+1}(z) \end{pmatrix} = \begin{pmatrix} z & -\bar\alpha_j \\ -z\alpha_j & 1 \end{pmatrix} \begin{pmatrix} \Phi_j(z) \\ \Phi^*_j(z) \end{pmatrix} \tag{3.13}$$

The 2×2 matrix in (3.13) has an inverse $z^{-1} \rho_j^{-2} \left(\begin{smallmatrix} 1 & \bar\alpha_j \\ \alpha_j z & z \end{smallmatrix}\right)$ where

$$\rho_j = (1 - |\alpha_j|^2)^{1/2} \tag{3.14}$$

Thus (3.14) implies the inverse Szegő recursions:

$$\Phi_j(z) = \rho_j^{-2} \frac{[\Phi_{j+1}(z) + \bar\alpha_j \Phi^*_{j+1}]}{z} \tag{3.15}$$

$$\Phi^*_j(z) = \rho_j^{-2} [\Phi^*_{j+1}(z) + \alpha_j \Phi_{j+1}(z)] \tag{3.16}$$

(3.8) implies

$$\alpha_N(d\mu) = \overline{-\Phi_N(0; d\mu)} = \overline{-\Phi_N(0; d\nu)} = \alpha_N(d\nu)$$

and thus, by (3.15) with $j = N-1$, we have (3.9) for $j = N-1$. By iterating this argument, we conclude that (3.9) and (3.10) hold.

We need only prove (3.11) and (3.12), assuming $c_0(d\mu) = c_0(d\nu) = 1$ since with $d\tilde\mu = d\mu / c_0(d\mu)$, we have $c_n(d\mu) = c_0(d\mu) c_n(d\tilde\mu)$ and $\|\Phi_j\|^2_{d\mu} = c_0(d\mu) \|\Phi_j\|^2_{d\tilde\mu}$. (3.12) is immediate from (2.11).

We prove (3.11) when $c_0(d\mu) = c_0(d\nu)$ by induction and noting $\langle 1, \Phi_k \rangle = 0$ yields a formula for c_k in terms of the coefficients of Φ_k and $c_0, c_1, \dots, c_{k-1}$. □

REMARK. The last paragraph of the proof shows that the α's determine the c's and proves that the map of μ to α is one-one.

To succinctly state this section's final result, we introduce

$$\varphi_N(z; d\mu) = \frac{\Phi_N(z; d\mu)}{\|\Phi_N\|} \tag{3.17}$$

the orthonormal polynomials. By (2.11),

$$\varphi_N(z; d\mu) = \prod_{j=0}^{N-1} \rho_j^{-1} c_0(d\mu)^{-1/2} \Phi_N(z; d\mu) \tag{3.18}$$

We define

$$\kappa_n = \left(\prod_{j=0}^{n-1} \rho_j^{-1}\right)(c_0(d\mu))^{-1/2} = \|\Phi_n\|^{-1} \tag{3.19}$$

and

$$\kappa_\infty = \lim_{n\to\infty} \|\Phi_n\|^{-1} \tag{3.20}$$

which exists (but it may be $+\infty$) by $\rho_j < 1$. The infinite product in (3.19), and so κ_∞, is finite if and only if

$$\kappa_\infty < \infty \Leftrightarrow \sum_{j=0}^{\infty} |\alpha_j|^2 < \infty \tag{3.21}$$

We also note the translation of (2.9)/(2.12) from Φ to φ:

$$\varphi_{n+1}(z) = \rho_n^{-1}(z\varphi_n(z) - \bar{\alpha}_n\varphi_n^*(z)) \tag{3.22}$$

$$\varphi_{n+1}^*(z) = \rho_n^{-1}(\varphi_n^*(z) - \alpha_n z\varphi_n^*(z)) \tag{3.23}$$

THEOREM 3.4. *Let $d\mu$ be a nontrivial probability measure on $\partial\mathbb{D}$. Define*

$$d\mu^{(N)} = \frac{d\theta}{2\pi|\varphi_N(e^{i\theta})|^2} \tag{3.24}$$

Then $d\mu^{(N)}$ is a probability measure on $\partial\mathbb{D}$ for which (3.1) holds. As $N \to \infty$, $d\mu^{(N)} \to d\mu$ weakly.

REMARK. We call measures of the form (3.24) *BS measures* and $d\mu^{(N)}$ the *BS approximation.*

PROOF. Since $d\mu^{(N)}$ is a multiple of $d\tilde{\mu}^{(N)}$, we have the bottom half of (3.1) by (3.5). Since (3.4) holds for $j = 0$, Theorem 3.3 and (3.10) imply the top half of (3.1).

Since $\Phi_N = \|\Phi_N\|_\mu \varphi_N$, we clearly have

$$\|\Phi_N\|_{\mu^{(N)}}^2 = \|\Phi_N\|_\mu^2$$

so, by (3.13), $c_0(d\mu^{(N)}) = c_0(d\mu) = 1$, that is, $d\mu_N$ is a probability measure.

By the above and (3.11), we have

$$c_j(d\mu^{(N)}) = c_j(d\mu) \qquad j = 0, 1, \dots, N \tag{3.25}$$

This and its complex conjugate implies that for any Laurent polynomial, f (polynomial in z and z^{-1}),

$$\lim_{N\to\infty} \int f(e^{i\theta})\, d\mu^{(N)} = \int f(e^{i\theta})\, d\mu \tag{3.26}$$

since the left side is equal to the right for N large. Since Laurent polynomials are dense in $C(\partial\mathbb{D})$, (3.26) holds for all f, that is, we have weak convergence. □

We note we have proven that

$$\alpha_j(d\mu^{(N)}) = \begin{cases} \alpha_j & j \le N-1 \\ 0 & j \ge N \end{cases} \tag{3.27}$$

The ideas of this section go back to Geronimus [**11**] and were rediscovered in [**9**] and [**7**]. In particular, the use of inverse recursion to prove Geronimus' theorem is taken from [**7**].

4. Szegő's Theorem

In this section, our main goal is to prove

THEOREM 4.1. *For any* $w \in L^1(\frac{d\theta}{2\pi})$,

$$\lim_{N\to\infty} \frac{1}{N} \log D_N(w) = \int \log w(\theta) \frac{d\theta}{2\pi} \tag{4.1}$$

REMARKS.

(1) Since $\log(x) \le x - 1$ and $w(\theta) \in L^1$, $\int \max(0, \log w(x)) \frac{d\theta}{2\pi} < \infty$, so the integral on the right side of (4.1) is either convergent or diverges to $-\infty$, in which case (4.1) says $D_n(w)^{1/n} \to 0$.

(2) This was conjectured by Pólya [**26**] and proven by the twenty-year old Szegő in 1915 [**30**]. Our proof here is essentially the one Szegő presented in [**31, 32**].

(3) The same result is true for the symbol $d\mu = w(\theta)\frac{d\theta}{2\pi} + d\mu_s$, that is, the limit is independent of $d\mu_s$. This extension was first proven by Verblunsky [**37**]. We will not prove this more general result here ([**28**] has four different proofs in Chapter 2) since it is peripheral to our main result.

The first half of the theorem is a simple use of Jensen's inequality:

PROPOSITION 4.2. *Let* $w = e^L$ *with* $w, L \in L^1$. *Then*

$$\|\Phi_n\|^2_{w\frac{d\theta}{2\pi}} \ge \exp\biggl(\int L(\theta) \frac{d\theta}{2\pi}\biggr) \tag{4.2}$$

In particular,

$$D_n(w) \ge \exp\biggl(\int (n+1) L(\theta) \frac{d\theta}{2\pi}\biggr) \tag{4.3}$$

PROOF. (4.3) follows from (4.2) and (2.4). To prove (4.2), we write

$$\begin{aligned}\|\Phi_n\|^2 = \|\Phi_n^*\|^2 &= \int \exp(2\log|\Phi_n^*(e^{i\theta})| + L(e^{i\theta})) \frac{d\theta}{2\pi} \\ &\ge \exp\biggl(\int [2\log|\Phi_n^*(e^{i\theta})| + L(e^{i\theta})]\biggr) \frac{d\theta}{2\pi} \end{aligned} \tag{4.4}$$

by Jensen's inequality.

By Theorem 3.1, $\log(\Phi_n^*(z))$ is analytic in $\mathbb{D}$, so

$$\begin{aligned}\int \log|\Phi_n^*(e^{i\theta})| \frac{d\theta}{2\pi} &= \operatorname{Re} \int \log(\Phi_n^*(e^{i\theta})) \frac{d\theta}{2\pi} \\ &= \log|\Phi_n^*(0)| = 0\end{aligned}$$

since Φ_n monic implies $\Phi_n^*(0) = 1$. □

The other half of the theorem depends on a variational principle noted by Szegő:

PROPOSITION 4.3. *We have for any* $w \in L^1(\partial\mathbb{D}, \frac{d\theta}{2\pi})$,

$$\lim_{n\to\infty} [D_n(w)]^{1/n} = \inf\biggl\{\int |f(e^{i\theta})|^2 w(\theta) \frac{d\theta}{2\pi} \biggm| f \in H^\infty(\mathbb{D});\ f(0) = 1\biggr\} \tag{4.5}$$

REMARK. $H^\infty(\mathbb{D})$ is the Hardy space of bounded analytic functions on $\mathbb{D}$. By general principles [**8, 27**], for $\frac{d\theta}{2\pi}$ a.e. $e^{i\theta} \in \partial\mathbb{D}$, $\lim_{r\uparrow 1} f(re^{i\theta})$ exists, and that is what we mean by $f(e^{i\theta})$ in (4.5).

PROOF. Since $\|\Phi_n^*\| = \|\Phi_n\|$, by (2.6),

$$\begin{aligned} \text{LHS of (4.5)} &= \lim_{n\to\infty} \|\Phi_n^*\|^2 \\ &= \inf_n \|\Phi_n^*\|^2 \end{aligned} \tag{4.6}$$

by (2.5). By the argument at the start of the proof of Theorem 2.3,

$$\Phi_n^* = \pi_n 1$$

where π_n is the projection in the space of polynomials, $\mathcal{P}_n$, of degree n onto the orthogonal component of the span of $z, z^2, z^3, \dots, z^n$. This span is $\{P \in \mathcal{P}_n \mid P(0) = 0\}$, so by standard geometry,

$$\|\Phi_n^*\|^2 = \inf\{\|P\|^2 \mid P \in \mathcal{P}_n;\ P(0) = 1\} \tag{4.7}$$

proving again that $\|\Phi_n^*\|^2$ is decreasing in n and proving (4.5) if H^∞ is replaced by the set of all polynomials.

To complete the proof, we need only show that for any $f \in H^\infty$ with $f(0) = 1$, there are polynomials $P_\ell(z)$ so that $P_\ell(0) = 1$ and

$$\int |P_\ell(e^{i\theta})|^2 w(\theta)\, \frac{d\theta}{2\pi} \to \int |f(e^{i\theta})|^2 w(\theta)\, \frac{d\theta}{2\pi} \tag{4.8}$$

If f is analytic in a neighborhood of $\mathbb{D}$, the Taylor approximations converge uniformly on $\bar{\mathbb{D}}$, so (4.8) holds. For general f, by the dominated convergence theorem,

$$\lim_{r\uparrow 1} \int |f(re^{i\theta})|^2 w(\theta)\, \frac{d\theta}{2\pi} = \int |f(e^{i\theta})|^2 w(\theta)\, \frac{d\theta}{2\pi}$$

so, by a two-step approximation, we find P_ℓ's so (4.8) holds. □

PROOF OF THEOREM 4.1. We will prove that for any $\varepsilon > 0$, there is an f in H^∞ with $f(0) = 1$ and

$$\int |f(e^{i\theta})|^2 w(\theta)\, \frac{d\theta}{2\pi} \le \exp\biggl(\int \log(w(\theta) + \varepsilon)\, \frac{d\theta}{2\pi}\biggr) \tag{4.9}$$

so taking $\varepsilon \downarrow 0$ yields the opposite inequality to (4.3).

Define

$$g(z) = \exp\biggl(-\int \log(w(\theta) + \varepsilon)\biggl(\frac{e^{i\theta} + z}{e^{i\theta} - z}\biggr) \frac{d\theta}{4\pi}\biggr) \tag{4.10}$$

and $f(z) = g(z)/g(0)$, so $f(0) = 1$. Moreover, $|g(z)| \le \varepsilon^{-1/2}$ by the fact that the Poisson kernel

$$P_r(\theta, \varphi) = \operatorname{Re}\biggl(\frac{e^{i\theta} + re^{i\varphi}}{e^{i\theta} - re^{i\varphi}}\biggr)$$

is nonnegative and $\int \frac{d\theta}{2\pi} P_r(\theta, \varphi) = 1$. By standard maximal function arguments (see **[27]**), $|g(e^{i\theta})| = |w(\theta) + \varepsilon|^{-1/2}$, so

$$\int |f(e^{i\theta})|^2 w(\theta)\, \frac{d\theta}{2\pi} \le g(0)^{-2} = \text{RHS of (4.9)}$$

□

5. The Szegő Function

When $w(\theta) = e^{L(\theta)}$ with $L \in L^1$, Szegő [**31, 32**] introduced a natural function, $D(z)$ on $\mathbb{D}$, which will play a critical role in several places below:

$$D(z) = \exp\biggl(\int \biggl(\frac{e^{i\theta}+z}{e^{i\theta}-z}\biggr) L(\theta)\, \frac{d\theta}{4\pi}\biggr) \tag{5.1}$$

Do not confuse D_n and $D(z)$. Both symbols are standard, but the objects are very different.

THEOREM 5.1.

(a) *If* (2.18) *holds, then for* $|z| < 1$,

$$D(z) = D(0) \exp\biggl(\sum_{k=1}^\infty \hat{L}_k z^k\biggr) \tag{5.2}$$

$$D(0) = \biggl[c_0\biggl(w\,\frac{d\theta}{2\pi}\biggr)\biggr]^{1/2} \prod_{n=0}^\infty (1-|\alpha_n|^2)^{1/2} \tag{5.3}$$

(b) $D(z)$ *lies in* $H^2(\mathbb{D})$.

(c) $\lim_{r\uparrow 1} D(re^{i\theta}) \equiv D(e^{i\theta})$ *exist for a.e.* θ *and*

$$|D(e^{i\theta})|^2 = w(\theta) \tag{5.4}$$

(d) D *is nonvanishing on* $\mathbb{D}$.

PROOF.

(a) We get (5.2) and

$$D(0) = \exp(\tfrac{1}{2}\, L_0) \tag{5.5}$$

from (5.1) and

$$\frac{e^{i\theta}+z}{e^{i\theta}-z} = 1 + 2\sum_{j=0}^\infty (e^{-i\theta} z)^n$$

uniformly in $e^{i\theta} \in \partial\mathbb{D}$ and $z \in \{|z| < r\}$. (5.3) then follows from (2.14), (4.1), and (5.5).

(b) Let $D^{(M)}(z)$ be given by (5.1) with $L(\theta)$ replaced by $\min(L(\theta), M)$. Then $D^{(M)} \in H^\infty$ and $|D^{(M)}(e^{i\theta})|^2 = \min(w(\theta), e^{2M})$,

$$\begin{aligned} \sup_{0<r<1} \int_0^{2\pi} |D^{(M)}(re^{i\theta})|^2\, \frac{d\theta}{2\pi} &= \int_0^{2\pi} \min(w(\theta), e^{2M})\, \frac{d\theta}{2\pi} \\ &\leq \int_0^{2\pi} w(\theta)\, \frac{d\theta}{2\pi} \end{aligned}$$

Thus, taking $M \to \infty$, we see $D \in H^2$.

(c) is immediate from properties of boundary values of the Poisson integral in (5.1).

(d) is trivial from (5.1).

□

The following simple but powerful L^2 calculation goes back to Szegő [**31, 32**] (it has a version when $d\mu_s \neq 0$; see [**28**]):

THEOREM 5.2. *Let* $d\mu = w\frac{d\theta}{2\pi}$ *where* $w = e^L$, $L \in L^1$. *Then, as* $n \to \infty$,

(i)

$$\int |D\varphi_n^*(e^{i\theta}) - 1|^2 \frac{d\theta}{2\pi} \to 0 \tag{5.6}$$

(ii)

$$\int |\varphi_n^*(e^{i\theta}) - D^{-1}(e^{i\theta})|^2 \, d\mu \to 0 \tag{5.7}$$

(iii)

$$\varphi_n^*(z) \to D(z)^{-1} \tag{5.8}$$

uniformly on compact subsets of $\mathbb{D}$.

PROOF.

(i) By (5.4), $\int |D\varphi_n^*|^2 \frac{d\theta}{2\pi} = \int |\varphi_n^*|^2 \, d\mu = 1$, so (5.6) is equivalent to

$$\operatorname{Re} \int \varphi_n^*(e^{i\theta}) D(e^{i\theta}) \frac{d\theta}{2\pi} \to 1 \tag{5.9}$$

$D\varphi_n^*$ is in H^2, so the Cauchy integral formula applies, that is,

$$\begin{aligned} \text{LHS of (5.9)} &= \varphi_n^*(0) D(0) \\ &= \kappa_n \kappa_\infty^{-1} \end{aligned} \tag{5.10}$$

since $\Phi_n^*(0) = 1$, $\varphi_n = \kappa_n \Phi_n$ and (5.3) and (3.19) imply $D(0) = \kappa_\infty^{-1}$. (5.10) implies (5.9).

(ii) is immediate from (5.6) and $d\mu = w\frac{d\theta}{2\pi} = |D|^2 \frac{d\theta}{2\pi}$ by (5.4).

(iii) (5.6) says $D\varphi_n^* \to 1$ in $\mathbb{H}^2$ and so, a fortiori, uniformly on compact subsets of $\mathbb{D}$.

□

We need to extend this result to a neighborhood of $\bar{\mathbb{D}}$ when L is real analytic. As a preliminary, we note:

LEMMA 5.3. *Let* $d\mu = w\frac{d\theta}{2\pi}$ *be a probability measure with* $\log w \in L^1(\frac{d\theta}{2\pi})$. *Then*

$$\alpha_n = -\kappa_\infty \int \overline{\Phi_{n+1}(e^{i\theta})} \, D(e^{i\theta})^{-1} \, d\mu(\theta) \tag{5.11}$$

PROOF. We will prove for $m \geq n+1$ that

$$\alpha_n = -\kappa_m \int \overline{\Phi_{n+1}(e^{i\theta})} \, \varphi_m^*(e^{i\theta}) \, d\mu(\theta) \tag{5.12}$$

from which (5.11) follows from (5.7).

φ_m^* is orthogonal to $\{z^\ell\}_{\ell=1}^m$, so if P is any polynomial of degree at most m with $P(0) = 0$,

$$\int \overline{P(e^{i\theta})} \, \varphi_m^*(e^{i\theta}) \, d\mu(\theta) = 0 \tag{5.13}$$

(5.12) follows from (5.13) by taking

$$P(z) = \bar{\alpha}_n \varphi_m^*(z) + \kappa_m \Phi_{n+1}(z)$$

which has $P(0) = \bar{\alpha}_n \kappa_m + \kappa_m(-\bar{\alpha}_n) = 0$. □

The following is due to Nevai-Totik [**25**]; it is needed in Section 11:

THEOREM 5.4. *Suppose that* $d\mu = e^L \frac{d\theta}{2\pi}$ *and* $e^{i\theta} \mapsto L(\theta)$ *is analytic in a neighborhood of* $\partial\mathbb{D}$. *Then* $\varphi_n^*(z) \to D(z)^{-1}$ *uniformly in a neighborhood of* $\bar{\mathbb{D}}$. *Moreover, the Verblunsky coefficients obey* $|\alpha_n| \le C_2 e^{-An/2}$ *for some* $A > 0$.

PROOF. Analyticity says $|\hat{L}_k| \le Ce^{-A|k|}$ for some $A > 0$. So, by (5.2), $D(z)$ is analytic and nonvanishing in a disk of radius e^A. In particular, if

$$D(z)^{-1} = \sum_{j=0}^{\infty} d_{j,-1} z^j \tag{5.14}$$

then

$$|d_{j,-1}| \le C_1 e^{-A|j|/2} \tag{5.15}$$

Plug (5.14) into (5.11) and note that

$$\int \overline{\Phi_{n+1}(e^{i\theta})}\, e^{ik\theta}\, d\mu(\theta) = 0$$

for $k = 0, 1, \dots, n$. Thus

$$\begin{aligned}
|\alpha_n| &\le \kappa_\infty \sum_{k=n+1}^{\infty} |d_{k,-1}| \left| \int \overline{\Phi_{n+1}(e^{i\theta})}\, e^{ik\theta}\, d\mu \right| \\
&\le \kappa_\infty \|\Phi_{n+1}\| \sum_{k=n+1}^{\infty} |d_{k,-1}| \\
&\le \kappa_\infty \sum_{k=n+1}^{\infty} |d_{k,-1}| \qquad (5.16)
\end{aligned}$$

since $\|\Phi_{n+1}\| \le 1$. So, by (5.15),

$$|\alpha_n| \le C_2 e^{-A|n|/2} \tag{5.17}$$

By (2.12) and $|\Phi_n^*(e^{i\theta})| = |\Phi_n(e^{i\theta})|$, we have

$$\begin{aligned}
\sup_\theta |\Phi_{n+1}^*(e^{i\theta})| &\le (1 + |\alpha_n|) \sup_\theta |\Phi_n^*(e^{i\theta})| \\
&\le \prod_{j=0}^{n} (1 + |\alpha_j|) \\
&\le \exp\biggl(\sum_{j=0}^{\infty} |\alpha_j| \biggr) = C_4 < \infty \qquad (5.18)
\end{aligned}$$

by iteration. $C_4 < \infty$ follows from (5.17). Since Φ_n^* is analytic, we get

$$\sup_{z \in \bar{\mathbb{D}}} |\Phi_n^*(z)| \le C_4 \tag{5.19}$$

and thus, since $\Phi_n^*(z) = z^n\, \overline{\Phi_n(1/\bar{z})}$, we get

$$z \in \mathbb{C} \backslash \mathbb{D} \Rightarrow |\Phi_n(z)| \le C_4 |z|^n \tag{5.20}$$

Returning to (2.12),

$$\sum_{n=0}^{\infty} |\Phi_{n+1}^*(z) - \Phi_n^*(z)| \leq \sum_{n=0}^{\infty} |\alpha_n|\,|\Phi_n(z)|$$
$$\leq C_4 C_2 \sum_{n=0}^{\infty} |ze^{-A/2}|^n$$

showing Φ_n^*, and so φ_n^*, converges uniformly in $\{z \mid |z| \leq e^{A/4}\}$. Since the limit is D^{-1} in $\mathbb{D}$, it is D^{-1} in this larger disk. □

Finally, in terms of D, we want to rewrite the second term in the Szegő asymptotic formula (1.4):

THEOREM 5.5. *Let $d\mu = e^{L(\theta)}\frac{d\theta}{2\pi}$ with $L \in L^1$. Let $\hat{L}_k$ be given by (1.3). Then*

$$\sum_{k=1}^{\infty} k|\hat{L}_k|^2 = \frac{1}{\pi}\int_{|z|\leq 1} |D(z)|^{-2}\left|\frac{\partial D}{\partial z}\right|^2 d^2z \tag{5.21}$$

where both sides can be infinite.

PROOF. (5.21) follows by taking $r \uparrow 1$ in

$$\sum_{k=1}^{\infty} k\,|\hat{L}_k|^2 r^{2k} = \frac{1}{\pi}\int_{|z|\leq r} |D(z)|^{-2}\left|\frac{\partial D}{\partial z}\right|^2 d^2z \tag{5.22}$$

(by using monotone convergence). To prove (5.22), note that by (5.2),

$$\log\left[\frac{D(z)}{D(0)}\right] = \sum_{k=1}^{\infty} \hat{L}_k z^k$$

converges uniformly in $|z| < R$ and that $|D|^{-2}|\frac{\partial D}{\partial z}|^2 = |\frac{\partial}{\partial z}\log D(z)|^2$. Thus (5.22) follows from

$$\frac{1}{\pi}\int_{|z|\leq r} \bar{z}^{k-1} z^{\ell-1}\, d^2z = k^{-1}\delta_{k\ell} r^{2k}$$

and

$$\frac{\partial}{\partial z}\log D(z) = \sum_{k=1}^{\infty} k\hat{L}_k z^{k-1}$$

□

6. Extending the Strong Szegő Theorem, Part I

With the Szegő function and Fatou's lemma, we have the tools for the Golinskii-Ibragimov [**12**] half of the extension theorem:

THEOREM 6.1 ([**12**]). *Suppose for any BS measure, $d\mu$, we know that*

$$G(d\mu) = \exp\left(\sum_{k=1}^{\infty} k|\hat{L}_k|^2\right) \tag{6.1}$$

Then for any measure $d\mu = e^L\frac{d\theta}{2\pi}$ with $L \in L^1$, we have

$$G(d\mu) \geq \exp\left(\sum_{k=1}^{\infty} k|\hat{L}_k|^2\right) \tag{6.2}$$

REMARK. In particular, $G < \infty \Rightarrow \sum k|\hat{L}_k|^2 < \infty$.

PROOF. Let $d\mu^{(N)}$ be the BS approximations to $d\mu$. Let $D^{(N)}(z)$ be the D function for $d\mu^{(N)}$ and let

$$M^{(N)} = \frac{1}{\pi} \int_{|z|\leq 1} |D^{(N)}(z)|^{-2} \left| \frac{\partial D^{(N)}}{\partial z} \right|^2 d^2 z \tag{6.3}$$

and similarly for M and D.

By (2.17) and (3.27),

$$G(d\mu^{(N)}) = \prod_{j=0}^{N-1} (1 - |\alpha_j|^2)^{-j-1}$$

so $G(d\mu^{(N)})$ is monotone increasing to $G(d\mu)$, that is,

$$G(d\mu) = \lim_{N\to\infty} G(d\mu^{(N)}) \tag{6.4}$$

On the other hand, since $\varphi_{N+j}(z, d\mu^{(N)}) = z^j \varphi_N(z, d\mu^{(N)})$, $\varphi^*_{N+j}(z, d\mu^{(N)}) = \varphi^*_N(z, d\mu^{(N)}) = \varphi^*_N(z, d\mu)$, so by (5.8),

$$D^{(N)}(z) = \varphi^*_N(z, d\mu)^{-1} \tag{6.5}$$

and thus, by (5.8) again,

$$D^{(N)}(z) \to D(z) \tag{6.6}$$

uniformly on compacts. By analyticity, the same thing is true for derivatives, so

$$|D(z)|^{-2} \left| \frac{\partial D}{\partial z} \right|^2 = \lim_{N\to\infty} |D^{(N)}(z)|^{-2} \left| \frac{\partial D^{(N)}}{\partial z} \right|^2$$

Thus, by Fatou's lemma,

$$M \leq \liminf M^{(N)} \tag{6.7}$$

By hypothesis,

$$G(d\mu^{(N)}) = \exp(M^{(N)}) \tag{6.8}$$

(6.8), (6.7), and (6.4) imply that

$$G(d\mu) \geq \exp(M)$$

which is (6.2). □

7. The Coulomb Gas Representation

In this section, we will provide an integral formula for D_n that will be critical in the next section. This formula appeared in Szegő's first paper [**30**] on asymptotics of Toeplitz determinants. He used it there for a minor technical purpose — and for us, too, it plays a relatively minor role. That said, it plays a central role in two proofs of the strong Szegő theorem and several applications.

While we will not pursue the Coulomb gas picture, if one uses $|z_j - z_k| = \exp(\log|z_j - z_k|)$, the formula we give for D_n says it is the partition function of a two-dimensional gas, and this point of view is the basis of Johansson's proof [**17**]. It also explains some interest in Toeplitz matrices in the physics literature [**20, 21, 10**].

Using Weyl's relation that Haar measure restricted to the classes of $\mathbb{U}(n+1)$, the group of $n \times n$ unitary matrices, is essentially $(2\pi)^{-n-1}[(n+1)!]^{-1} \prod_{k<j} |e^{i\theta_k} - e^{i\theta_j}|^2 \, d\theta_0 \dots d\theta_n$ with $\{e^{i\theta_j}\}_{j=0}^n$ the eigenvalues of $U \in \mathbb{U}(n+1)$, one can use the formula below to rewrite D_n as an integral over $\mathbb{U}(n+1)$. This is both the starting point of the Bump-Diaconis [**6**] proof and of the many applications of Toeplitz matrices in the theory of random matrices [**24**].

It is, of course, well-known that $\prod_{k<j}(z_k - z_j)$ is a Vandermonde determinant. Expanding two such products, we get

$$\left| \prod_{0 \le k < j \le n} (z_k - z_j) \right|^2 = \sum_{\pi,\sigma \in \Sigma_{n+1}} (-1)^\pi (-1)^\sigma \prod_{j=0}^n \bar{z}_j^{\pi(j)} z_j^{\sigma(j)} \tag{7.1}$$

where Σ_{n+1} is the permutations of $\{0, \dots, n\}$ to itself.

Setting $z_j = e^{i\theta_j}$ and integrating $d\mu(\theta_0) \dots d\mu(\theta_n)$, we get that

$$\int |\pi(z_k - z_j)|^2 \, d\mu(\theta_0) \dots d\mu(\theta_n) = \sum_{\pi,\sigma \in \Sigma_{n+1}} (-1)^\pi (-1)^\sigma \prod_{j=0}^n T_{\pi(j)\sigma(j)} \tag{7.2}$$

where $T_{k\ell}$ are the matrix elements of the $(n+1) \times (n+1)$ Toeplitz matrix. Now

$$\prod_{j=0}^n T_{\pi(j)\sigma(j)} = \prod_{j=0}^n T_{j(\sigma\pi^{-1})(j)}$$

and $(-1)^\pi (-1)^\sigma = (-1)^{\sigma\pi^{-1}}$. Thus, summing over σ for π fixed and then over π, we see the right side of (7.2) is $(n+1)! D_n(d\mu)$. Specializing to $d\mu = e^L \frac{d\theta}{2\pi}$, we have proven

THEOREM 7.1 (Coulomb Gas Representation for D_n). *Let e^L, $L \in L^1(\partial\mathbb{D}, \frac{d\theta}{2\pi})$. Then with $z_k = e^{i\theta_k}$,*

$$D_n\left(e^L \frac{d\theta}{2\pi}\right) = [(n+1)!]^{-1} \int_{\partial\mathbb{D}^{n+1}} \left| \prod_{0 \le k < j \le n} (z_k - z_j) \right|^2 e^{\sum_{j=0}^n L(\theta_j)} \prod_{j=0}^n \frac{d\theta_j}{2\pi}$$

8. Extending the Strong Szegő Theorem, Part II

The Coulomb representation and Fatou's lemma are the tools for Johansson's half of the extension theorem. A measure of the form $e^L \frac{d\theta}{2\pi}$ where L is a real Laurent polynomial (i.e., $\sum_{k=-n}^n \hat{L}_k e^{ik\theta}$ with $\hat{L}_{-k} = \bar{\hat{L}}_k$) is called a *GI measure* (after its early use in [**12**]). If $d\mu = e^L \frac{d\theta}{2\pi}$ with $L \in L^1$, we define the *GI approximations*, $d\mu_{(N)}$, by

$$d\mu_{(N)} = \exp\left(\sum_{|k| \le N} \hat{L}_k e^{ik\theta} \right) \frac{d\theta}{2\pi} \equiv \exp(L_{(N)}(\theta)) \frac{d\theta}{2\pi} \tag{8.1}$$

THEOREM 8.1 ([**17**]). *Suppose for any GI measure, $d\mu$, we know that*

$$G(d\mu) = \exp\left(\sum_{k=1}^\infty k |\hat{L}_k|^2 \right) \tag{8.2}$$

Then for any measure $d\mu = e^L \frac{d\theta}{2\pi}$ with $L \in L^1$, we have

$$G(d\mu) \le \exp\left(\sum_{k=1}^\infty k |\hat{L}_k|^2 \right) \tag{8.3}$$

REMARK. In particular, if $\sum_{k=1}^\infty k|\hat L_k|^2 < \infty$, then $G(d\mu) < \infty$.

PROOF. Without loss (since we can multiply $d\mu$ by a constant), suppose $\hat L_0 = 0$. Given $d\mu = e^L \frac{d\theta}{2\pi}$, let $d\mu_{(N)}$ be its GI approximations and let

$$C_{(N),m} = [(n+1)!]^{-1} \int_{\mathbb{D}^{m+1}} \prod_{0\le j<k\le m} |z_j - z_k|^2 e^{\sum_{j=0}^m L_{(N)}(\theta_j)} \prod_{j=0}^m \frac{d\theta_j}{2\pi} \tag{8.4}$$

so, of course, $C_{(N),m} = D_m(d\mu_{(N)})$ by Theorem 7.1. Let C_m be the integral with $L_{(N)}$ replaced by L.

There is nothing to prove in (8.3) if $\sum_{k=1}^\infty k|\hat L_k|^2 = \infty$, so suppose the sum is finite. Thus, $L_{(N)} \to L$ in L^2, and so, we can find a subsequence $L_{(N_j)} \to L$ pointwise. By Fatou's lemma, for each m,

$$C_m \le \liminf_{j\to\infty} C_{(N_j),m} \tag{8.5}$$

By Theorem 7.1, (2.16), $\hat L_0 = 0$, and the assumption for GI measures,

$$\begin{aligned} C_{(N_j),m} &= G_m(d\mu_{(N_j)}) \\ &\le G(d\mu_{(N_j)}) \\ &= \exp\biggl(\sum_{k=1}^{N_j} k|\hat L_k|^2\biggr) \\ &\le \exp\biggl(\sum_{k=1}^{\infty} k|\hat L_k|^2\biggr) \end{aligned}$$

By (8.5) and $\hat L_0 = 0$,

$$G_m(d\mu) = C_m \le \exp\biggl(\sum_{k=1}^\infty k|\hat L_k|^2\biggr)$$

Taking $m \to \infty$ yields (8.3). □

Theorems 6.1 and 8.1 reduce the proof of the sharp form of the strong Szegő theorem to proving the result for BS and GI measures. In both cases, $d\mu = e^L \frac{d\theta}{2\pi}$ with $e^{i\theta} \mapsto L(\theta)$ analytic in a neighborhood of $\partial\mathbb{D}$. In the last three sections, we will prove the strong Szegő theorem in that case.

9. The CD Formula

If u_1, u_2 solve $-u_j'' + Vu_j = \lambda_j u_j$ with $u_j(0) = 0$, then $(\bar\lambda_1 - \lambda_2)\bar u_1 u_2 = (\bar u_1 u_2' - u_2 \bar u_1')'$, so

$$(\bar\lambda_1 - \lambda_2) \int_0^a \overline{u_1(x)}\, u_2(x)\, dx = \bar u_1(a) u_2'(a) - \bar u_1'(a) u_2(a) \tag{9.1}$$

The Christoffel-Darboux formula is just the analog of this Wronskian relation that many undergraduates learn!

THEOREM 9.1 (CD Formula). *Let*

$$K_n(z,\zeta) = \sum_{j=0}^n \overline{\varphi_j(\zeta)}\, \varphi_j(z) \tag{9.2}$$

Then

$$K_n(z,\zeta) = \frac{\overline{\varphi_{n+1}^*(\zeta)}\,\varphi_{n+1}^*(z) - \overline{\varphi_{n+1}(\zeta)}\,\varphi_{n+1}(z)}{1-\bar{\zeta}z} \tag{9.3}$$

$$= \frac{\overline{\varphi_n^*(\zeta)}\,\varphi_n^*(\zeta) - z\bar{\zeta}\,\overline{\varphi_n(\zeta)}\,\varphi_n(z)}{1-\bar{\zeta}z} \tag{9.4}$$

PROOF. By using (3.22)/(3.23) and their conjugates to write φ_{n+1} and φ_{n+1}^* in terms of φ_n and φ_n^*, we get

$$\overline{\varphi_{n+1}^*(\zeta)}\,\varphi_{n+1}^*(z) - \overline{\varphi_{n+1}(\zeta)}\,\varphi_{n+1}(z) = \overline{\varphi_n^*(\zeta)}\,\varphi_n^*(z) - z\bar{\zeta}\,\overline{\varphi_n(\zeta)}\,\varphi_n(z) \tag{9.5}$$

where we used $\rho_n^{-2}(1-|\alpha_n|^2)=1$, and that the $\varphi^*\varphi$ cross terms cancel.

First of all, (9.5) says that (9.3) is equivalent to (9.4). In addition, if we let $Q_n(z,\zeta)$ be what we know is the common value of the right sides of (9.3) and (9.4), then subtracting (9.3) for $n-1$ from (9.4) for n, we see that

$$\begin{aligned} Q_n(z,\zeta) - Q_{n-1}(z,\zeta) &= (1-z\bar{\zeta})\left[\frac{\overline{\varphi_n(\zeta)}\,\varphi_n(z)}{1-z\bar{\zeta}}\right] \\ &= K_n(z,\zeta) - K_{n-1}(z,\zeta) \end{aligned}$$

so we need only prove $Q_0(z,\zeta) = K_0(z,\zeta)$. Since (9.4) for $n=0$ is 1 and that is $K_0(z,\zeta)$, (9.3) is proven. □

COROLLARY 9.2. *For* $e^{i\theta}\in\partial\mathbb{D}$,

$$K_n(e^{i\theta},e^{i\theta}) \equiv \sum_{j=0}^{n} |\varphi_j(e^{i\theta})|^2 = -\frac{\partial}{\partial r}\,|\varphi_{n+1}^*(re^{i\theta})|^2\bigg|_{r=1} + (n+1)|\varphi_{n+1}^*(e^{i\theta})|^2 \tag{9.6}$$

PROOF. In (9.3), take $z=\zeta=re^{i\theta}$ and take $r\uparrow 1$. If we note $|\varphi_{n+1}(re^{i\theta})|^2 = r^{2n+2}|\varphi_{n+1}^*(r^{-1}e^{i\theta})|^2$, we get two terms: the one from the $-(r^{2n+2}-1)/(1-r^2)$ gives the $(n+1)|\varphi_{n+1}^*(e^{i\theta})|^2$ term in (9.6), and the other terms give the derivative in (9.6). □

The CD formula for OPUC is due to Szegő [**33**]. Our proof is taken from Golinskii [**13**].

REMARK. Integrating both sides of (9.6) yields

$$\int \frac{\partial}{\partial r}\,|\varphi_{n+1}^*(re^{i\theta})|^2\bigg|_{r=1} d\mu(\theta) = 1 \tag{9.7}$$

10. The Feynman-Hellman Theorem for Toeplitz Determinants

Here we want to begin with a simple but useful formula for $\frac{\partial}{\partial\lambda}\|\Phi_n\|_{d\mu_\lambda}^2$, where the measure $d\mu_\lambda$ depends smoothly on λ. This appeared in [**29**], strongly motivated by closely related ideas of Baik et al. [**2**].

THEOREM 10.1. *Let* $d\mu_\lambda$ *be a family of positive measures on* $\partial\mathbb{D}$ *which are* C^1 *in* λ *in the sense that* $\lambda\mapsto c_n(d\mu_\lambda)$ *is* C^1 *for each* n. *Then* $\|\Phi_n(\,\cdot\,,d\mu_\lambda)\|_{\mu_\lambda}$ *is* C^1 *and*

$$\frac{d}{d\lambda}\log\|\Phi_n\|_{\mu_\lambda}^2 = \int |\varphi_n(e^{i\theta},d\mu_\lambda)|^2\,\frac{d\mu_\lambda(\theta)}{d\lambda} \tag{10.1}$$

PROOF. Since $\frac{d}{d\lambda}\log\|\Phi_n\|^2 = (\frac{d}{d\lambda}\|\Phi_n\|^2)/\|\Phi_n\|^2$, it suffices to prove that

$$\frac{d}{d\lambda}\,\|\Phi_n\|^2_{\mu_\lambda} = \int |\Phi_n(e^{i\theta}, d\mu_\lambda)|^2\, \frac{d\mu_\lambda(\theta)}{d\lambda} \tag{10.2}$$

Since

$$\|\Phi_\lambda\|^2 = \int \overline{\Phi_n(e^{i\theta}, d\mu_\lambda)}\,\Phi_n(e^{i\theta}, d\mu_\lambda)\, d\mu_\lambda \tag{10.3}$$

its derivative is a sum of three terms, of which one term is the right side of (10.2) and the other two are

$$\int \overline{\frac{\partial}{\partial\lambda}\,\Phi_n(e^{i\theta}, d\mu_\lambda)}\,\Phi_n(e^{i\theta}, d\mu_\lambda)\, d\mu_\lambda \tag{10.4}$$

and its conjugate.

But, for all λ, Φ_n is monic, so $\frac{\partial}{\partial\lambda}\Phi_n$ is a polynomial of degree at most $n-1$ and so orthogonal to Φ_n. It follows that the term in (10.4) is zero. □

REMARK. Those familiar with the Feynman-Hellman theorem [**35**] will see a metaphysical link to this proof.

We want to do two things with (10.1). First, we sum several logs in

$$\log D_n(d\mu) = \sum_{j=0}^{n} \log\|\Phi_j\|^2$$

(by (2.4)), and second, we make an explicit choice of $d\mu_\lambda$:

THEOREM 10.2. *Let w_t be a family of L^1 functions, C^1 in t so $\log w_t$ is also C^1 in t, and $w_0 = 1$. Then]*

$$\begin{aligned}\log D_n\left(w_1\,\frac{d\theta}{2\pi}\right) &= (n+1)\log(\|\Phi_{n+1}\|^2_{t=1}) \\ &\quad - \int_0^1 dt \int \left[\frac{d}{dt}\,(\log w_t)\right]\left(\frac{\partial}{\partial r}\,|\varphi^*_{n+1}(re^{i\theta}; d\mu_t)|^2\bigg|_{r=1}\right) d\mu_t(\theta)\end{aligned} \tag{10.5}$$

PROOF. We have that

$$\frac{d}{dt}\,w_t\frac{d\theta}{2\pi} = \frac{w_t'}{w_t}\,d\mu_t = \frac{d}{dt}\,(\log w_t)\,d\mu_t$$

Using this, (2.4), (10.1), and the definition (9.2) of K, we get

$$\frac{d}{dt}\,\log D_n\left(w_t\,\frac{d\theta}{2\pi}\right) = \int \left[\frac{d}{dt}\,(\log w_t)\right] K(e^{i\theta}, e^{i\theta}; d\mu_t)\, d\mu_t \tag{10.6}$$

Now use (9.6) to get two terms. If we integrate dt, the first term gives the integral in (10.5). The second term can be integrated using (10.1) to give the first term in (10.5). □

11. Completion of the Proof

We complete the proof, following [**29**], by showing:

THEOREM 11.1. *Let $d\mu(\theta) = e^{L(\theta)}\frac{d\theta}{2\pi}$ where $e^{i\theta} \mapsto L(\theta)$ is analytic in a neighborhood of $\partial\mathbb{D}$. Then*

$$\lim_{n\to\infty}\,[\log D_n(d\mu) - (n+1)\hat{L}_0] = \sum_{k=1}^{\infty} k|\hat{L}_k|^2 \tag{11.1}$$

PROOF. Without loss, we suppose $\int d\mu(\theta) = 1$. We claim first that (under the analyticity assumption)

$$\lim_{n\to\infty} (n+1)[\log \|\Phi_{n+1}\|^2 - \hat{L}_0] = 0 \tag{11.2}$$

For in terms of the Verblunsky coefficients,

$$\|\Phi_{n+1}\|^2 = \prod_{j=0}^{n} (1 - |\alpha_j|^2)$$

(by (2.11)), while (4.1) and (2.14) say that

$$e^{\hat{L}_0} = \prod_{j=0}^{\infty} (1 - |\alpha_j|^2)$$

Thus

$$\begin{aligned} \|\Phi_{n+1}\|^2 e^{-\hat{L}_0} &= \prod_{j=n+1}^{\infty} (1 - |\alpha_j|^2)^{-1} \\ &= 1 + O(e^{-An/2}) \end{aligned}$$

by (5.17). This proves (11.2).

Thus, by (10.5), (11.1) is equivalent to there being a choice w_t with

$$-\int_0^1 dt \int \left(\frac{d}{dt} \log w_t\right) \frac{\partial}{\partial r} |\varphi_{n+1}^*(re^{i\theta}; d\mu_t)|^2 \bigg|_{r=1} d\mu_t(\theta) \to \sum_{k=1}^{\infty} k|\hat{L}_k|^2 \tag{11.3}$$

Define for $0 \le t \le 1$,

$$c_t = \log\left[\int e^{tL(\theta)} \frac{d\theta}{2\pi}\right] \tag{11.4}$$

and

$$w_t(\theta) = \exp(tL(\theta) - c_t) \tag{11.5}$$

so $\int d\mu_t = 1$ also. Thus, the Szegő function for μ_t is given by

$$\log D_t(z) = t \log D(z) - \tfrac{1}{2} c_t \tag{11.6}$$

$$= t \sum_{k=1}^{\infty} z^k \hat{L}_k + \tfrac{1}{2} t\hat{L}_0 - \tfrac{1}{2} c_t \tag{11.7}$$

It follows that $D_t(z)$ is analytic in a t-independent disk, and then the Verblunsky coefficients obey

$$|\alpha_n(d\mu_t)| \le Ce^{-An/2}$$

uniformly in $t \in [0,1]$, and then, from the proof of Theorem 5.4, that $\varphi_n^*(z, d\mu_t) \to D_t(z)^{-1}$ uniformly in $t \in [0,1]$ and z in $\{z \mid |z| \le e^{A/4}\}$. Thus, the $\frac{\partial}{\partial r}$ derivative in the integral on the left side of (11.3) converges to a $\frac{\partial}{\partial r}$ derivative of D_t.

We also write

$$\frac{\partial}{\partial r} |\varphi_{n+1}^*|^2 = 2|\varphi_{n+1}^*|^2 \operatorname{Re} \frac{\partial}{\partial r} \log(\varphi_{n+1}^*) \to -2|D_t|^{-2} \operatorname{Re}\left(t \frac{\partial}{\partial r} \log D\right) \tag{11.8}$$

and we can use $|D_t|^{-2}\, d\mu_t = \frac{d\theta}{2\pi}$.

By (11.5),

$$\frac{d}{dt}\log w_t = L(\theta) - \frac{d}{dt}c_t \tag{11.9}$$

Since the $\frac{d}{dt}c_t$ is θ-independent and by using (9.7), we see the $\frac{d}{dt}c_t$ can be integrated to $c_1 - c_0 = 0$. Thus, in (11.3) we can replace $\frac{d}{dt}\log w_t$ by $L(\theta)$. The result is that

$$\text{LHS of (11.3)} \to \int_0^1 dt \int 2t\, L(\theta)\, \text{Re}\left[\frac{\partial}{\partial r}\log D\right]\frac{d\theta}{2\pi} \tag{11.10}$$

The only t-dependence is the $2t$ and $\int_0^1 2t\,dt = 1$, so

$$\begin{aligned}\text{RHS of (11.10)} &= \int\Big[\sum_{k=-\infty}^{\infty}\hat{L}_k e^{ik\theta}\Big]\,\text{Re}\Big(\sum_{k=1}^{\infty}k\hat{L}_k e^{ik\theta}\Big)\frac{d\theta}{2\pi}\\ &= \int\Big(\sum_{k=-\infty}^{\infty}\hat{L}_k e^{ik\theta}\Big)\Big[\tfrac{1}{2}\sum_{k=-\infty}^{\infty}|k|\hat{L}_k e^{ik\theta}\Big]\frac{d\theta}{2\pi}\\ &= \tfrac{1}{2}\sum_{k=-\infty}^{\infty}|k|\bar{\hat{L}}_k\hat{L}_k = \sum_{1}^{\infty}k|\hat{L}_k|^2\end{aligned}$$

as claimed. □

References

1. F.V. Atkinson, *Discrete and Continuous Boundary Problems*, Academic Press, New York, 1964.
2. J. Baik, P. Deift, K. McLaughlin, and X. Zhou, in preparation.
3. M. Bakonyi and T. Constantinescu, *Schur's Algorithm and Several Applications*, Pitman Research Notes in Math. **261**, Longman, Essex, U.K., 1992.
4. G. Baxter, *A convergence equivalence related to polynomials orthogonal on the unit circle*, Trans. Amer. Math. Soc. **99** (1961), 471–487.
5. A. Böttcher, *The Onsager formula, the Fisher-Hartwig conjecture, and their influence on research in Toeplitz operators*, J. Statist. Phys. **78** (1995), 575–584.
6. D. Bump and P. Diaconis, *Toeplitz minors*, J. Combin. Theory Ser. A **97** (2002), no. 2, 252–271.
7. P. Delsarte, Y.V. Genin, and Y.G. Kamp, *Orthogonal polynomial matrices on the unit circle*, IEEE Trans. Circuits and Systems **CAS-25** (1978), 149–160.
8. P.L. Duren, *Theory of H^p Spaces*, Pure and Applied Math., Vol. 38, Academic Press, New York-London, 1970.
9. T. Erdélyi, P. Nevai, J. Zhang, and J. Geronimo, *A simple proof of "Favard's theorem" on the unit circle*, Atti Sem. Mat. Fis. Univ. Modena **39** (1991), 551–556.
10. M.E. Fisher and R.E. Hartwig, *Toeplitz determinants: Some applications, theorems and conjectures*, Stochastic Processes in Chemical Physics (K.E. Shuler, ed.), Advances in Chemical Physics, Vol. 15, Wiley Interscience, New York, 1969, pp. 333–353.
11. J. Geronimus, *On the trigonometric moment problem*, Ann. of Math. (2) **47** (1946), 742–761.
12. B.L. Golinskii and I.A. Ibragimov, *On Szegő's limit theorem*, Math. USSR Izv. **5** (1971), 421–444.
13. L. Golinskii, *Orthogonal polynomials on the unit circle, Szegő difference equations and spectral theory of unitary matrices*, second Doctoral thesis, Kharkov, 2003.
14. V. Guillemin and K. Okikiolu, *Spectral asymptotics of Toeplitz operators on Zoll manifolds*, J. Funct. Anal. **146** (1997), no. 2, 496–516.
15. K. Hoffman, *Analytic functions and logmodular Banach algebras*, Acta Math. **108** (1962), 271–317.
16. I.A. Ibragimov, *A theorem of Gabor Szegő*, Mat. Zametki **3** (1968), 693–702. [Russian]

17. K. Johansson, *On Szegő's asymptotic formula for Toeplitz determinants and generalizations*, Bull. Sci. Math. (2) **112** (1988), no. 3, 257–304.
18. T. Kailath, *A view of three decades of linear filtering theory*, IEEE Trans. Inform. Theory **IT-20** (1974), 146–181.
19. H.J. Landau, *Maximum entropy and the moment problem*, Bull. Amer. Math. Soc. **16** (1987), 47–77.
20. A. Lenard, *Momentum distribution in the ground state of the one-dimensional system of impenetrable Bosons*, J. Mathematical Phys. **5** (1964), 930–943.
21. ———, *Some remarks on large Toeplitz determinants*, Pacific J. Math. **42** (1972), 137–145.
22. G. Lumer, *Analytic functions and Dirichlet problem*, Bull. Amer. Math. Soc. **70** (1964), 98–104.
23. B.M. McCoy, *Introductory remarks to Szegő's paper "On certain Hermitian forms associated with the Fourier series of a positive function"*, Gabor Szegő: Collected Papers, Volume 1, 1915–1927, (R. Askey, ed.), Birkhäuser, Boston, 1982, pp. 47–51.
24. M.L. Mehta, *Random Matrices and the Statistical Theory of Energy Levels*, Academic Press, New York, 1967.
25. P. Nevai and V. Totik, *Orthogonal polynomials and their zeros*, Acta Sci. Math. (Szeged) **53** (1989), 99–104.
26. G. Pólya, L'Intermédiaire des Mathématiciens **21** (1914), S. 27.
27. W. Rudin, *Real and Complex Analysis*, 3rd edition, McGraw-Hill, New York, 1987.
28. B. Simon, *Orthogonal Polynomials on the Unit Circle. Part 1. Classical Theory*, Amer. Math. Soc. Colloq. Publ., Vol. 54, Providence, RI, 2005.
29. ———, *Orthogonal Polynomials on the Unit Circle. Part 2. Spectral Theory*, Amer. Math. Soc. Colloq. Publ., Vol. 54, Providence, RI, 2005.
30. G. Szegő, *Ein Grenzwertsatz über die Toeplitzschen Determinanten einer reelen positiven Funktion*, Math. Ann. **76** (1915), 490–503.
31. ———, *Beiträge zur Theorie der Teoplitzschen Formen, I*, Math. Z. **6** (1920), 167–202.
32. ———, *Beiträge zur Theorie der Teoplitzschen Formen, II*, Math. Z. **9** (1921), 167–190.
33. ———, *Orthogonal Polynomials*, Amer. Math. Soc. Colloq. Publ., Vol. 23, Providence, RI, 1939; 3rd edition, 1967.
34. ———, *On certain Hermitian forms associated with the Fourier series of a positive function*, Comm. Sém. Math. Univ. Lund **1952** (1952), Tome Supplementaire, 228–238.
35. W. Thirring, *Quantum Mechanics of Atoms and Molecules*, Springer-Verlag, New York, 1981.
36. S. Verblunsky, *On positive harmonic functions: A contribution to the algebra of Fourier series*, Proc. London Math. Soc. **38** (1935), 125–157.
37. ———, *On positive harmonic functions (second part)*, Proc. London Math. Soc. **40** (1936), 290–320.
38. J. Wermer, *Dirichlet algebras*, Duke Math. J. **27** (1960), 373–381.
39. H. Widom, *Eigenvalue distribution theorems for certain homogeneous spaces*, J. Funct. Anal. **32** (1979), no. 2, 139–147.

MATHEMATICS 253-37, CALIFORNIA INSTITUTE OF TECHNOLOGY, PASADENA, CA 91125, U.S.A.
E-mail address: `bsimon@caltech.edu`

Titles in This Series

387 **Michael Entov, Yehuda Pinchover, and Michah Sageev, Editors,** Geometry, spectral theory, groups, and dynamics, 2005

386 **Yasuyuki Kachi, S. B. Mulay, and Pavlos Tzermias, Editors,** Recent progress in arithmetic and algebraic geometry, 2005

385 **Sergiy Kolyada, Yuri Manin, and Thomas Ward, Editors,** Algebraic and topological dynamics, 2005

384 **B. Diarra, A. Escassut, A. K. Katsaras, and L. Narici, Editors,** Ultrametric functional analysis, 2005

383 **Z.-C. Shi, Z. Chen, T. Tang, and D. Yu, Editors,** Recent advances in adaptive computation, 2005

382 **Mark Agranovsky, Lavi Karp, and David Shoikhet, Editors,** Complex analysis and dynamical systems II, 2005

381 **David Evans, Jeffrey J. Holt, Chris Jones, Karen Klintworth, Brian Parshall, Olivier Pfister, and Harold N. Ward, Editors,** Coding theory and quantum computing, 2005

380 **Andreas Blass and Yi Zhang, Editors,** Logic and its applications, 2005

379 **Dominic P. Clemence and Guoqing Tang, Editors,** Mathematical studies in nonlinear wave propagation, 2005

378 **Alexandre V. Borovik, Editor,** Groups, languages, algorithms, 2005

377 **G. L. Litvinov and V. P. Maslov, Editors,** Idempotent mathematics and mathematical physics, 2005

376 **José A. de la Peña, Ernesto Vallejo, and Natig Atakishiyev, Editors,** Algebraic structures and their representations, 2005

375 **Joseph Lipman, Suresh Nayak, and Pramathanath Sastry,** Variance and duality for cousin complexes on formal schemes, 2005

374 **Alexander Barvinok, Matthias Beck, Christian Haase, Bruce Reznick, and Volkmar Welker, Editors,** Integer points in polyhedra—geometry, number theory, algebra, optimization, 2005

373 **O. Costin, M. D. Kruskal, and A. Macintyre, Editors,** Analyzable functions and applications, 2005

372 **José Burillo, Sean Cleary, Murray Elder, Jennifer Taback, and Enric Ventura, Editors,** Geometric methods in group theory, 2005

371 **Gui-Qiang Chen, George Gasper, and Joseph Jerome, Editors,** Nonlinear partial differential equations and related analysis, 2005

370 **Pietro Poggi-Corradini, Editor,** The p-harmonic equation and recent advances in analysis, 2005

369 **Jaime Gutierrez, Vladimir Shpilrain, and Jie-Tai Yu, Editors,** Affine algebraic geometry, 2005

368 **Sagun Chanillo, Paulo D. Cordaro, Nicholas Hanges, Jorge Hounie, and Abdelhamid Meziani, Editors,** Geometric analysis of PDE and several complex variables, 2005

367 **Shu-Cheng Chang, Bennett Chow, Sun-Chin Chu, and Chang-Shou Lin, Editors,** Geometric evolution equations, 2005

366 **Bernhelm Booß-Bavnbek, Gerd Grubb, and Krzysztof P. Wojciechowski, Editors,** Spectral geometry of manifolds with boundary and decompositon of manifolds, 2005

365 **Robert S. Doran and Richard V. Kadison, Editors,** Operator algebras, quantization, and non-commutative geometry, 2004

364 **Mark Agranovsky, Lavi Karp, David Shoikhet, and Lawrence Zalcman, Editors,** Complex analysis and dynamical systems, 2004

363 **Anthony To-Ming Lau and Volker Runde, Editors,** Banach algebras and their applications, 2004

362 **Carlos Concha, Raul Manasevich, Gunther Uhlmann, and Michael S. Vogelius, Editors,** Partial differential equations and inverse problems, 2004

361 **Ali Enayat and Roman Kossak, Editors,** Nonstandard models of arithmetic and set theory, 2004

360 **Alexei G. Myasnikov and Vladimir Shpilrain, Editors,** Group theory, statistics, and cryptography, 2004

359 **S. Dostoglou and P. Ehrlich, Editors,** Advances in differential geometry and general relativity, 2004

358 **David Burns, Christian Popescu, Jonathan Sands, and David Solomon, Editors,** Stark's Conjectures: Recent work and new directions, 2004

357 **John Neuberger, Editor,** Variational methods: open problems, recent progress, and numerical algorithms, 2004

356 **Idris Assani, Editor,** Chapel Hill ergodic theory workshops, 2004

355 **William Abikoff and Andrew Haas, Editors,** In the tradition of Ahlfors and Bers, III, 2004

354 **Terence Gaffney and Maria Aparecida Soares Ruas, Editors,** Real and complex singularities, 2004

353 **M. C. Carvalho and J. F. Rodrigues, Editors,** Recent advances in the theory and applications of mass transport, 2004

352 **Marek Kubale, Editor,** Graph colorings, 2004

351 **George Yin and Qing Zhang, Editors,** Mathematics of finance, 2004

350 **Abbas Bahri, Sergiu Klainerman, and Michael Vogelius, Editors,** Noncompact problems at the intersection of geometry, analysis, and topology, 2004

349 **Alexandre V. Borovik and Alexei G. Myasnikov, Editors,** Computational and experimental group theory, 2004

348 **Hiroshi Isozaki, Editor,** Inverse problems and spectral theory, 2004

347 **Motoko Kotani, Tomoyuki Shirai, and Toshikazu Sunada, Editors,** Discrete geometric analysis, 2004

346 **Paul Goerss and Stewart Priddy, Editors,** Homotopy theory: Relations with algebraic geometry, group cohomology, and algebraic K-theory, 2004

345 **Christopher Heil, Palle E. T. Jorgensen, and David R. Larson, Editors,** Wavelets, frames and operator theory, 2004

344 **Ricardo Baeza, John S. Hsia, Bill Jacob, and Alexander Prestel, Editors,** Algebraic and arithmetic theory of quadratic forms, 2004

343 **N. Sthanumoorthy and Kailash C. Misra, Editors,** Kac-Moody Lie algebras and related topics, 2004

342 **János Pach, Editor,** Towards a theory of geometric graphs, 2004

341 **Hugo Arizmendi, Carlos Bosch, and Lourdes Palacios, Editors,** Topological algebras and their applications, 2004

340 **Rafael del Río and Carlos Villegas-Blas, Editors,** Spectral theory of Schrödinger operators, 2004

339 **Peter Kuchment, Editor,** Waves in periodic and random media, 2003

338 **Pascal Auscher, Thierry Coulhon, and Alexander Grigor'yan, Editors,** Heat kernels and analysis on manifolds, graphs, and metric spaces, 2003

For a complete list of titles in this series, visit the AMS Bookstore at **www.ams.org/bookstore/**.